THE OXFORD HISTORY
OF ENGLAND

Edited by G. N. CLARK

THE OXFORD HISTORY OF ENGLAND

Edited by G. N. CLARK

Provost of Oriel College, Oxford

(*The titles of some volumes are provisional*)

THE REIGN OF
ELIZABETH
1558–1603

By

J. B. BLACK

Burnett-Fletcher Professor of History
University of Aberdeen

OXFORD
AT THE CLARENDON PRESS

Oxford University Press, Amen House, London E.C. 4

GLASGOW NEW YORK TORONTO MELBOURNE WELLINGTON
BOMBAY CALCUTTA MADRAS CAPE TOWN

Geoffrey Cumberlege, Publisher to the University

FIRST PUBLISHED 1936
REPRINTED 1937, 1945, 1949

PRINTED IN GREAT BRITAIN
AT THE UNIVERSITY PRESS, OXFORD
BY CHARLES BATEY, PRINTER TO THE UNIVERSITY

PREFACE

THE purpose of this book is to consider, so far as the space allows, all aspects of the reign of Elizabeth, giving more emphasis than usual to social and cultural as distinct from political affairs. Obviously the arrangement of the matter cannot be decided simply on the chronological principle; for no historian, however skilled in the art of narrative, could hope to carry forward the whole burden of fact from year to year without involving himself and his readers in confusion. Moreover, a strict adherence to chronology would make it impossible to treat any one subject consecutively and intelligibly. On the other hand, the substitution of a topical for a chronological treatment, although it might simplify the issues at stake, would have the effect of sacrificing unity, and would leave the book in the form of a series of essays, more or less disjointed, which is equally to be deprecated. As a working compromise between these two opposite methods, we have adopted the plan of arranging the chapters in rough chronological sequence, and at the same time keeping each chapter focused on the particular topic with which it deals.

While this is, broadly, the plan of the volume, it has to be modified in one respect, in order to accommodate the discussion of certain matters which have a central importance for the period, but cannot be treated adequately in the general body of the narrative. In books of this kind, non-political questions are frequently relegated to the end, or at least the latter part, as if they were in the nature of addenda. The reader will find that we have incorporated them in the central portion of the volume, side by side with discussions of the catholic and puritan problems, and of the constitution. In short, the scheme as a whole works out thus. First come four chapters dealing with the Religious Settlement, England and France, Mary Stuart and the Succession, and the critical years 1568–75, which follow each other in approximately chronological order. Then this method is frankly abandoned in favour of specialized studies of the catholic and puritan challenges to the establishment, of the working of the constitution, and of the economic, social, literary, artistic, scientific, and cultural features of the age. In chapter ix, which deals with England's relations with the Netherlands, the method employed in the first four chapters is resumed and continued through chapter x (the Execution of Mary Stuart, and the Armada) and chapter xi (the Last Years of the Reign). The final chapter deals with the Irish problem, not because Ireland is important only at the close of the reign, but because this is the only point in the narrative where it can be discussed comprehensively.

In regard to the matter contained in the book, it need hardly be said that a completely objective account of events in Elizabeth's reign, however desirable it may be in theory, cannot in actual fact be written. From time to time one treads on embers of controversies that still flicker with a baleful light as soon as they are disturbed. And there are questions on which historians will probably be divided to the end of time, for the simple reason that their points of view differ so greatly that the gulf between them is unbridgeable. Of such a nature is the question of Mary Stuart's responsibility for the fate that eventually overtook her. No catholic historian will accept the view that Elizabeth, or her minister Walsingham, was not animated by a deadly hatred of the Stuart queen, whose claim to the English throne was the *fons et origo* of most of the troubles of the reign. Similarly it may be doubted whether a well-informed Scottish historian, working intelligently over the state papers now available for the reign of James VI, could come to any other conclusion than that English policy, after Morton's fall, was directed towards keeping Scotland in a continual welter unless and until those who controlled affairs in Edinburgh were 'at England's devotion'. And who would venture to suggest that a Spaniard, a Frenchman, or even a Netherlander could be expected to take any but a hostile view of the inner working of the same policy in regard to his own country? In the present volume we have been compelled to observe events predominantly through English eyes, or, to be more correct, through the eyes of the English government—to look upon the kings of France and Spain, the popes, Mary Stuart, and James VI as 'problems' to be solved rather than as beings entitled to a separate and sympathetic consideration. But the writer is aware, and the reader ought also to be aware, that there is another point of view which must be taken into account before we begin to speak of objective history in the proper sense of the words. An attempt has been made to keep this in mind while writing the book, but the paramount necessity of placing the reader at the standpoint of the queen and her ministers has prevented a rigorous following out of the principle.

In accordance with the general rules governing the series of which this volume forms a part, notes and references have been very sparingly used, only in fact when for one reason or another some elucidation of the text seemed to be necessary, or a ruling had to be given on a disputed point, or it was deemed advisable to lighten the narrative of its heavy burden of fact. Spelling has been modernized for the convenience of the reader, without, it is hoped, any violence being done to the sense. The sources are indicated in the bibliography, which aims at giving a fairly representative selection of authorities, but is not to be regarded as exhaustive on any topic. The researcher will naturally turn to Conyers Read's *Bibliography of British History: the Tudor Period* for further guidance in his work.

To sum up one's indebtedness to other writers is never easy, and to particularize it would perhaps be invidious. Nevertheless I should be singularly ungrateful and neglectful were I not to put on record how much I owe to the labours of Queen Elizabeth's most recent biographer, Professor J. E. Neale, whose researches in English political and constitutional history of the later sixteenth century have placed all students of the period under a lasting obligation. The extent of my debt to his various writings is by no means covered by the references in the text.

It remains for me to thank heartily those who have helped me in the later stages of my task. I am profoundly grateful to the editor, Professor G. N. Clark, for his unfailing courtesy and for the many suggestions he made on points both of matter and of style. His critical acumen and wide knowledge saved me from more errors of omission as well as of commission than I care to remember. To Mr. J. M. Henderson, lecturer in British History in the University of Aberdeen, I am also greatly indebted for the care and discrimination with which he read the proofs, for much fruitful interchange of opinion, and for the numerous occasions when his eye detected ambiguities which I myself failed to discover owing to too great familiarity with the written word. Both have in different ways contributed materially to the readableness and, I dare hope also, to the usefulness of the book.

J. B. B.

KING'S COLLEGE, ABERDEEN,
January 1936.

TABLE OF CONTENTS

III. MARY STUART AND THE SUCCESSION

IV. YEARS OF CRISIS: 1568–75

VII. THE EXPANSION OF ENGLAND, AND THE ECONOMIC AND SOCIAL REVOLUTION

VIII. LITERATURE, ART, AND THOUGHT

IX. ELIZABETH AND THE NETHERLANDS

X. THE EXECUTION OF MARY STUART:
THE SPANISH ARMADA

XI. THE ELIZABETHAN STATE AT WAR:
LAST YEARS OF THE REIGN

XII. THE IRISH PROBLEM

XIII. THE END

LIST OF MAPS *(at end)*

GENEALOGICAL TABLES *(at end)*

I

THE QUEEN AND THE RELIGIOUS SETTLEMENT

FEW rulers have impressed themselves so forcibly on the
memory and imagination of the English race as Queen
Elizabeth. It may be admitted, of course, that much of the
lustre surrounding her name is adventitious; that her reputation as a queen, like that of Louis XIV, lies bathed in the
reflected glory of the people she governed. Even so, however,
'Elizabethan England' was, in a very real sense, Elizabeth's
England. She it was who nursed it into being, and by her
wisdom made possible its amazing development. Her characteristic virtues and defects, her sympathies and antipathies, her
very whims and caprices are writ large across its political
firmament. She inspired its patriotism, its pageantry, its
heroisms, stimulated its poetry, and shaped its destiny. And
when she died she left behind her a kingdom that had won
a commanding position among the great powers of Europe.

On 17 November 1558, when she succeeded her half-sister,
Mary, on the throne, there was no glimmering of the splendid
future in store for her. England was 'ragged and torn with
misgovernment'; the treasury was empty; the principal for-
tresses of Portsmouth and Berwick were falling into ruin; the
country was bare of munitions; and a huge debt of more than
£266,000 had to be liquidated, part of which was owing to
foreign creditors, and charged with a 'biting' interest. The
French king, Henry II, bestrode the realm like a Colossus, with
one foot in Calais and the other in Edinburgh; and the alliance
with Spain had played itself out with the loss of England's last
continental possession. The prospect was rendered gloomier
by the uncertainty of the young queen's title, and the probability
that the Stuart claim, barred by Henry VIII, would be revived
by the French dauphiness, Mary Stuart. So insecure did the
state of affairs appear that few expected the new régime to last,
and many calculated on its speedy downfall. To one competent
observer, at least, it looked as if the two great continental
monarchies, taking advantage of the defenceless state of the
country, would turn it into a second Piedmont.

As for the queen, what contemporaries saw was a tall, 'comely
rather than handsome' woman of twenty-five summers, with

fair hair, 'fine' eyes, and a delicate 'olive' complexion. While, in some respects, she took after her mother, Anne Boleyn, her temperament and bearing were those of her imperious father, in whom, we are told, she 'gloried'. Close beneath her winsome, debonair exterior lay the *terribilità* of the Tudor 'lion'. Her character, however, was largely an unknown quantity, and her capacity for government altogether untried. Few probably realized, and time alone would show, that this slip of a woman, ignorant as yet of the technique of statesmanship, had already graduated in the hard school of experience, and was, in all essential respects, mistress of her destiny.

Wisdom, indeed, had come early to Elizabeth. Fate had deprived her of a mother[1] at the tender age of two years and eight months; but, as a sort of compensation for the loss, had developed in her a remarkable precocity of intellect. She became an observant, introspective child, apt to learn both from books and from life. Growing up, as she did, in a world of swift political changes and sudden reverses of fortune, when the fragility of happiness, the insecurity of power, and the uncertainty of all human affairs were the recurrent themes of philosopher and poet, it would have been strange if this intelligent offspring of Henry VIII had not pondered deeply on these things. But it was experience rather than books that disillusioned her of any tendency she might have had to romanticism, and converted her plastic and unformed mind into the calculating machine it afterwards became. She ripened quickly, almost too quickly to preserve that balance between emotion and restraint which is the glory of true womanhood. At the age of fifteen, an unsavoury but harmless love-affair with Admiral Seymour, which cost the admiral his life and involved the princess herself in public disgrace, had given her a first bitter taste of the power of scandal, and shown her the importance of keeping a tight hold over her natural impulses. It was, in fact, the fiery crucible in which all that remained of irresponsible childhood was remorselessly burnt to dross and ashes: out of it came a woman with a purpose, schooled to self-repression, prudence, and mistrust. Following upon this crude awakening came the five momentous years of her sister's reign, when, as heir presumptive to the throne, she became willy nilly the centre of every plot against the existing régime, and had

[1] Anne Boleyn was executed on 19 May 1536.

to pick her steps with care between the calculated malice of her enemies and the pitfalls dug by her friends. For Elizabeth it was probably the most trying time in her life: her stamina was tested to the uttermost. For weeks she lay a prisoner in the Tower for alleged complicity in the Wyatt rebellion, a prey to gnawing anxiety lest this gloomy keep, which had been for so many the antechamber to death, might be for her also the end of all things. But she escaped disaster by her own innate shrewdness and by the lack of convincing evidence against her. Danger had sharpened her wits: she had learnt to screen her thoughts from others, to prevaricate, to dissimulate, to deceive, and to overcome difficulties less by vanquishing than by circumventing them. Circumstances had bred in her a hard, self-regarding type of mind, not particularly sensitive to fine issues, nor open in its acceptance of life, but strong in the grain and pliant as steel. She was wise with this world's wisdom— resourceful, self-reliant, cautious, and morally courageous in moments of stress. But she had lived too long in an atmosphere of plot and intrigue to cultivate the virtue of magnanimity—it was a luxury she could ill afford; and suspicion was a second nature to her. It cannot be said, therefore, that, as she stood on the threshold of her reign, she was a particularly heroic figure, or that her personal character, abstruse and recondite as it was, would be likely to provide material for the lyric or the tragic muse. In short, she was like one who had been 'saved, yet so as by fire': the marks of the scorching she had received, during the most impressionable period of her life, remained with her—a psychological inheritance which the passage of time would never alter. Of religious feeling, in the ordinary sense of the word, she probably had little. Her cold, entirely humanist outlook, nourished by classical study, kept her apart from the deeper spiritual currents of her time. Moreover she had seen too much of the ravages of fanaticism, both protestant and catholic, to set any store by the dogmatic formularies of either side. The only religious faith she can be said to have held with any degree of conviction was a belief in an over-ruling Providence—the refuge of all distressed human beings. Her culture, on the other hand, was considerable. In an age that could boast of such feminine prodigies as Jane Grey, Mildred Cecil, and the accomplished daughters of Sir Thomas More, Elizabeth was notable for her learning. Her schoolmaster,

Roger Ascham, tells us that she could speak French and Italian fluently, Latin readily and well, and Greek moderately. Above all, she had realized, while still quite young, that the stability of her throne would depend upon the success with which she interpreted the national aspirations and gave them articulation—the very point, be it noted, on which her sister had blundered so badly. Whatever mistakes she might make at the commencement of her reign, there would at least be no repetition of 'Spanish' rule and 'Rome' rule, which had been linked together in the public mind as the causes of England's troubles at home and abroad. The future policy of the country would be based upon its vital needs. Thus did Elizabeth strike the key-note to a new age when she resolved to identify herself with her people, and become, in fact as well as in name, the 'most English woman in England'.[1]

The accession was a memorable occasion in the civic annals of London; and London in those days was like Paris in the days of the French revolution—the heart and brain of a kingdom. Something of what people felt may be gathered from the glowing exordium with which Holinshed begins his chronicle of the reign:

'After all the stormy, tempestuous, and blustering windy weather of Queen Mary was overblown, the darksome clouds of discomfort dispersed, the palpable fogs and mist of the most intolerable misery consumed, and the dashing showers of persecution overpast: it pleased God to send England a calm and quiet season, a clear and lovely sunshine, a quitsest[2] from former broils of a turbulent estate, and a world of blessings by good Queen Elizabeth.'

From the first day of her arrival in the capital, 23 November, to her coronation day, 15 January, the young queen revelled in the enthusiastic loyalty of her subjects, feasting their eyes with equipages through the city and on the Thames. The popular rejoicing reached a climax on the eve of the coronation, when the glittering royal procession wound its way from the Tower to Westminster, stopping from time to time, so that the queen might admire the splendid pageants set up by the Londoners in honour of the occasion, and make suitable

---

[1] The best account of Elizabeth's youth is L. Wiesener's *La Jeunesse d'Élisabeth d'Angleterre*, 1878, trans. C. M. Yonge, 2 vols., 1879; but cf. J. E. Neale, *Elizabeth*, 1934, chaps. i, ii.

[2] i.e. release, a corruption of *quietus est*.

responses to their greetings. There could be no mistaking the meaning of the pageantry: tableau after tableau was charged with significant references to past miseries and future hopes. There was the pageant of the Roses in Gracechurch street; the pageant of the 'Respublica Ruinosa' and the 'Respublica Bene Instituta' in Cheapside; and the pageant of Deborah in Ludgate and Fleet street. The allusion to Deborah, 'judge and restorer of Israel', was merely one of the many allusions that day to religion, for London was overwhelmingly protestant. At Temple Bar, where a chorus of children sang a farewell to the queen as she left the city, stress was laid on the establishment of truth and the suppression of error; and it was observed that when religion was mentioned Elizabeth lifted up her hands and eyes to heaven, expressing the wish that all around should respond—Amen! More striking, perhaps, than the popular rejoicing was the affability of the queen: she entered into the spirit of the celebrations with a frankness that disconcerted at least one foreign spectator, who thought that she exceeded the bounds of gravity and decorum. But there was method in this wooing of the people. Elizabeth's first care was to show that she had their interests at heart, and thereby to recover for the Crown the favour her sister had so recklessly squandered.

Meanwhile signs of an impending revolution in national policy were beginning to be visible in the political world; and one of the first to notice them was the count de Feria, Philip II's ambassador, who had been dispatched from Brussels shortly before Mary's death to present the king's felicitations to Elizabeth. With a confidence strangely out of keeping with the circumstances of the time, this proud Spaniard had journeyed down to Hatfield to inform her that her succession was assured, and that she owed her good fortune to the kind services of his master. To his chagrin he was firmly but politely told that in this matter her gratitude was due solely to her people. The privy council treated him even more perfunctorily. 'They received me', he commented, 'as a man who came accredited with the bulls of a dead pope'; and again: 'They are all as ungrateful to your Majesty as if they had never received anything from your hands.' Worse still, Spain's representative soon found that he was denied the courtesy of a room in the palace, after the queen took up residence at Whitehall; nor could he get any reliable information as to what was going on behind the scenes, for every one shunned him 'as if I were the devil'. He concluded angrily that the realm was in confusion, and that the best way to deal with it was 'sword in hand'. This cavalier treatment of de Feria was more or less inevitable: it was a gesture to the world that English policy was henceforth to be modelled on English interests and not as the Spaniard should dictate.

In regard to religion changes were also imminent, but here the signs were more difficult to discern. This was due to the fact that the political question of the queen's personal and dynastic security took precedence over all other matters. Doubtless if the throne could have been secured by means of a concordat with Rome, involving concessions on both sides, this would have been the method employed. The queen herself, despite outward friendliness to protestantism, had no vested interest in the movement, nor was she in favour—rather the reverse—of the prevailing Calvinism of the time. But the character of the reigning pope, Paul IV,[1] and the fear of French machinations at Rome in support of Mary Stuart, rendered such a plan impossible, even if it were seriously considered. On the other hand, so radical a step as a decisive breach with the papacy was too dangerous to be contemplated in the present military and financial weakness of the kingdom. Extreme caution was the only feasible line to follow. Matters must not be brought to a crisis until the new government was properly in the saddle, and until a religious policy had been formulated that would secure the allegiance of as many of the queen's subjects as possible.

The preliminary moves were by no means reassuring to catholics, but moderate enough not to arouse their resentment. Elizabeth chose as her principal secretary Sir William Cecil— an eminently safe if not heroic figure, who had already served under Somerset and Northumberland in a similar capacity, and had so cleverly dissimulated his protestant leanings during Mary's reign as to be employed by her in an important embassy to the Netherlands. Cecil belonged to the 'politique' group of Englishmen, to whom might be applied the epithet—*non ex quercu*

---

[1] 'Fully conscious of his own dignity, he regarded princes not as his sons but as his subjects. . . . He told the ambassadors that the place of kings was at the feet of the pope, from whom they should receive their laws as his pupils. . . . The utterances of his volcanic nature were as sudden as the eruptions of Vesuvius.' (L. Pastor, *History of the Popes*, vol. xiv, ch. iii, pp. 69–70.)

sed ex salice orti. His greatness as an administrator still lay in the future, when he became, if not the brain, at least the regulative balance in the Elizabethan government for more than thirty years: so that in selecting him to be her chief co-operator at the beginning of the reign the queen showed great discrimination. Even the Spanish ambassador could find nothing worse to say of him than that, so far as he could discover, Cecil was a man of intelligence and virtue; indeed he thought it worth while to recommend him for a Spanish pension, although there was no likelihood of such a bribe being accepted. Cecil was sworn in by the queen with the following words: 'This judgment I have of you that you will not be corrupted with any manner of gift, and that you will be faithful to the state, and that without respect of my private will you will give me that counsel that you think best.' To the best of our knowledge the secretary preserved the letter as well as the spirit of his pledge until his death in 1598. In the choice of her council Elizabeth was no less discriminating. Discarding the majority of her predecessor's advisers, all staunch catholics, she retained only those who were likely to prove pliable; and to these she added seven of her own choice, all of whom were protestants, and some related to herself by marriage. The new board could hardly be described as a revolutionary junta, and de Feria exaggerated when he remarked that the kingdom was 'entirely in the hands of heretics'; but he was right to this extent that the really significant part of the council was protestant.

In regard to policy, again, the note of caution was especially prominent, not unmingled with a touch of dissimulation. Thus, for example, the first important proclamation of the reign forbade, under severe penalties, 'all manner our subjects of every degree' to attempt of their own authority any alteration of the established order of religion. Ostensibly the purpose was to quieten the catholics, but observant critics, reading between the lines, came to the conclusion that, as the proclamation bound only subjects, it reserved to the Crown the right to make any innovation it pleased, and was really an announcement to that effect.[1] Similarly the Marian ambassador at Rome, Sir

---

[1] 'It is true that in the proclamation by the queen "that no one was to dare (of his own authority) . . . to alter the present state of religion", the phrase "of his own authority" is construed to imply that the queen, at her own time, will herself give the authority.' (Michiel Surian to the Doge and Senate, 10 Dec. 1558: *Venetian Calendar,* 1557–8.)

Edward Carne, was not only allowed for the time being to remain at his post, but instructed to inform the pope that an important mission would shortly be sent to him from England, although such a mission was inherently improbable from the very first. It was also observed that the royal title which the queen affected omitted all reference to the supreme headship, Elizabeth being content to style herself simply 'Queen of England, France, and Ireland, etc.' But the use of the 'etc.' left her the loop-hole to restore the obnoxious phrase if she saw fit to do so. It was no ordinary brain that devised so clever a subterfuge.[1]

All these measures might be described as anticipatory of coming change; but they were of a negative rather than a positive nature; and the queen steadfastly refused to communicate to any one what use she intended to make of them. Yet, in spite of her caution, Elizabeth was not quite consistent in her behaviour, and many indications were given long before parliament met of how the land lay at Court. The first divine to preach at Paul's Cross—a regular Court appointment—was a known protestant, William Bill; and when Bishop Christopherson of Chichester dared to refute him, on the following Sunday, he was reprimanded and confined to his house. In December, when the requiems for Mary and Charles V were celebrated, in strict accordance with the old ritual, Bishop White, who preached the sermon at the former, ventured to refer to coming changes in religion. For this he was punished in the same way as Christopherson. On Christmas day, however, came a marked change in the queen's demeanour. She instructed the officiating bishop at the royal chapel, Oglethorpe of Carlisle, to omit the elevation of the host from the service of the Mass; and when he refused to mutilate the rubric she left the chapel as soon as the reading of the Gospel was concluded. Meanwhile the return of the 'wolves from Germany'—the Marian exiles for conscience' sake—became the signal for an outburst of theological polemics, little to the queen's liking; and scurrilous attacks were made by pamphleteers against the hierarchy. To still the uproar, a proclamation was issued on 28 December, forbidding all preaching, but permitting the use of the Gospel, the Epistle, and the Ten Commandments—which

---

[1] The 'etc.' had indeed been used by Mary until Wyatt's rebellion; but with the opposite intention, i.e. as a means of dropping the supreme headship.

were to be recited in English. By this time the catholics were becoming thoroughly alarmed: so that it was difficult to find a cleric willing to officiate at the coronation ceremony on 15 January. Eventually Oglethorpe consented, and the queen took the oath in the traditional manner, being proclaimed 'Defender of the True, Ancient, Catholic Faith'. But at the subsequent service in the Abbey she repeated her tactics of Christmas, left before the elevation of the host, and refused to receive the communion which was given in one kind, according to the old rite. The final scene of this interim period occurred a few days later, on the occasion of the state opening of parliament, when the queen was met on her way to the house by the abbot of Westminster and his monks bearing lighted tapers. 'Away with these torches,' she cried, 'for we see very well.'

It was now beyond dispute that great alterations in religion were about to be submitted to the verdict of the nation's representatives.[1] For weeks committees had been busy drafting bills and preparing business, and when parliament met, on 25 January, the government was ready for the fray. The lord keeper, Sir Nicholas Bacon, in his opening speech, explained that 'her Majesty's desire was to secure and unite the people of the realm in one uniform order to the glory of God and to general tranquillity'; and for this purpose she required them to avoid 'contumelious and opprobrious words, as heretic, schismatic, and papist, as causes of displeasure and malice, enemies to concord and unity, the very marks they were now to shoot at'. On 4 February Carne was recalled from Rome, his presence there being no longer necessary; and on the 9th a bill for annexing the supremacy of the church to the Crown was laid before the commons. The great controversy had begun.

A certain amount of criticism used to be expended on the composition of this first parliament of the reign, on the assumption that the validity of the reformation settlement depends upon the integrity of the legislative body that carried it through. It must be remembered that a Tudor parliament was the product of a very narrow franchise, and of an electoral system that was subject to frequent government manipulation. It could be packed by circulating lists of approved candidates

---

[1] 'The catholics are very fearful of the measures to be taken in this parliament.' (De Feria to the king, 31 Jan. 1559: *Span. Cal.*, 1558–67.)

to the sheriffs with appropriate instructions. But there is no reason to suppose that the elections of 1558-9 were different in any essential respect from previous elections. The bulk of the members 'returned' in January 1559 had already sat in the last parliament of Mary's reign. All that can justly be said is that the queen took the ordinary precautions permissible in her day for facilitating the passage of government measures. There was no abnormal packing. The question is really of relatively small importance; for, if the conduct of Tudor parliaments be examined, it will be found that, despite government interference, they had a mind and will of their own, which they sometimes asserted with disconcerting emphasis. Mary could not move her parliaments to redistribute the confiscated church lands, nor could Somerset touch the land question without endangering his position. Packing a parliament, therefore, did not mean making it a slavish servant of the Crown.

In point of fact discussion was prolific in the parliament of 1559, in spite of the strength of the government's supporters. Harding, the catholic apologist, writing a few years after the event, asserts that 'many learned in parliament' spoke against the proposed changes; and there is plenty of evidence from other sources to corroborate him. For example, the bill to secure uniformity in the church service created wide divergencies of opinion, and in the last resort only passed by the narrow margin of three votes. It has been pointed out, also, that there were no less than three supremacy bills introduced before the act took its final shape and won the approval of both houses. Not only so: the last draft of the measure shows, according to Maitland,[1] traces of extensive erasures and interlineations, which can only be explained on the assumption that the clerk of the house of commons was kept busy with his pen as the discussion proceeded. Naturally the strongest opposition was encountered in the house of lords, where the forces of the old order were entrenched. But even here the full conservative weight of the catholics was not brought into play, because there were no less than ten sees unrepresented through death or illness and the carelessness of 'the accursed cardinal',[2] and the dissolution of the monasteries had deprived the abbots of their seats; while several catholic peers were either absent or voted, by an inconceivable slack-

[1] F. W. Maitland, *Collected Papers*, iii, pp. 185-204.
[2] i.e. Reginald Pole, who died on 19 Nov. 1558.

ness, through protestant proxies. But such as it was, the opposition in the upper house, particularly the episcopate, presented an unbroken phalanx to the measures sent up from the commons, and succeeded to some degree in modifying the resultant legislation in the conservative interest. Altogether three months elapsed before the two key statutes of the Reformation became law—a fact that seems to show, apart from all other indications, that discussion was not suppressed nor the action of the government unduly precipitate.

A more difficult question now emerges, to which, unfortunately, it is impossible to give a precise answer. Granting that parliament was not inordinately packed and that debate was not 'gagged', how far may it be said that the voice of the nation's representatives expressed the conscious will of the nation? Simpson in his *Life of Campion* caustically describes the anglican settlement as 'originally an abasement of religion before *a supposititious public opinion*[1] for the convenience of statesmen'. Some colour for this view is lent by the statements of the Spanish ambassador, de Feria, and of Dr. Sanders, the catholic apologist and refugee, both of whom were eyewitnesses and had some opportunity of acquiring information as to the state of public opinion. Both assert categorically that the catholics were greatly in excess of the protestants. De Feria places the non-catholic part of the population at one-third;[2] Sanders at one per cent.[3]

It will be noted, however, that there is a very great discrepancy between the estimates; and the reason for this is plain: there was, as yet, no means of ascertaining the relative strength of parties, nor can it be said that parties, as such, existed. In January 1559 there was no definite issue before the country regarding the religious question, for the government had, as we have seen, carefully refrained from divulging its programme.

[1] The italics are ours.
[2] *Span. Cal.*, 1558–67, p. 39.
[3] 'The English common people consist of farmers, shepherds, and artisans. The two former are catholics. Of the others none are schismatic except those who have sedentary occupations, as weavers and shoemakers and some idle people about the Court. The remote parts of the kingdom are still very averse from heresy, as Wales, Devon, Westmorland, Cumberland, and Northumberland. As the cities in England are few and small, and as there is no heresy in the country, nor even in the remoter cities, the firm opinion of those capable of judging is that hardly one per cent. of the English people is infected.' (*Cath. Rec. Soc.* I: Report to Cardinal Morone.)

Twenty years were still to elapse before public opinion, under the pressure of government action, jesuit and seminarist propaganda, and puritan agitation, began to take shape and cleave England into two passionately opposed camps. At the beginning of the reign the credal conflict was entirely subsidiary. While, therefore, it is true to say that the vast bulk of the nation were untouched by any desire to revolt from the old faith, it is equally true to affirm that they were not moved by any marked desire to defend it. The whole dispute about England's relations to Rome was beyond the lay mind. We must also remember that this was the fourth religious settlement within the space of less than a generation; and during these momentous years England had allowed herself to be bandied about from the caesaropapalism of Henry VIII to the protestantism of Edward VI and back again to the catholicism of Mary with singular indifference: showing that collective thinking on the issues involved was unformed, or that ecclesiastical questions did not deeply interest the country. Possibly also the memory of recent fanaticism, both catholic and protestant, had caused a general languishing of religious enthusiasm similar to that which prevailed during the Restoration period. All three explanations are probably true.

The only organized body capable of expressing a considered opinion on the religious policy of the government was the clergy; but their co-operation was not invited, and they were given no opportunity of influencing the course of events except, as we have seen, through parliamentary channels. Convocation, the official organ of the church, was deliberately ignored. Nor were the returned protestants better treated in this respect than their opponents. On the contrary, they were rather worse off; because being without cohesion or organization, and prohibited from preaching by the law, they could make no use of the stores of learning they had acquired abroad, and had to look on passively while the catholics freely voiced their views in convocation and defended the old order in the house of lords. Thus, on 28 February, convocation under Bonner's presidency passed a series of articles reaffirming its belief in transubstantiation, the sacrifice of the Mass, the papal supremacy, and condemning the usurped right of the laity to decide matters affecting the faith, sacraments, and church discipline. Obviously the balance seemed to be tilted unfairly against the reformers.

a Bread and wine into Body and Blood

It was to remedy this grievance as well as to afford the reformers a chance of fleshing their weapons on their opponents that the government planned a conference between both parties shortly before Easter. On 31 March a debate was staged at Westminster Hall by royal command, at which the rival theologians might discuss in public certain of the questions already pronounced upon by convocation. There was no intention, be it noted, of giving the debate anything more than an academic character: it cannot, for example, be compared with the celebrated colloquy of Poissy which the French government arranged two years later; nor was it in any sense a free interchange of opinion. No sooner were the proceedings opened than the disputants began to wrangle over the method of presenting their arguments, the catholics alleging, with some justification, that they had not been correctly informed beforehand of the arrangements. Sir Nicholas Bacon, who acted as president, ruled that, in refusing to abide by the conditions laid down, they had shown contempt of the Crown, and forthwith dissolved the assembly. A few days later two of the ringleaders, bishops White of Winchester and Watson of Lincoln, were sent to the Tower, and heavy fines were inflicted on the others. The only notable result of the conference, therefore, was to deprive the catholic cause of its most redoubtable champions and to weaken the episcopal defence in the house of lords.

The work of parliament, which received the queen's approval on 8 May, may now be conveniently summarized. By the Act of Supremacy the whole of Mary's reactionary legislation was swept away, the anti-papal statutes of Henry VIII revived in all essential points, and the supreme power over the national church vested for ever in the Crown. At the same time the Act of Uniformity restored the second Edwardian prayer book, slightly modified, as the directory of public worship. Both of these acts carried their penal codes. Refusal to take the oath of supremacy was punishable with loss of office, and any attempt to maintain by 'writing, printing, teaching, preaching, express words, deed or act' the authority of a foreign prince, prelate, or potentate, within her majesty's dominions, exposed the offender, on the third committal, to death for high treason. The oath was compulsory for all the clergy, judges, justices, mayors, royal officials, and persons taking orders or receiving degrees at the universities. The provisions of the Act of

Uniformity, on the other hand, applied to the entire community, clerical and lay alike: clerical offenders against the prayer book being liable, on the third committal of the offence, to imprisonment for life, and the laity to a fine of 12d. for every absence from church.

When Elizabeth sanctioned the Act of Supremacy she took care to point out that her position was substantially the same as her father's under the earlier supremacy statutes. She claimed, as he did, that the power vested in the Crown was merely a jurisdictional authority over the clergy (*potestas jurisdictionis*), or, as the Thirty-nine Articles afterwards expressed it, the right 'to rule all estates and degrees . . . whether they be ecclesiastical or temporal, and restrain with the civil sword the stubborn and evil doers'. This would seem to imply only a temporal authority over ecclesiastical persons akin to that exercised over the laity. Such a statement, however, is a gloss upon the statute rather than an exposition of its content. Although the supremacy did not transfer to the Crown the spiritual authority in the church (*potestas ordinis*), that is the right to minister the rites of religion, which can only spring from consecration, it did transfer, as the subsequent publication of the royal Injunctions clearly shows, the right to make ordinances touching spiritual matters. Some of the queen's counsellors went so far as to maintain that she had power under the law equal to that of the pope or the archbishop of Canterbury, extending even to the articles of the faith. There is, then, a marked discrepancy between the gloss put by the queen on the statute and the powers it actually conferred or was believed to confer. Why was this so? There seems to be only one reason, namely, a desire on the part of the government to quieten public opinion on the catholic side, which strongly resented anything in the nature of a regal supremacy by a lay person over the church. A jurisdictional authority, merely temporal in character, would not carry the same obnoxious implications. For the same reason also the title accorded to Henry VIII of 'the only supreme head in earth of the church of England', which was followed without variation by Edward VI, was altered in the Elizabethan statute to 'the only supreme governor of this realm, as well in all spiritual and ecclesiastical things or causes as temporal'. This qualitative difference in the wording, be it noted, sacrificed nothing of the substance of power, but it was intended to soften the impact of the measure on the catholic conscience, and to make the transference of ecclesiastical power to the Crown as little obtrusive as possible.[1]

The same spirit of compromise is noticeable in the prayer book. The petition to be delivered from the tyranny of the bishop of Rome and all his detestable enormities was struck out of the litany; the rubric declaring that by kneeling at the sacrament no adoration was intended to any corporal presence of Christ was expunged; and in the delivery of the sacrament to communicants the Zwinglian wording of the second Edwardian compilation was tempered by the addition of the catholic wording of the first. These were undoubtedly concessions to those who were attached to the old order; and it may be questioned whether, in view of the queen's moderation, the average Englishman was conscious at first of any marked change in the ministration of religion beyond, perhaps, the use of English in the service in place of Latin.

One other important act remains to be noted, that confirming the queen's title. Technically Elizabeth's position was somewhat dubious. By the succession statute (35 Henry VIII) she was the lawful inheritor of the crown after Mary; but an earlier statute (28 Henry VIII) had declared her 'preclosed, excluded, and barred to the claim', and this statute was still unrepealed. Her half-sister Mary had been placed in exactly the same position, but had taken steps to have herself legitimated by act of parliament. It was therefore a moot point whether Elizabeth should follow her example, or simply let sleeping dogs lie, and ground her claim entirely on the statute of 1544. Bacon, who was her legal adviser in the matter, decided on the latter course; and an act was passed establishing the royal title on Henry VIII's later statute and the queen's descent from the blood royal of England. Thus the question of Elizabeth's 'legitimacy' was ignored and continued to trouble her for many years to come.

After the passing of the Acts of Supremacy and Uniformity, the government lost no time in putting them into execution.

[1] For the philosophical basis of the supremacy see Richard Hooker, *Ecclesiastical Polity*, bk. viii; and for its legal implications, Sir William Holdsworth, *History of English Law*, i. 589–91; cf. also W. P. M. Kennedy, *Elizabethan Episcopal Administration*, I, ch. viii, and F. Makower, *Constitutional History of the Church of England*, pp. 251–9.

The bishops as protagonists of the old order were the first to be brought to the test; and with a single exception—Kitchen of Llandaff—they refused to have anything to do with the supremacy oath or the prayer book. Dogged determination to stand by their guns at all hazards was perhaps to be expected of serious-minded men who had fallen once already and been pardoned: any other course of action would have argued an inconceivable weakness of character. Consequently during the summer they were deprived of office; but since it was not the policy of the government to make martyrs of them and thereby weaken its own position, they were not subjected to the full rigour of the law. Some were detained in a condition of semi-imprisonment in the custody of the new state bishops who took their place; others were given their liberty conditionally upon reporting themselves to the council from time to time; eight were sent to the Tower; and one at least, Bonner, for whom the queen manifested an unqualified dislike, was sent to the Marshalsea, where he subsequently died. Heath, archbishop of York and chancellor of the kingdom in the previous reign, was allowed to retire to his estates in Surrey, and seems to have remained on excellent terms with Elizabeth until his death. Likewise also fared the aged Tunstall of Durham, whose adherence to the settlement the government was particularly anxious to secure. 'The recovery of such a man', said Cecil, 'would have furthered the common affairs of the realm very much.' But Tunstall was not to be won over: having come through the whole stormy period since the first attack on the church by Henry VIII, he was proof against cajolery, and remained a stalwart defender of the ancient church to the end.[1]

The courage and unanimity of the episcopate was not, however, supported by a similar attention to conscience and principle on the part of the lower ranks of the clergy. On the contrary their surrender to the government in the day of adversity was so complete that even catholic writers describe it as a *débâcle*. The government commissions that began their perambulation of the country in August returned to London in October with the comfortable feeling that 'the ranks of the papists have fallen almost of their own accord'. It is impossible,

[1] For the subsequent history of the deposed bishops see Bridgett Knox, *The True History of the Catholic Hierarchy deposed by Elizabeth* (1889), and G. E. Phillips, *The Extinction of the Ancient Hierarchy* (1906).

of course, to prove this statistically, for the simple reason that we neither know the number of priests then in England, nor is there any accurate record as to the number of subscriptions to the oath. All estimates, therefore, are the result of ingenious and elaborate guessing; and they are so discrepant that little profit can come from considering them.[1] The one indubitable fact is that the Marian pastorate in overwhelming numbers passed over into the service of the establishment without a murmur. The small minority who resisted and were deprived of their livings, or resigned their charges for conscience' sake, were either reabsorbed into the general body of the laity, or became the 'lurking' priests of later days; or they left the country for a more agreeable domicile in the Netherlands and other friendly parts.[2] It must be added, however, that the success of the government, great as it was, did not mean the total extinction of the old church. True, it disappeared as a visible organized body; but the loyalty of the priests who took the oath demanded by the government was often qualified by time-serving, and a watchful readiness to revive the past whenever fortune's wheel should take another turn. The situation in 1559 was far from irremediable. The queen might marry a catholic prince, or she might die prematurely like her sister; and who was to say that in either of these contingencies the new religious settlement would survive? Impermanence had been the lot of the previous religious revolutions: impermanence might also be the fate of this one. Hence many who accepted Elizabeth's innovations and paid lip-service to the new liturgy were buoyed up with the hope that a day would come when the old religion would again be restored to favour.

The surrender of the pastorate, however, was substantial enough to ensure an overwhelming acceptance by the laity of

---

[1] The following figures show the extraordinary discrepancy in the estimates. Camden (*Annales*) places the number of livings at 9,400 and the non-jurors at 175. H. N. Birt (*The Elizabethan Religious Settlement*) reduces the former figure to 8,000 and increases the latter to 700. J. H. Pollen (*The English Catholics in the Reign of Queen Elizabeth*) follows Birt fairly closely, but scales down the non-jurors to 600. Both writers, however, reckon between 1,000 and 2,000 unexplained 'disappearances' which may have been due to non-acceptance of the religious changes. A. O. Meyer (*England and the Catholic Church under Queen Elizabeth*) is content with the remark that out of about 8,000 clergy only 200 or 300 were deprived during the first six years of the reign. The question cannot be settled by the data available at present.

[2] According to Pollen (op. cit.) there were 68 refugee English priests in Flanders in 1566.

C

the religious changes imposed upon them. The success of the government, of course, was by no means uniform or complete, but varied greatly in different parts of the country. It soon became apparent, for example, that whereas the south, south-east, and midlands found little difficulty in adjusting themselves to the new conditions, the north and north-west harboured a good deal of 'recusancy'. It was here that the so-called 'massing priests' were to find a congenial outlet for their activities. Plying their profession precariously on the hills, or, more securely, in the houses of the nobility and gentry, they contrived to supply the needs of a spiritually backward population, whose attach-ment to the catholic faith was nevertheless the deepest thing in their lives. When the new bishops came into office and took stock of their dioceses, the reports they sent to the council of the difficulties and problems confronting them were dismal in the extreme. Pilkington of Durham, whose sphere of influence in-cluded the pestilent border region, the haunt of 'arrant knaves and stark thieves', complained that he was 'fighting wild beasts with Paul at Ephesus', and that the cleansing of his diocese was more than the labour of a Hercules. At Carlisle matters were no better; for persistent rumours of change were being blown about the countryside by priests and 'evilly disposed persons'; while the cathedral church was in a state of decay, and all the prebendaries save one were 'ignorant papists or monks'. The noblemen of Westmorland, on the other hand, repressed pro-testantism among their peasantry by threatening eviction. Lan-cashire was a veritable hot-bed of 'rampant popery'. Even in Yorkshire the reformed faith could not be regarded as secure. It should not be concluded, however, that the north was rebel-lious. The fact that eleven years of the reign passed without civil strife beyond Trent is a sufficiently valid testimony to the non-existence of anything more than passive disaffection.

Between the comparatively small minority who resisted the law and became recusants, and the great majority who con-tinued from a sense of use and wont to attend the parish churches, there was a considerable number of the laity who found it necessary to compromise with their consciences in deciding their attitude to the new religious régime. Naaman bowed in the house of Rimmon while his heart was in the temple of the true God: so too, it was argued, might a true catholic be present, from a bodily point of view, at the heretical

services of the anglican church, and yet remain a faithful
adherent of the Roman creed. This silent compromise of conscience was in conflict with a strict interpretation of the canon
law, which left no such loop-hole of escape for the tortured
conscience of the individual. The pope, the committee of the
council of Trent which considered the matter, and prominent
English catholics like Dr. Sanders and William Allen, were
unanimous in their verdict that attendance at heretical services
was a heinous sin. Yet in spite of it a sort of dual loyalty was
conceived to be possible by many catholics in England, and
became the source of much anxiety to all who had the recovery
of the country for catholicism deeply at heart. On the strength
of it some went to the anglican service, but left before the end;
others attended morning and evening prayer, but scrupled at
the communion; and others again received the 'Lord's Supper'
in accordance with the new ritual, but afterwards enjoyed the
'Lord's Body' in secret to satisfy their consciences. In other
words, the distinction was drawn between inner belief and outward conformity. In the confusion of the time all manner of
irregularities prevailed. Sometimes the same clergyman was
both anglican minister and catholic priest in one, balancing his
functions between the formal and the real needs of his flock.
But the government showed no inclination at first to inquire
into these doubtful proceedings: it was satisfied if the letter of
the law was obeyed: the private opinion of the subject was a
matter that lay outside its province. This was in keeping with
the queen's remark that she did not wish to 'make a window
into men's souls' nor to force their consciences. Time and the
slow moral attrition of repeated church attendance might be
relied upon to convert unwilling obedience into a real conformity.

It has been said also, with some justification, that the substitution of English for Latin in the service was as acceptable to
catholics as to protestants, and that this may have helped forward the settlement. The demand for the bible in the vernacular was certainly common to both catholics and protestants;
and we know that it was with difficulty that those catholics
who had possessed themselves of English bibles in Edward's
reign could be got to give them up under Mary. Nor must one
forget that one of the principal achievements of the catholic
seminary at Rheims was the translation of the New Testament

into English. The use of the vernacular in the anglican service may, therefore, have acted as an attractive force on the side of the reformation.

It is clear, however, that when the second parliament of the reign met, in 1563, the government was by no means satisfied with the progress of the settlement. Many influential people who were in a position to hinder the establishment of the church had been able to escape the oath of supremacy, either because they were not included in the classes specifically required to take it, or because the officials entrusted with its administration were lax in the discharge of their duty. The bishops, too, were much handicapped in the enforcement of the secular penalties of excommunication, their principal weapon against the froward, by the lack of support from the sheriffs on whom they depended. The only way to remedy these defects was to amend the law. Consequently when parliament assembled in 1563 two supplementary acts were passed. The first, an act for the assurance of the queen's power, increased the penalties for upholding the authority of the pope, and swept several new classes into the circle of those to whom the oath of supremacy was applicable: e.g. members of the house of commons, schoolmasters, lawyers and barristers, sheriffs, and 'all persons whatsoever who have or shall be admitted to any ministry or office belonging to the common law or any other law within the realm'. At the same time the administration of the oath was placed in the hands of the archbishops and the bishops: the lord chancellor was authorized to issue similar powers to any one he considered fit to be entrusted with it; and the court of king's bench was made the general supervisory authority. In the second, an act for the better execution of the writ *de excommunicato capiendo*, the censure of the church was strengthened by penalizing officials who stood in the way of its enforcement.

Even so, however, the irregularities could not be eradicated. It is observable that in most cases the episcopal reports on the condition of the northern dioceses take the form of complaints against the hostile or doubtful attitude of the justices of the peace. In 1564, when, for the first time, an official inquiry was made into the matter, it was found that practically half of the justices in the country were either unfavourable or openly adverse to the policy of the government; and in the north the proportion of the 'evilly disposed' to the loyal was much greater.

Clearly the private opinions and affinities of the magistracy
were a serious obstacle to the successful establishment of the
church. Nor was it possible for the government, in view of its
own moderate religious policy and the nature of Tudor local
administration, to purge the roll of the justiceship too drastically.
As late as 1578 complaints were rife that owing to the slackness
of the justices in divers counties 'not only is God dishonoured
and the laws infringed, but very evil example given to the
common sort of people'.

Let us now take a wider view. The religious revolution introduced by the legislation of 1559, however carefully prepared
and cautiously carried out, was not a thing that could be confined in its effects to England alone: it was bound to create
widespread concern throughout the length and breadth of
catholic Europe. In particular, the pope as guardian of the
spiritual integrity of christendom could not remain passive while
a kingdom that had been reunited so recently to the Holy See
again proclaimed its independence. Theoretically his course of
action was clear enough. The law of the church in the sixteenth
century, as in the thirteenth, prescribed that temporal rulers
who were guilty of maintaining heresy in their dominions
should be proceeded against by ecclesiastical censure and deprivation.[1] If, therefore, the English question had been susceptible of treatment purely on grounds of justice and religion, the
punishment of the queen ought to have followed as soon as plain
evidence of the fact convicted her of malevolent intentions
towards Rome. It is noteworthy that Paul IV was ready to
move in this direction as early as the spring of 1559. But in
point of fact the pope was not free to act on spiritual grounds
alone. Besides launching the ecclesiastical thunderbolt, he had
to consider how it might be made effective. Expediency, practicability, and the convenience of the secular power on whom
would fall the burden of carrying out the will of the church had
to be consulted; to pronounce sentence and find no suitable
means of executing it would reduce disciplinary action to a mere
absurdity. Moreover, there was the mortifying fact to be taken
into account that in the sixteenth century princes no longer
stood in the same dread of papal bulls as formerly. Henry
VIII's excommunication had shown that the foundations of

[1] Pollen, p. 49.

secular sovereignty were too secure to be shaken by a discharge of the papal artillery. It may be doubted, too, whether the papacy was not now afraid of its own weapon, and inclined to hesitate before using it, however great the provocation. But the principal danger, so far as the pope was concerned, lay not so much in the promulgation of the bull of excommunication as in its execution.

In virtue of his position as 'King Catholic' and mightiest catholic monarch in the world, Philip II of Spain was the predestined champion of the old order. Of his anxiety to save catholicism in England from destruction there is plenty of evidence in the documents. But worldly cares and the conviction that a conflict with England would redound to his own disadvantage in the struggle for power in Europe made him unwilling to sponsor violent measures against the English queen. Consequently his constant advice to the pope was to delay extreme action and trust to the diplomatic means which he himself could employ through his ambassador in London. Thus while the English catholics living in exile on the Continent, the imprisoned bishops in England, and many in high position in the catholic church were in favour of immediate corrective measures regardless of consequences, Spanish influence at Rome never ceased to counsel patience and restraint. This conflict of opinion may serve to explain the wavering and uncertain attitude of the papacy towards Elizabeth during the early years of the reign, and that curious mixture of menace and suavity, of energy followed by supineness, that prevented both Paul IV and Pius IV from taking the only decisive step capable of retrieving to some extent the downfall of the old religion in England.

A contributory cause of the indecision of the popes might perhaps be discovered in the prevalent misconception at Rome and elsewhere as to the real state of affairs in Elizabeth's dominions. The general tenor of reports reaching the Holy See from interested catholics was that things were not so bad as they appeared to be: that the English people were overwhelmingly loyal to the ancient faith: that the religious revolution was the work of a handful of greedy nobles, apostate priests, and adventurers who, having possessed themselves of power and poisoned the queen's mind, had imposed their will on the country; but that Elizabeth herself was not hostile to the church, and might even be induced to restore it, if only she were freed from the evil advice of her councillors. Spain as well as Rome was the victim of this hallucination; and such was the queen's ability to dissimulate her true sentiments that she succeeded in keeping up the convenient illusion for a considerable time. Discreet 'asides' in conversation with the Spanish ambassador, the use of candles in her private chapel, affectation of innocence on points of doctrine, &c., were all pressed into the service of her artful make-believe.

Nevertheless, in spite of all countervailing influences, no less than three major efforts, and several subsidiary, were made by Pius IV during the years between 1559 and 1565 to pave the way for a catholic revival in the country. In the spring of 1560 he dispatched Parpaglia, abbot of San Salvatore, with a letter to the queen adjuring her to return to the bosom of the church. If this envoy did not reach England but found himself stopped by the Spanish authorities at Brussels, the blame for the failure rests not with the pope but with the king of Spain, who represented that the time was unpropitious. Parpaglia was therefore recalled to Rome at the end of the year.[1] The following spring, after the bull announcing the early summons of a general council (the council of Trent) had been published at Rome, another papal nuncio, the Abbot Martinengo, was sent to England, with instructions to invite the queen to send representatives. But again the effort failed. In common with the German protestants, England refused to have anything to do with the council, and the envoy was denied access to the realm (May 1561). The reasons for this drastic decision—which, incidentally, was largely the work of Cecil and Bacon—were that the queen had not been consulted as to the summoning of the council, that the council would not be 'free, pious, and christian', and that the pope was seeking at the very moment of the invitation to stir up trouble for the government in England and Ireland. It was not, therefore, through Spanish influence that Martinengo was turned away from our shores: on the contrary, the Spanish ambassador, De Quadra, did what he could to win acceptance for him; but after the rebuff was administered, Spain played an important part in warding off

[1] Parpaglia was a bad choice: he was regarded in Spanish circles as a partisan of France; only eighteen months before, he had been expelled from Flanders for espionage; and, as a friend of Pole, he was certain to be disliked by Elizabeth.

the retaliatory blow of the pope, which in normal circumstances would have been a bull of excommunication. Nor was this all. After the council met at Trent, and the question of the queen's excommunication was submitted to it in a memorandum from the English catholic refugees at Louvain (June 1563), it was through the influence exerted by the Habsburg rulers, the Emperor Ferdinand and Philip II, that nothing was done to give effect to the petition. They pointed out that such a step was dangerous to the peace of Europe and therefore to the hope of re-establishing credal unity. After these outstanding failures Pius contented himself with indirect attempts to restore relations with England by means of individuals like the cardinal of Ferrara, Thomas Sackville, and the Italian Gurone Bertano.[1] But the queen resisted all advances.

It will be apparent, then, that if catholicism in England succumbed to the government the explanation must be sought for not only in a lack of morale among the Marian pastorate, but also in the impotence of their friends abroad to render effective aid when it was most needed. But Philip II must bear by far the heaviest share of the responsibility for what happened. Although he claimed to be the patron and protector of the English catholics, his attitude was that of an obstructor rather than of a helper. He would neither allow others to lend assistance nor himself intervene materially to remedy the situation. The only attempt at friendly intervention came not from him but from his kinsman Ferdinand, who, at the pope's request, indited a letter to Elizabeth, in 1563, urging her to modify her laws in favour of the catholics, to the extent at least of granting them freedom of worship in specified places. Needless to say, this appeal also met with a firm, albeit courteous, refusal.

We have seen how the religious settlement was reached, and have discussed its nature and success: it remains to be seen how the church was reconstituted.

By November 1559 the last of the Marian bishops had been removed from office and the filling of the vacancies began.[2] On 17 December Matthew Parker was consecrated and installed as

---

[1] On the subject of these missions see C. G. Bayne, *Anglo-Papal Relations, 1558–66*, and Pastor, XVI. vii.

[2] The installations were not complete until 1562: London, Ely, Bangor, Worcester (21 Dec. 1559); Salisbury, Lincoln, St. David's, St. Asaph's, Rochester, Bath and Wells, Lichfield and Coventry, Exeter, Norwich (1560); York, Winchester, Peterborough, Carlisle, Durham, Chester (1561); Gloucester (1562).

primate at Canterbury. Elizabeth and Cecil had long marked
out Parker for the onerous and responsible task of piloting the
half-manned ship into port, but it was with considerable reluctance that he would consent to undertake the office. He was
over fifty years old, he said; he suffered periodically from the
quartan ague; his voice was somewhat decayed; he was lame
of a fall from horseback. Above all, being a man of backward
and retiring disposition, he pleaded that he could never hope
to influence the world by his personality, though he might hope
to do so by his pen. He therefore asked to be allowed to remain
in obscurity. But Elizabeth and Cecil were imperative, and
Parker, betraying all the shrinking diffidence of the scholar, was
dragged into the theatre of public life. Perhaps the government
knew their man better than he knew himself; the archbishop
was indeed a remarkable man.[1] Although cut out by nature
for the university and the school rather than public affairs, he
did not lack courage nor practicality of mind. By all accounts,
too, he was a model of wisdom and piety and generosity. 'How
prudent, how grave, how learned he was,' wrote one person,
'that one would think that Cato or Quintus Fabius lived again
in him.' These were the qualities the government wanted. A
church founded on a compromise, whose path was to be a *via
media*, required a pilot whose mind and character were resolute,
albeit opposed to all fanaticism and recklessness. Parker had
an excellent tradition behind him. He had been one of the little
society, 'little Germany' as it was called, which gathered round
Bilney at Cambridge in the earlier days of the Renaissance; he
was a thorough-going believer in schools and universities and
learning; and he had personally known the great protagonists
of the English Reformation, Cranmer, Bucer, Latimer, and
Ridley. Above all he had no taint of presbytery, which marked
most of the prominent protestants in 1559, who had spent years
on the Continent.

The archbishop's task was one of colossal difficulty. The
situation as revealed by the ecclesiastical commissioners and the
new bishops was chaotic in the extreme. Many parishes had no
clergyman; and out of the few who administered the sacraments

---

[1] He had already had considerable administrative experience as master of
Corpus Christi College, Cambridge (1544), and vice-chancellor of Cambridge
(1545, re-elected 1549). He had also been chaplain to Anne Boleyn (1535) and
to Henry VIII (1537), and dean of Lincoln (1552).

there was scarcely one who was both able and willing to preach the word of God. 'Incredible ignorance and superstition' prevailed among the people. Doubtless the desolation which had settled down on the church is partly to be explained by the deprivations or voluntary withdrawals which had removed the better and more conscientious clergy; but there were many causes. The universities, which, in the last resort, maintained the level of intelligence among both clergy and people, had suffered practically continuous decay since Henry VIII's time. The confused revolutions of the preceding generation had depraved the tone of religious life everywhere. Moreover, the church had failed to keep pace with the growth of the population and the shifting of the centres of population. Cities, market towns, and boroughs where the greatest concourse of people was, writes a contemporary, were destitute of learned men, and many were lapsing into pure heathenism. To some extent, also, the problem was the creation of the protestant conscience. Under a catholic hierarchy the so-called 'superstition' of the people, the 'rank ignorance' of the clergy, the lack of 'reverence' for church buildings and the other 'evils' which bulked so large in the reports of the new episcopate, might have caused little stir. It was a condition into which the catholic church fell from time to time, to be rescued by a reactionary wave of reform whenever the public conscience was deeply touched. But with the establishment of protestantism these things took on a new colour; for the protestantizing of the church meant not merely a change in the personnel of the pastorate, but also a change in the religious attitude of the community. The new service with its greater inwardness of appeal, which took the place of the ritual of the Mass and the more outward and symbolic observances of catholicism, demanded, if not an increased intelligence on the part of both clergy and laity, at least a new point of view. And this was the root of the matter to men like Parker. The queen might be satisfied with outward conformity and leave the conscience alone, but to her bishops the problem presented itself in exactly the opposite form. If protestantism was to have a foothold in the country it was necessary to alter the religious outlook of the nation. This, briefly, was the work the new governors of the church had taken in hand. But owing to the indifference of the queen, its accomplishment was slow.

The establishment of the church was completed, so far as

Elizabeth was concerned, by the issue of the royal Injunctions of 1559. For any subsequent effort to give it a real corporate and self-conscious existence, the archbishop had to depend upon his own authority and assume the responsibility: it was sufficient for the political purposes of the government that the papal control should be broken and the superstitious practices of the middle ages abrogated. The consequence was that for some years there was neither a distinctive doctrine nor an effective discipline in the anglican church. At first the only legal standard in matters of belief was the subscription to the oath of supremacy by the clergy and church attendance by the laity. The Act of Supremacy, it is true, prescribed the Scriptures, and the decrees of the first four general councils, as the test for heresy; but this was the common heritage of all christians, and cannot be compared with the Confession of Augsburg and the other elaborate formularies that gave individuality and cohesion to the protestant churches abroad. The *Book of Homilies* published by Parker in 1562, for the use of the pulpit, carefully refrained from mentioning points of doctrine and confined itself to moral precepts concerning the heinousness of rebellion, the reverence due to church buildings, excess of apparel, and other matters of a socio-political character. It was not until Convocation passed the Thirty-nine Articles in 1563 that the church acquired a definite body of doctrine; and even then, the fact that the Articles were not given legal sanction for another eight years, i.e. after the papal bull made the breach with Rome irreparable, would seem to indicate that the government was averse to any statement of the church's position which might hamper such moves as the political situation demanded.

On the disciplinary side, too, the reconstruction of the church was accompanied by great difficulties. Parker found that his freedom to exercise authority was weakened by the lack of support from the queen, and by open or secret obstruction from certain of her councillors. It was only after extensive disorders had broken out, too glaring to be passed over, that he ventured to issue his *Book of Advertisements* (1566); and he was careful to temper his regulations with the remark that they were merely for the preservation of order and decency in the church service. Thanks to his efforts, however, there gradually emerged a national church with a distinctive service, a distinctive creed, and a distinctive clergy.

Meanwhile, in 1562, Bishop John Jewel of Salisbury published his *Apology* in order to explain and justify the principles of the settlement. This celebrated work, although it lacked the balance and repose of Hooker's greater apology a generation later, remained the basis of the anglican reply to catholic controversialists throughout the period. Jewel took his stand on the theory already embedded in the Acts of Supremacy and Uniformity, and contended that the archetype of the anglican church was the christian church as it existed prior to the establishment of the universal power of the Roman bishop, and of the elaborate ritualistic accretions of the middle ages. 'We have planted no new religion,' he wrote, 'but only have renewed the old that was undoubtedly founded and used by the apostles of Christ and other holy fathers in the primitive church, and of this long late time, by means of the multitude of your traditions and vanities hath been drowned.' The importance of the argument lay in the fact that it boldly appealed to the church of the first six centuries as the standard of Christian belief and practice. On this firm historical foundation the Anglican defence rested until the 'judicious' Hooker still further strengthened it by introducing the appeal to pure reason. Meanwhile the *Apology* found its way, either in its original Latin garb or in translations, into every European country, and gained widespread recognition as one of the most notable books of its time.[1]

The coming into force of the anti-papal legislation, and the diocesan visitations of the ecclesiastical commissioners, were accompanied, particularly in London and adjacent parts, by an ebullition of public feeling against the church; and a considerable amount of looting, burning, and destruction took place. In the autumn of 1559 'papist gear' of all kinds—roods, crucifixes, images of Mary and John, copes, altar-cloths—were pillaged from the churches, piled up in the streets, and given to the flames in what has been called a protestant Bartholomew. Possibly some degree of violence was unavoidable on the occasion; but the appetite for destruction once whetted did not confine itself to 'things savouring of superstition', whose abolition the queen sanctioned by her Injunctions. It passed into an

---

[1] The publication of the *Apology* led to a furious controversy (1564–7), no fewer than forty volumes of catholic polemic being written in reply by those who had settled in the Netherlands. The revolt in the Low Countries extinguished the controversy.

insensate *furor* of vandalism, in which valuable monuments and family memorials were defaced. To meet this contingency a proclamation was issued (Sept. 1560) prohibiting the destruction of monuments of antiquity set up in the churches for memory and not for superstition. But much irreparable mischief had been done by mob violence before it could be arrested. Patrons of cures appear to have been no less rapacious. Responsible churchmen like Bishop Jewel speak of them as 'latrones', robbers, rather than patrons, and complaints were frequent that valuable church property disappeared, to say nothing of the plunder of bells and lead roofing. Even the attitude of the court savoured of vandalism, though in a more subtle way. Elizabeth took over the most eligible of the 'lands, manors, tenements, and castles' belonging to the bishoprics in exchange for the much less valuable 'tithes and parsonages impropriate' which had been annexed by the Crown at the dissolution of the abbeys.

Many protestations were addressed to the queen to prove that the step was both iniquitous and dangerous. Would it not lead directly to the decay of learning? Were not bishoprics rewards for studious men? Did not the bishops voluntarily maintain scholars at the universities out of their revenues, and how could they do so if they were cramped? Besides, a married clergy needed a larger income for purposes of hospitality and the education of their families than the old celibate prelates. But however good the arguments, the installation of the primate and the consecration of the bishops elect was delayed until the lands had been surveyed and the exchange carried out.

The effect of all this was incalculable. Destruction and desecration of sacred things brought about a general growth of irreverence in both the destroyers and those who watched the disappearance of the symbols they held dear; and it is questionable if the public mind recovered from the shock to which it was subjected. 'It cannot cause surprise', says a well-known writer, 'that to many people, and those the best, even the abuses of the old system were more dear than the reforms of the new'.[1]

[1] W. H. Frere, *The English Church in the Reigns of Elizabeth and James I.*

II

ENGLAND AND FRANCE

MEANWHILE Elizabeth had made her first experiments in foreign policy with a boldness that hardly corresponded with her slender resources.

For some weeks previous to her accession commissioners representing France, Spain, Savoy, and England had been engaged, at Cercamp, a Flemish village near the French border, in winding up the last of the great wars between the house of Valois and the house of Habsburg. War-weariness, financial exhaustion, and indecisive results rendered it comparatively easy for the two chief belligerents, France and Spain, to settle their differences on the basis of a mutual withdrawal from conquered territory, and a return to the *status quo ante bellum*. But the fate of Calais, which England had lost to France in the course of the struggle,[1] still hung in the balance, delaying indefinitely the conclusion of a general peace. It was a matter that admitted of only one rational solution, namely, the formal cession of the town to its captors; for the time had come when England must cut herself adrift from territorial commitments on the Continent, and launch out on her true destiny as an island kingdom. The wounded pride of a defeated nation, however, would not brook such a solution. The loss of Calais, over whose walls the English flag had floated for two hundred years, whose gates once bore the haughty inscription:

> Then shall the Frenchman Calais win
> When iron and lead like cork shall swim—

had created as much excitement in London as, let us say, the loss of a place like Gibraltar might be expected to produce at the present day. So much glamorous sentiment had attached itself to this 'brightest jewel in the English crown', as the saying then went, that its recovery by diplomatic means would have done much to mitigate the gloom that surrounded the death-bed of Mary Tudor. But France was adamant: under no circumstances would she include Calais in the general programme of restitutions; rather than surrender it, said the French king, Henry II, he would hazard his crown. So Mary

[1] 7 Jan. 1558.

went to her grave with Calais written on her heart; and Elizabeth found herself confronted with a bad debt, which prudence suggested should be abandoned at the earliest possible moment, but the recovery of which might well prove to be a splendid augury for the success of the new reign, an earnest of better things to come. Naturally she resolved to make the recovery of Calais the prime object of her diplomacy, and a proof that the vindication of the national honour was her deepest concern. She could hardly have acted otherwise.

But the quest was hopeless from the start. The French were in effective occupation; they had won the place in fair fight in a legitimate war; its acquisition completed the territorial unity of the kingdom, and ended once for all the medieval anomaly of an English settlement in the midst of a community homogeneously French. Nothing, therefore, would induce them to reconsider their decision, although they were willing enough to fall in with any verbal subterfuge whereby England's humiliation might be softened. The duke of Guise, whose prowess had resulted in the original capture of the town, only voiced the opinion of his fellow countrymen when he stated that he would rather sacrifice a hundred thousand lives than evacuate it. Against this stubborn attitude all juridical and canonical arguments beat in vain. Nor could King Philip II of Spain, in whose interest England had entered the war, prevail upon France to abate one jot of her demand. His desire to see the English again in possession of Calais was undeniable.[1] He was in honour bound to do what he could for his unfortunate ally and fully cognizant of his obligation; it was also important that a place of such strategic value for his dominions, commanding, as it did, the sea route between Spain and the Netherlands, should be in friendly hands. Besides, there was the over-riding consideration that Spanish *Weltpolitik* and the maintenance of the existing balance of power in Europe required the continuance of Anglo-Spanish co-operation. But Philip was not a free agent. His treasury was empty and his need for peace very great.[2] He was worried with the many distractions incidental to a ruler whose empire reached from the North Sea to

---

[1] 'As the English entered into this war for our cause the treaties which bound them to do so also bind us not to treat without them, and we are determined to fulfil this obligation and to conclude no peace without their consent.' (The king to de Feria, 28 Dec. 1558, *Span. Cal.* i.)

[2] A. de Ruble, *Le Traité de Cateau-Cambrésis*, pp. 8, 9.

the Mediterranean, from the confines of Germany to Peru. The Turkish fleet had made an onslaught on Minorca; the Moors had inflicted a serious reverse on the Spanish forces in Morocco; the plague of heresy was making headway in the Netherlands and even in Spain itself—in short the position, as the Spaniard saw it, was so desperate that ruin stared him in the face if the peace negotiations miscarried and war broke out again in northern Europe. All things considered, then, King Philip could not give England unconditional and unlimited support in her demand for Calais; nor, indeed, was he prepared to continue his support unless he had ample assurance that the change of government in London would not mean a change in the direction of English foreign policy. If, however, he could feel sure that England under the new régime would remain, as before, the client state of Spain: if, better still, Elizabeth would marry him and keep her hands off the catholic religion, he would do his utmost to give her satisfaction in regard to Calais; otherwise he would be compelled to let matters take their course. These conditions were perhaps reasonable enough from the Spanish point of view; but any hope the king may have had of their acceptance was rudely dashed by the incalculable audacity of Elizabeth. Politely but firmly she turned aside his offer of marriage and pursued her preparations for the religious revolution in England uninterruptedly. Thus Philip's interest in Calais gradually faded. In March 1559, after the conference had resumed its meetings at Cateau-Cambrésis, he announced the transference of his matrimonial offer from Elizabeth of England to Elizabeth of Valois, eldest daughter of Henry II— a plain intimation that the conclusion of a Franco-Spanish peace was not far off.

There was now nothing left for Elizabeth to do but to get the best terms she could from France; and so, on 2 April, she compromised on the following conditions. Calais was to remain in the possession of the French for eight years, at the expiry of which it was to revert to England, under penalty for default of half a million crowns. The French undertook to send hostages as a token of their honourable intentions, and the question of ultimate sovereignty was deferred to the end of the probationary period. It was also stipulated—an important reservation, as we shall see—that the fulfilment of the bargain should be dependent upon mutual abstention from acts of aggression in the meantime, and that Scotland, as France's ally, should be in-cluded in the protective guarantee. Under this transparently face-saving device was Calais virtually ceded to France—not in intention but in effect; for no one doubted that the French government would find a pretext, under the clause of reserva-tion, for retaining the place when the allotted period ran out. On the other hand, events were soon to show that Elizabeth, for her part, did not hold herself bound by the treaty, but would attempt to recover Calais by force or stratagem as soon as a fair opportunity offered. The morality of governments in relation to international obligations was proverbially low in the six-teenth century, and treaties were regarded as binding only in so far and for so long as they subserved the interests of the states that made them. Of such a nature was the Anglo-French section of the treaty of Cateau-Cambrésis. Mutual distrust, rancour, and unfulfilled hopes dogged it like a shadow.

Elizabeth had good reason to be suspicious of the French. From the first, their attitude to her accession had been one of ill-concealed hostility. Already the dauphiness had quartered the English insignia on her shield along with those of France and Scotland; and more than once during the discussions at Cateau-Cambrésis the malevolence of the French delegates had broken through the thin crust of diplomatic suavity. Rumours were also current that Henry II was using his influence at Rome to draw the papal thunders against the daughter of Anne Boleyn, and that preparations were in train for an invasion of England. The tension was probably increased by the pub-lication of the bull, *Cum ex apostolatus*, on 16 February; for although Paul IV did not mention any ruler by name, he declared that all sovereigns who supported heresy in their dominions fell from their right by the mere fact of their heresy. Circumstances therefore seemed to point to the likelihood of a French crusade on behalf of Mary Stuart, and de Feria thought it necessary to warn his master on the subject, on 11 April. But Henry II was too circumspect a prince to risk a fresh European conflagration on the morrow of a definitive peace settlement. So long as he lived the Stuart claim to the English Crown would be no more than a tacit threat.

In July, however, fate took an unexpected hand in the game. The king died suddenly of a lance-thrust at a tournament held in honour of his daughter Elizabeth's marriage to Philip II; and the crown passed to the dauphin, Francis, and his consort, Mary Stuart. The house of Guise was now dominant in France. The queen's uncles, the duke of Guise and the cardinal of Lorraine, formed along with the chancellor a secret cabinet for the direction of national policy.[1] All offices in the patronage of the Crown were at their disposal, and were filled with their partisans. In Scotland another member of the same family, Mary of Guise, ruled in trust for her daughter. Is it surprising that Elizabeth should view with grave concern the amassing of power by so formidable a group, whose enmity to herself was the common talk of Europe, and who put family aggrandizement before all other considerations? What was to hinder them, now that they had the resources of both France and Scotland at their feet, from seeking to establish a Guise ascendancy over England as well? The lure was great: their niece's claim would furnish a colourable pretext for the enterprise; and Scotland would make an admirable *point d'appui*.

A French intervention in Scotland was indeed overdue, if the protectorate established by Henry II was to be maintained; for the country was in a state of appalling disorder, and administration had become difficult. The Scots, whose love for their 'auld' ally had never been a positive sentiment nourished by community of culture, but an artificially created affection resting on the negative basis of hatred to England, had given way to a francophobia. The Frenchman was too much in evidence everywhere—in the high seat of government, in society, in the army; and everything seemed to point to his making himself master of the country, to the destruction of its independence. In many respects the position tallied with that which obtained in England shortly before Elizabeth's accession. Just as the London populace cold-shouldered the Spaniard in the streets, so the Scots rabbled the bodyguard of M. d'Oysel in the streets of Edinburgh. The government of the regent was watched with suspicion and distrust, and her efforts to enforce the laws was met with the cry of outraged liberties. 'And now that it is a question of my determination to see justice take a straight-

---

[1] 'Before the "Council of Affairs" assembles, the Cardinal of Lorraine, the duke of Guise, with the chancellor meet first, either in the king's chamber or in that of the queen mother, where they discuss privately all matters of the greatest importance, without communicating their deliberations to others.' (*Venetian Cal.*, 1558–80, Tiepolo to the Doge and Senate, 16 July 1559.)

forward course, and they find me a little severe,' wrote Mary of Guise to her brother the duke, 'they will not endure it, and say that these laws are the laws of the French.' Racial antagonism, however, was of small moment compared with the vast social and political upheaval caused by the growth of protestantism. After a decade of propaganda, persecutions, and martyrdoms, the followers of the new doctrines had secured the support of an influential section of the nobility, and banded themselves together in a covenant for the defence of the gospel, under the title of the 'Congregation'. In 1559 the movement was greatly strengthened by the fiery apparition of John Knox, who brought with him to Scotland a militant Calvinism excelling in severity anything that had hitherto appeared in Europe. The whole structure on which the religious, political, and civil life of society had rested for centuries was suddenly shaken to its foundations, and a fury of destruction, popular in origin, laid waste ancient historic churches and abbeys at Perth, Scone, Stirling, St. Andrews, and Edinburgh. At first the regent, actuated by the desire to save the French alliance from ship-wreck amid the clash of religious passions, had tried to steer a middle course between the reformers and the catholic church; but the inflammatory preaching of Knox, and the vandalism of the 'rascal multitude' which it provoked, caused her to use her power on the side of the existing order, and to arraign the reformers as rebels. Although she had no option in the matter, the step was fraught with the gravest consequences.

If the catholic church in Scotland had been worthier of support, or if the secular arm had been longer and stronger, matters might have ended differently. But in reality the rotten ecclesiastical fabric, long since known to be tottering and in need of drastic reform, merited most of the censures showered on it by Knox and his party; and the resources of the regent were totally inadequate for its defence. Hence the reformers boldly defied her authority, and sought to paralyse any coercive measures she might take against them by recourse to arms. Safety, as Knox said, consisted in numbers. Nay, he went farther, and in the name of a most revolutionary political philosophy, deduced from the Scriptures as a precept of Almighty God, proclaimed that a wicked ruler setting herself up against the godly might with divine sanction be deposed. Being also an idolator—as all worshippers at Mass appeared to him to be—she might lawfully be put to death; for did not the Scriptures plainly command that idolators should die the death?[1] Thus a holy war was begun against France and the catholic church, and protestantism became identified with patriotism. The die was cast when the Congregation, after vainly attempting to arrest the military precautions of the government by a peremptory demand to stop the fortification of Leith, formally declared the regent deposed at the market cross of Edinburgh (Oct. 1559).

In the ordinary course of events, and granted free access to the port of Leith, a powerful military state like France should have had little difficulty in suppressing the rebellion. Ill-armed peasantry and burghers, without experience in war, were no match for infantry trained in the tradition of Bayard and Gaston de Foix. Herein, precisely, lay Elizabeth's peril. She could not help the Scots openly because she was pledged to a policy of non-intervention by the treaty of Cateau-Cambrésis; and yet, if the French conquered them, the next step might be the invasion of England. Clearly something would have to be done to checkmate France before this danger materialized. Subsidies and munitions might, for example, be smuggled into Scotland from Berwick: the earl of Arran might be brought back from France, in order to give the rebellion a show of legality and a titular head: an army might be mobilized in the northern English counties, as a precautionary measure, and to lend moral support. All these things Elizabeth did. But the situation in Scotland grew steadily worse as the autumn wore on, and the subsidies were as good as wasted. Not only were the Scots defeated in the field, but defeat greatly damped the fiery ardour which, in the first flush, had brought men to the standard in such numbers that it 'rained steel bonnets'. Sorrowfully Knox had to admit that the arm of the flesh was a failure. 'The whole multitude standen in such doubt', he wrote to Cecil, 'they cannot tell to which party they shall incline.' 'For God's sake', wailed Cecil to Sir Ralph Sadler at Berwick, 'comfort them to stand fast.'

The military weakness of the Scots was bad enough, but not so serious as their lack of a navy. The North Sea in winter might be a fickle friend to the French, as they found to their

---

[1] Knox's 'murderous syllogism'. He relied on the Old Testament denunciations of the Amalekites, and Elijah's slaughter of the prophets of Baal.

cost in December, when two expeditions bound for Leith with men, guns, and stores, were blown back and wrecked, only a few stragglers reaching the haven in safety. Nevertheless, it was confidently stated in France that by Candlemas there would be an army of ten thousand on Scottish soil; and in any case these disasters did not greatly hinder the plans of the garrison at Leith. On Christmas Eve, a detachment crossed the Forth to Burntisland, and set out on a rapid march to St. Andrews accompanied by their ships. The Congregation, whose numbers had been much thinned, owing to the season of the year, could do nothing to arrest their progress. Obviously the time had come for Elizabeth, if she wished to save the Scottish cause from destruction, to throw off the mask of neutrality with which she had been vainly trying to delude the regent.

For weeks she had been watching the darkening prospect with an anxious eye, fully aware of the extensive military preparations in France and the bellicose utterances of the Guises, but reluctant to commit herself to war because of doubtful reports from the Netherlands as to Spain's attitude, and her own unpreparedness. However, she had mobilized 4,000 men beyond Trent for service on the border under the duke of Norfolk and Lord Grey de Wilton, and—more important still— had instructed Admiral Winter to take a fleet to the Forth and 'impeach' the French in whatever way he could, provided he acted on his own responsibility. Then at length, on 24 December, the privy council resolved upon open intervention in Scotland, and instructions were sent to Norfolk on the following day. He was to proceed in aid of the Scots with the utmost care and deliberation, and not to unleash the army under Grey's command until he was reasonably certain that the regent would not agree to the removal of the French troops by negotiation, and that the combined forces of England and Scotland were sufficient for the enterprise. Five days later, these orders were countermanded, the duke being informed that he must limit his action to sending secret succours—captains, powder, shot, and guns—into Scotland, and to rely mainly on Winter and the fleet.

Meanwhile the admiral, who left Queenborough on 27 December, was slowly making his way northward through heavy weather to strike the first blow against the French. It took him three weeks to reach Berwick; but on 23 January he was safely in the Forth. Fortified by words of encouragement

from Norfolk, and more detailed instructions, he sailed boldly
for the narrow passage between Queensferry and Burntisland,
which formed the line of communication between Leith and
the French army in Fife. The effect was immediate. Hardly
had his guns begun to speak at Inchkeith than the campaign
on the north side of the Forth came to an abrupt termination.
Seeing the English in command of the sea, and fearful of being
marooned in Fife, the French abandoned their ordnance, set
their stores in a blaze, and retreated swiftly by the circuitous
route of Dunfermline and Stirling to their base on the southern
side of the firth.

By this time Norfolk and Grey were getting into position on
the border; but the formalities of a treaty arrangement with
the Scots delayed action for a month. Eventually, on 27
February, an offensive and defensive alliance was made between
Norfolk and the Scottish lords, whereby Elizabeth took over the
protection of Scotland, its civil and religious liberties, so long
as the marriage of the queen of Scots and the French king
endured, and one month longer; the Scots undertaking, for
their part, to aid England with 2,000 foot and 1,000 horse in
the event of an invasion by France. The object of the alliance
was not to abrogate the sovereign rights of Queen Mary—these
were safeguarded by the treaty—but to prevent the overthrow
of Scottish independence by the French.

Elizabeth, however, did not yet see her way clear to give the
treaty its military expression. She had to keep a wary eye on
the Netherlands, where the Spanish troops employed in the
late war were still under arms, and the viceroy, Margaret of
Parma, was urging the king to use them for the preservation of
European peace.[1] At the same time Granvelle, president of the
Spanish *consulta* at Brussels, took the English ambassador, Sir
Thomas Challoner, to task for his mistress's rashness in provoking a war for which she was utterly unprepared. 'Is it not
strange', he asked, 'that ye believe the world knoweth not nor
seeth your weakness? . . . Is there one fortress or hold in all
England, that is able for one day to endure the breath of a
cannon?' 'The cock may scrape so long in the dunghill', said
another Spaniard in high position, 'till at last he discovereth
the knife to cut his own throat.' Challoner was much perturbed by what he heard, but more so by the hostile chatter in

---

[1] The Spanish forces were not recalled to Spain until the autumn of 1560.

Spanish military circles, where intervention was regarded as certain. He therefore wrote to Cecil: 'Whiles two cocks fight, beware the still cock, looking on and taking breath, when he seeth them tired, do not set upon both.' Very different, however, was the tale which Sir Thomas Gresham had to tell. Gresham was Elizabeth's financial agent in the Low Countries, and had his finger on the pulse of the merchant community, as well as access by his extensive system of espionage to the innermost secrets of the *consulta*. He therefore knew just how much or how little importance to attach to rumours of Spanish bellicosity. He had discovered among other things that the royal finances were so chaotic, and the royal credit so low, that it was morally certain King Philip could not raise a finger against England; that all the gold of the Indies would not suffice to extinguish his mountainous debts; and that the Flemings who lived by commerce with England would not tolerate a war between the two countries.

Gresham's exploits at this time were literally amazing. Keeping an open house and a bounteous table, he dined the great financiers of Antwerp, the Fuggers and the Schetzes:[1] 'dried up' the Bourse by the loans he contracted, and 'robbed the city of Antwerp of all its fine gold and silver': bribed the customs officers of the port, especially the chief searcher, and was able by their connivance to snap his fingers at proclamations prohibiting the export of arms and munitions. Thus, under the very eyes of the government, consignments of the precious metals and shiploads of ordnance, arquebuses, morions, corselets, pikes, dags, and gunpowder left the quays at Antwerp for the Tower of London. By the aid of his accomplices he even plundered the royal arsenal at Malines. 'I will confess here', he wrote to Cecil, 'that by trickery I have drawn 2,000 corselets from the royal arsenal at Malines and they are now in England. You will understand that their disappearance has created no small stir among the officers.'

Meanwhile, very heartening news was coming in from Sir Nicholas Throckmorton, the English ambassador in Paris, who began to allude, as early as 4 February, to 'factions in religion springing up everywhere' in France, and urged the queen to 'beat the iron while it is hot'. 'Many wise men', he wrote, 'agree that now is the time and good occasion offered to abate

[1] Gaspard Schetz was *trésorier-général des finances* at Brussels.

the French pride.' About the same time Christopher Mundt transmitted startling news to Cecil from Strassburg about a conspiracy hatched in France against the Guises, of which, he remarked, 'some of the first rank in France are cognisant'. Then came the actual outbreak of the Tumult of Amboise, on 16 and 17 March, the attempted capture or assassination of the duke and cardinal of Guise, and the temporary paralysis of the government. Four days later Throckmorton was jubilant. 'The queen may now see', he wrote, 'what advantage she has over the French, who are so troubled that they know not where to levy any force. . . . It will be to her advantage to have gained somewhat of the French before the king of Spain interposes himself in these matters.'

The chance was not to be missed. On 28 March word was sent to Lord Grey to set the English army of 6,500 in motion from Berwick, and to join the Scots at Prestonpans. Thus, when a messenger[1] arrived from King Philip on 5 April, with urgent representations that an armistice should be declared and war avoided, he was confronted with a *fait accompli*: Grey had effected a junction with the Scottish army, and was on his way to the siege of Leith. The war of the insignia had begun.

It is not necessary to follow the course of the campaign. It was by no means glorious, except, perhaps, to the defeated French. The besiegers were numerous enough to contain the enemy in their entrenchments, but insufficient to carry the place by assault. Nor did they agree at first as to the objects of the war. The Scots demanded the deposition of Mary of Guise, the evacuation of Scotland by the French, and the transference of the administration to a committee of themselves. Elizabeth was interested only in the ulterior motives of the French, and the defence of England; and if she could have security by treaty she was prepared even to aid the regent in reducing her rebellious subjects to order. It was not until Mary of Guise, buoyed up by false reports of coming succours from France, refused to recognize the Anglo-Scottish alliance, or to consider a settlement, that the aims of the allies became focused on the same object—the expulsion of the French bag and baggage. Even so, however, events moved with leaden foot, and the siege was grossly mismanaged. The grand assault on 8 May ended

---

[1] De Glajon, a Fleming by birth and friendly to England, who secretly advised Cecil to continue hostilities.

disastrously for the attackers, from inadequate preparation and lack of proper siege-tackle, and was not repeated. In the end it was not military defeat but sheer starvation and want of support from France that caused the gallant garrison to capitulate. Throckmorton had been a true prophet.

The peace negotiations, which were begun at Edinburgh while the siege was still in progress, were much delayed by Elizabeth's insistence on the restoration of Calais to England and an indemnity for her losses; but eventually, on 6 July 1560, after threatening to break off the conference and reinforce Grey for another assault on Leith, Cecil compelled the French to surrender on all essential points. The treaty of Edinburgh, as it was called, solemnly reaffirmed the provisions of the previous treaty, pledged both France and England to a policy of non-interference in Scotland, handed over the government to a Scottish council of twelve (of whom five were to be chosen by the lords of parliament and seven by Queen Mary), and provided for the evacuation of the kingdom by the French and the destruction of their fortifications at Leith, Dunbar, and Eye-mouth. No mention was made of religion or of the Anglo-Scottish alliance, both subjects being regarded as too burning to be touched. In regard to the question of the insignia, however, it was agreed that Francis and Mary should abstain 'for all time coming' from using the style and arms of England and Ireland.

These were substantial results to have achieved after so indecisive a war. They would have to be ratified later by the king and queen of France; but for the moment Elizabeth might rest assured that she was safe from a French attack through Scotland. Moreover, the reformation was now securely established north of the Tweed, and a bond of common interest forged between the two kingdoms, of incalculable use to herself and the Scottish reformers in the future. Yet in spite of it all the queen was far from contented with the settlement. She bitterly upbraided Cecil for his failure to secure the town of Calais and the indemnity, and on his return to London treated him with studied disfavour, even to the extent of leaving him out of pocket for his expenses. Worse still, he learned, to his surprise, that she had no intention of following up her victory and consolidating the favourable position she held with the Scottish lords. In vain he argued that the 'union of hearts' brought about by the war should be strengthened by a policy of lavish money grants to men like Lord James Stuart, the earl of Glencairn, the Lords Maxwell and Hume, and the lairds of Lethington and Grange. In vain, also, Throckmorton pointed out, in dispatch after dispatch from Paris, that £20,000 yearly was not too much to pay for 'the assured amity of Scotland'. Neither his advice, nor his reports that France was still bellicose and unlikely to ratify the treaty, made any impression on the stubborn refusal of the queen to spend another penny on Scotland: it had cost her enough already (£133,886 4s.), and the short-distance loans contracted by Gresham would have to be repaid at an early date. Likewise she received the proffered hand of friendship from Edinburgh with a cold indifference; and when the 'lords of the parliament of Scotland' sent an embassy urging her 'to join herself in marriage with the earl of Arran' as the most effective way of establishing the security of both kingdoms, and of making herself 'the strongest prince in christendom upon the seas', she placidly remarked that while she prized the amity of Scotland she thought the two countries could work together for common political ends quite as effectively without marriage as with it. Nay, she went farther and seized the occasion to tell the embassy that the conduct of the Scottish parliament in carrying out a thorough reformation of the church and religion, since the close of the war, was jeopardizing the ratification of the treaty by the French government.

This was a dangerous line to follow; but fortune favoured Elizabeth. In December (1560) Francis II died—'of a rotten ear', said the unsympathetic Knox: the Guises fell from power: the widowed Mary Stuart was thrust into the background of French politics; and the queen mother, Catherine de Médicis, assumed the regency during the minority of the new king, Charles IX. To Throckmorton these unexpected events seemed like a crowning mercy from Heaven. 'The queen', he wrote, 'has now cause to thank God for so well providing for her security by taking away the late king.' Up to a point he was right. The new French government was not disposed to reopen the war with England, nor to fight the battle of the Guises and their niece; and within a few months Elizabeth was able to boast that 'save for manners' sake she needed no ambassador at the court of France'. But if Throckmorton had been able to pierce a little farther into the mysterious ways of providence he would not have been so optimistic—as we shall see presently.

Meanwhile let us look at what was happening behind the scenery of the political pageant. If Cecil was perturbed at the nonchalant way in which Elizabeth gambled away the amity of Scotland, he was still more perturbed, on his return from Edinburgh, to find her sunk in amorous dalliance with one who bore the most tainted name among the aristocracy of England. After solemnly assuring her first parliament that she intended to live and die a virgin, and, in the spirit of this assurance, rejecting several suitors much higher in rank—including the archduke of Austria and Eric prince of Sweden—she had apparently become infatuated with the charms of Lord Robert Dudley, son of the arch-traitor Northumberland, who perished on the scaffold in Mary's reign.[1] It may be assumed as practically certain that Elizabeth entertained the hope of marrying Dudley as soon as the way was clear: that is, as soon as he was freed from Amy Robsart, who had been his wife for eleven years. Moreover, it was believed by many that considerations of propriety, morality, or the public interest would not deter Lord Robert from using fraud or violence to attain his end, if either promised a chance of success; and rumours were afloat as early as November 1559 that he intended to poison his wife. It was noted also that he became an assiduous cultivator of the Spanish ambassador, de Quadra: a sign, presumably, that if public opinion countered him, he would carry through his project with the help of Spain. Thus the potential danger to the state was twofold. If Dudley killed his wife, or procured her death, or even if she died in circumstances that gave rise to a suspicion of violence, the queen could not escape a share of the obloquy; and if she married him thereafter, under Spanish auspices, it could only be at the price of far-reaching concessions, tantamount to a complete reversal of all she had hitherto achieved, and the renewal of Spanish influence in England.

[1] The precise nature of the liaison with Lord Robert cannot be determined by the recorded facts of history. Men of the world, taking their stand on the statements of the Spanish ambassador, will probably always believe that the queen was Dudley's mistress, in the full sense of the word. But the evidence in favour of such a view comes, it will be observed, from a hostile source, and cannot, therefore, be accepted *simpliciter*; while, on the other hand, we have Elizabeth's own assurance, given to her council at a solemn moment (Nov. 1562), when she was seriously ill of the small-pox and believed herself to be dying, that 'nothing unseemly' had ever passed between them.

Such, then, was the posture of affairs when Cecil returned from his peace labours in Scotland. He found Lord Robert in the seat of power, the queen listless and apathetic in matters of business, and de Quadra a prime favourite at court. In fact, the stage was set for the amazing drama about to burst on the world.

On 6 September 1560 the secretary, pocketing his pride, approached the Spaniard with a doleful tale of coming ruin to the state, and besought him to remonstrate with the queen on her misconduct. At the same time he communicated the startling information that a plot for the murder of Amy Robsart by poison was actually in train, and that she was already reported to be ill, although he knew that at the moment she was quite well. The very next day, on her return from hunting, Elizabeth herself told de Quadra, in confidence, that Lord Robert's wife was dead or nearly so—a strangely significant remark if Cecil was right about the state of her health. Twenty-four hours later Lady Dudley was found dead, with her neck broken, at the bottom of a stair in Cumnor Place, her husband's Berkshire residence. The news of the occurrence was not published at Windsor until the 11th, when it was represented to be the result of an accident: she had broken her neck, said Elizabeth to the Spanish ambassador: she must have fallen down a stair. If this was the official version of the tragedy, it was also the view of the coroner's jury, who returned a verdict of death by misadventure, and acquitted Lord Robert of all responsibility in the matter. But public opinion, cognizant of the long intrigue leading up to the event, refused to accept the accident theory. Before this unconventional tribunal Dudley stood condemned, and the queen was regarded as consentient to the crime.

It stands to reason that we can never know the truth about the Cumnor tragedy. It may have been a case of suicide—Amy Robsart had plenty to grieve and distract her; but, in the opinion of the most competent historian[1] who has investigated the facts, so far as they are available, the presumption is strong

---

[1] E. Bekker, *Elisabeth und Leicester* (*Giessener Studien*, v, 1890). Bekker's conclusion is that Amy Robsart was probably murdered by poison, and then her neck was broken and the body placed at the foot of the stairs, so as to give the appearance of death by a fall. Froude's comment on the episode is as follows: 'The conclusion seems inevitable, that although Dudley was innocent of a direct participation in the crime, the unhappy lady was sacrificed to his ambition. She was murdered by persons who hoped to profit by his elevation to the throne.'

that she was deliberately murdered; and, if this be so, the candid critic might reasonably conclude that Elizabeth was just as much a party to it as Mary Stuart was to the Darnley murder. The only material difference between the two cases is that Mary married Bothwell after the murder, and Elizabeth did not marry Dudley; but this was due to a difference of temperament, not of intention.

Everywhere in Europe the worst construction was put upon the deed. Either it was believed that Dudley had killed his wife in order to marry the queen, or that he was already secretly married to her. Throckmorton was so affronted with the abominable reports circulating in Paris that 'every hair on my head stareth and my ears glow to hear'. 'The queen your mistress', said the Spanish ambassador to him, 'doth show that she hath honour but for a few in her realm, for no man will advise her to leave her folly.' In Germany, where the protestant princes had begun to build their hopes on England, consternation was the prevailing note.

Now although Elizabeth might—and did—affect an attitude of bland indifference, she could not be altogether impervious to the shafts of criticism that struck her from all quarters. She realized that she had committed a *faux pas*. The marriage with Lord Robert was definitely given up, and Cecil was called in to help in the extrication of the queen and her lover from the consequences of their conduct. Once Cecil was back at the helm, the ship of state, which had been momentarily deflected from its true course by gusts of passion, was guided into safe and charted waters; de Quadra was ousted from his favoured position at court; and the queen never again allowed her private impulses to endanger the safety of the state. Marriage-negotiations in the future were entrusted to statesmen.

Meanwhile, as a protective measure against possible aggression from abroad, Elizabeth had begun her quest for allies on the Continent. During the recent war, English agents had been busy in the chief heretical centres of Europe, pleading for union and co-operation with England; and Throckmorton had shown by his dealings with the huguenots, at the time of the Amboise conspiracy, how effective the plan could be, if it were pursued with secrecy and resolution. 'As the king of Spain advances his greatness', he wrote, 'by countenancing papists in his own and

other countries, so it will be to the queen's advantage to sustain
the protestants in her own realm and abroad.' As if to emphasize the need for some such *rapprochement*, continuous reports
were reaching England from various sources of projected
leagues among the catholic powers for the destruction of the
reformed faith. Generally speaking, there was no solid foundation of fact behind these reports—merely vague surmises, fears,
and suspicions engendered by the perfervid imagination of a
world oppressed by a nightmare of uncertainty and widespread
espionage. Nevertheless it was as easy to believe in the existence
of hostile combinations among catholics as it was difficult to
disprove them; and protestants everywhere lived under the
demoralizing conviction that a day was surely approaching
when a bloody deluge would break over them and submerge
their religion for ever. Germany was a factory of rumours.
Situated in the heart of Europe, and being the principal re-
cruiting-ground for mercenaries, it became, in effect, a gigantic
whispering gallery, where the military secrets of the catholic
powers were overheard, speculated upon, and distorted out of recognition. Throckmorton heard some of the whispering in Paris,
and, having a *flair* for anything in the nature of a plot against
the reformation, passed on the information dutifully to London.

However unreal these alleged catholic leagues may have been,
there was no unreality about Elizabeth's efforts to knit up the
protestant communities of Europe in a common policy of defence. The relationships she entered into with them cannot be
dignified by the name of alliances, since they were not established with sovereign independent authorities, and as a rule
carried no binding obligations; but they had much the same
effect as alliances, for they linked up the protestant communities with a secret network of intelligences and promises of mutual
support, which stimulated and strengthened the hands of each.
Ostensibly the basis of co-operation was the defence of the
Gospel—it was so understood by the lutherans and huguenots;
but, to Elizabeth, the religious aspect was quite subsidiary.
What she wanted of her friends abroad was that they should
enable her to shift the strategic frontier of England into Germany, France, and the Netherlands.

It was difficult for the queen to reconcile such a policy with
her avowed hatred of rebellion and solicitous regard for the
rights of sovereignty. Probably she never troubled to rationalize

her position in the matter. But the representatives of foreign states, whose interests were damaged, remonstrated bitterly. 'This woman', wrote de Quadra, the Spanish ambassador in London, 'desires to make use of religion in order to excite rebellion in the whole world'; and again, 'If she had the power to-day she would sow heresy broadcast in all your Majesty's dominions, and set them in a blaze without compunction.' Events were soon to show that the Spaniard was substantially right in his prognostications, and that although the object was self-protection rather than aggression, the friendly relations Elizabeth had set up with foreign protestants might easily be used as an instrument of offence as well as defence. It was France, not Spain, however, that was to feel the first blow.

In the spring of 1562 the French nation was moving rapidly towards one of the gravest crises in its history. Religious fanaticism, temporarily suppressed after the Tumult of Amboise, was now rampant again, playing havoc with social order, disorganizing administration, and even threatening to obliterate from men's minds all sense of political obligation. The authority of the Crown had sunk to its lowest ebb for a century. After a vain attempt to avert bloodshed by the *Édit de Janvier* (1561), granting a wide measure of toleration to the huguenots, Catherine de Médicis saw the rival religious groups fall to blows in various parts of the kingdom. Plunder and sacrilege by the huguenots led to murderous retaliation by the catholics. In the south, free companies (*pieds nus*) were formed, and hired out their services to whichever side offered the more attractive rewards. Pitched battles became the order of the day: the highways became unsafe: trade languished: husbandry was neglected; and the very foundations of civil society seemed to be shaken like a vessel in a storm.[1]

Meanwhile the duke of Guise, observing from his retirement in Lorraine the rising tide of religious passion, decided that the hour had struck for him to play his familiar role of catholic champion. Setting out at the head of his retainers, late in February 1562, he directed his steps towards Paris, permitting his men by the way to flesh their weapons on a huguenot

---

[1] For a description of the condition of France see Félix Rocquain, *La France et Rome pendant les Guerres de Religion*; also the memoirs of Gaspard de Saulx-Tavannes, Michael de Castelnau, Blaise de Montluc, &c., in the *Collection des mémoires*, ed. by J. F. Michaud and J. J. F. Poujoulat.

congregation in the village of Vassy. Early in March he entered the capital to the plaudits of the masses, the incarnation of militant catholicism. Followed in quick succession the flight of the huguenot members of the council, the reconstruction of the government under a triumvirate consisting of Guise, the constable Montmorency, and the marshal St. André—all staunch catholics, and the coercion of the king and queen mother into acceptance of the *coup d'état*.

On 31 March Throckmorton, who had watched the course of events closely, informed Elizabeth that the huguenot leaders, the Prince de Condé and the admiral Coligny, were in arms, and that civil war was imminent. Early in April, in defiance of the royal command to lay down their arms, they seized Orleans and issued an apologia for their action, calling upon all good subjects to resist the aggression of the house of Guise and rescue the tottering throne of France. At the same time appeals for help were sent to England and Germany, and an effort was made to rally the whole of protestant Europe in defence of the Reformation.

How did Elizabeth respond to the appeal? In the first place she was fully aware of the issues at stake in France, and, generally speaking, she knew what would follow if the Guise faction succeeded in annihilating its opponents, or at all events she could make a pretty shrewd guess. Not only would the reformed cause in Europe receive a blow the effects of which might prove to be fatal, but the triumphant Guises, in collusion with Spain, whose interest in the suppression of heresy was notorious, would inaugurate a great crusade for the faith, and on the crest of this movement rise to the political domination of the Continent. Then would come England's turn. Mary Stuart would perhaps marry Don Carlos, the heir to the Spanish throne, the English catholics would rise in support, Spain would launch an armada, and Elizabeth would lose her crown. In other words, the flank attack through Scotland, which had proved a miserable failure in 1560, would be retrieved in 1562 by a gigantic frontal attack on England, supported this time by both France and Spain. If these surmises were anywhere near the truth, clearly the huguenots must not be allowed to quail for lack of support. Their cause was in large measure England's cause: their defeat would be tantamount to the removal of the strongest bastion in Elizabeth's system of defence. But in what form should the assistance be given? Would moral support be enough? Would it suffice for the queen to act through the ordinary diplomatic channels and offer her services as mediator between the court and Condé? The idea pleased her because it entailed no expenditure of blood and treasure, did not commit her to the objectionable policy of aiding rebellion, and, if successful, would greatly enhance her prestige. Moreover, a pacification of France by means of a compromise would have the special advantage of preserving the huguenot faction intact and more than ever dependent on her good graces: whereas war, if it did not actually result in the annihilation of the huguenots at the hands of Guise and his supporters, would certainly weaken them. Accordingly, at the end of April 1562, Sir Henry Sidney was dispatched from London to join Throckmorton in urging the French government to accept an offer of mediation from Elizabeth. But Catherine de Médicis was suspicious— suspicious of some ulterior motive at work behind the scenes, and especially suspicious of Throckmorton, who was a bosom friend of the huguenot leaders. She had not forgotten the conspiracy of Amboise; and her dislike of the ambassador was fully sustained by the Parisians, who vented their hatred of the accursed Englishman by shooting arquebuses into his house and molesting his servants in the streets. The queen mother had no compunction, therefore, in giving Elizabeth's offer a blunt and uncompromising refusal. She would deal with Condé herself and in her own way.

Meanwhile the fertile brain of the English ambassador had been busy turning over a scheme whereby the turmoil in France might be used to the great advantage of England. As early as 17 April he divulged his plan in a letter to Elizabeth. 'It may happen', he wrote, 'that in these garboils some occasion may be offered as that again you may be brought into possession of Calais, or some port of consequence on this side.' It was a subtle suggestion by one who thoroughly understood the *Realpolitik* of his mistress, and, like the seed that fell on good soil, it took root and germinated at once. To Cecil, with whom he could afford to be more explicit, he wrote in more detail, explaining possible ways and means of bringing about the desired consummation. 'Our friends the protestants', he commented, 'must be so handled and dandled, that in case the duke of Guise, the constable, the marshal St. André, and that sect, bring the king of Spain into France and give him possession of some places and forts, then the protestants for their defence, or for their desire of revenge, or affection for the queen, may be moved to give us possession of Calais, Dieppe, or Newhaven,[1] perhaps all three.' Here indeed was a way in which assistance to the huguenots, necessary now in any case, could be combined with an excellent stroke of business. But would the huguenot leaders be willing to make their appeal for help—the necessary preliminary—in the terms required by the queen for the realization of the Throckmorton scheme? Would they accept succour from England at the price of a betrayal of their country? Their need for money was patent and urgent, and England alone could supply it in the necessary amount; but so far there was no indication that troops were wanted, still less was there any likelihood that a port would be conceded. So the queen drew tight her purse-strings, and resolved to 'creep into the matter' with caution. In June Armigil Wade, who had been sent to sound the pro-huguenot governor of Dieppe as to the reception of English assistance, reported that the time was not yet ripe for such a move and that it would be better to wait until the pinch of adversity had begun to operate. As a matter of fact negotiations between the court and Condé were still going on, and on 25 June it was rumoured that a settlement had been reached— a rumour soon dissipated when it was learnt that the huguenot demands included not only the removal of Guise and his party from the council but the re-publication of the Edict of January.

In the meantime the campaign was on foot in France for the settlement of the quarrel by force. The duke of Guise and the constable were on the way to Orleans and the Loire, while the dukes of Aumâle and Bouillon were moving into Normandy to cut off communications between England and the huguenot centres in the north. Early in August Cecil responded by a naval demonstration off the Norman ports. But still Condé and Coligny would not speak the words of concession that would bring England to their rescue; and at the beginning of September Elizabeth, falling back on her old policy of mediation, sent Sir Thomas Smith to tender her services to Catherine de Médicis. By this time, however, the huguenot cause, despite certain financial aids from the German protestants, was in desperate straits, and its leaders were in a mood to grasp at any

[1] i.e. Le Havre.

conditions. The royal army against them was daily increasing: the Rhinegrave was on the march to Paris with substantial German catholic levies: Spanish forces were concentrating in the neighbourhood of the Pyrenees; and papal troops were at Avignon. The only way out of the peril was to yield to Elizabeth and throw themselves on her mercy. About the middle of September, accordingly, the vidame de Chartres and La Haye were sent over to London with a blank sheet already signed by their principals and instructions to procure the most reasonable terms they could. The queen was now able to dictate the conditions of her intervention and to drive, incidentally, one of the hardest bargains of her life. On 20 September 1562 the secret treaty of Richmond was signed. Its terms were few. Le Havre was to be garrisoned by 3,000 English troops and held as a pledge until Calais was surrendered; a further 3,000 men were to be placed at Condé's disposal for the defence of Dieppe and Rouen in case of necessity; and a sum of 140,000 crowns was to be paid as a loan into the huguenot war chest—repayable by the huguenots when the English were in possession of Calais.

From the French point of view it would be difficult to imagine a more humiliating contract. The huguenot clergy, who represented a kind of protestant cosmopolitanism, might gloss it over with the flattering formula that it was for the glory of God and the good of religion; but in reality the cession to England of the mouth of the Seine was simply a throw-back to the bad old days when France, a prey to the strife between Burgundians and Armagnacs, fell under the domination of the Lancastrians. Even the huguenot leaders must have had their qualms when they learned the terms of the treaty, for they asserted later that their plenipotentiaries had misunderstood their instructions; but it was not until the time came for the fulfilment of the bargain, some months later, that they fully grasped the baseness of the concessions they had made. For the moment the heat of passion and the urgency of the cause thrust reflection into the background. Elizabeth, of course, was perfectly aware of the machiavellian character of her diplomatic victory. In a proclamation issued immediately the treaty was signed, she expatiated on the justice of her intervention in France. It was in defence of Condé 'who requires nothing but the maintenance of the honour of God, and repose of the realm, and the liberty of the king', and the occupation of the coastal places by English troops was merely to preserve them from falling into the hands of those 'who have advanced themselves in force beyond the authority of the king'. Furthermore, she laid stress on the religious aspect of the struggle she was about to enter upon: the cause (i.e. of Guise) had become 'a manifest enterprise not by teaching but by the sword to force men's consciences', and the quarrel would spread to all other countries, being really part and parcel of a great catholic league for the suppression of the protestant faith. To Philip of Spain, however, she could not present the issue in this fashion with any hope of winning his sympathy. As the Most Catholic King he could not be expected to give his blessing to a war for the defence of heresy. So on 20 September Elizabeth explained her intentions to him quite frankly. She was at war, she said, not against the king of France but against the house of Guise: with the king she intended to live in peace; but she must safeguard her right to Calais, 'which she manifestly sees they do not mean to deliver', and as soon as this object was accomplished she would withdraw her forces. Moreover, since Philip had been in part at least responsible for the loss of the seaport, and presumably regarded it as a debt to be wiped out, she ventured to solicit his help in its recovery.

This was hardly Philip's intention. He had hoped, subject to French approval, to intervene in the struggle against heresy and rebellion, on the side of the Crown; and his offer of help had been provisionally accepted by the queen mother. Spanish troops were therefore moved up to the French frontier in July, and orders were sent to the duchess of Parma to prepare a Netherlands contingent. Here the assistance of Spain to Charles IX came to a dead stop; for not a man could be sent from the Low Countries, and Philip soon found himself in much the same position as the king he was so anxious to help. The Netherlands were, in fact, seething with an incipient revolt against Spanish rule and the religious policy which it practised. As Thomas Windebank put it in a letter to Cecil: 'If the Low Countries of Flanders had a head, the likelihood is of such troubles as was at the enterprise of Amboise, and so forth greater to come.' Moreover, the sympathies of the Flemings—thanks to the activities of Gresham and others, who followed Throckmorton's tactics in France—were as strongly on the side of England in regard to the huguenot crisis as they had been when the Franco-Scottish crisis occurred three years before. 'All men, saving the papists', remarked Windebank, 'do wish the Queen's Majesty to make her profit, thinking it should be the profit of all christendom.' Gresham corroborated, and on 1 August he informed Cecil that 'now is the time (they say here) to recover those pieces we have lost of late in France, or better pieces'.

Elizabeth was thus in the happy position of being able to dispatch her expeditionary force to Normandy without any possibility of interruption from the side of Spain. Nay, the position of affairs in the Netherlands became so alarming in the autumn that Philip was compelled to execute a rapid *volte-face*. In December the duke of Alva, speaking in his master's name, informed Challoner in Madrid that he was in favour of the English right to Calais, and that the king would willingly mediate with the French for its restoration. To Granvelle, in Brussels, he wrote at the same time that the only way to stop the English from helping the huguenots and creating trouble in the Netherlands was to let them have Calais.

Meanwhile the war in France was going badly for the huguenots. In spite of the gallant help of Captain Leighton and an English company, the royalists had captured Rouen on 26 October, thus breaking the chief link between the huguenot forces and the Norman ports; and in December they completely defeated Condé and Coligny at the battle of Dreux, taking the former prisoner and driving the latter with the remnant of the army back on the Loire. By this time the occupation of Le Havre by the English under Sir Adrian Poynings and the earl of Warwick had excited the liveliest concern in catholic circles. Rather than see the country dismembered, the queen mother now approached Condé with the most enticing inducements to disown his perfidious ally, who had taken the opportunity of domestic troubles to seize one of the most important seaports of the kingdom. Circumstances favoured her appeal—Navarre, the senior member of the house of Bourbon, was dead, killed at the siege of Rouen; the constable was a prisoner of the huguenots; and the duke of Guise was soon to be removed by the stroke of a fanatical huguenot.[1] Why should not Condé clasp the proffered hand of friendship extended by the court, and join with the government in expelling the impudent foreigner? Already wavering in his attachment to the English alliance,

[1] February 1563.

which had profited him nothing, and seeing the way open for a general conciliation, Condé succumbed, and in March 1563 signed the peace of Amboise, thereby gaining toleration for the huguenot religion, and political recognition for himself and his coadjutor Coligny.

Elizabeth, whose information was perfect—thanks to the plentiful communications of Smith and Throckmorton—was greatly perturbed at this trend of events; and Condé was not allowed to make his surrender to the court without many caustic reminders of his treaty obligations. Tormented by his conscience over Le Havre, and urged on by the feeling that duty as well as interest left him only one line of policy to follow, the unfortunate prince prevaricated, excused himself, urged the queen to moderate her terms, made eloquent but futile suggestions about a possible marriage between her and the French king; but in the end it was clear he would do nothing to implement his promise with regard to Calais. While the unedifying wrangle was in progress, a great movement of national feeling began to surge through France. Grievances that had seemed irremediable, divisions that apparently cut deeper than life itself, were salved as by the stroke of an enchanter's wand. The intervention of the foreigner had re-established the moral unity of the nation. 'D'ici à Bayonne', exclaimed Montmorency, 'tout crie: Vive France!'[1]

The defence of Le Havre which followed is one of the many minor episodes of heroism and tragedy that crowd the reign of Elizabeth. All through the winter and spring (1562-3) Warwick and his little garrison of 5,000 men had toiled unceasingly at the fortifications, sleeping in their harness, badly provided with arms and munitions, lacking food, drink, and clothing, but in excellent spirits and able to hold the attacking forces of the Rhinegrave at a distance. 'This is as noble a garrison', wrote its gallant commander, 'as ever served a prince. They fight like Hectors, labour like slaves, are worse fed than peasants, and are poorer than common beggars.' But the hand of fate was soon to descend upon them with a crushing blow. On 7 June Warwick noted in a communication to the council that a strange disease had come amongst them 'whereof nine died this morning and many before very suddenly'. By the

---

[1] A good account of the whole episode is given in the duc d'Aumâle's *Histoire des Princes de Condé*, tome i.

third week of the month this new and unforeseen enemy, more terrible than 'all the cannon of France', was carrying off two hundred a week. At the end of June the mortality more than doubled itself, and Warwick was reduced to desperation; half his men were out of action, either dead or dying. Recruits were hurried across from England to make good the wastage; but they were raw, untrained louts, unacquainted with the use of firearms, hastily dispatched without leaders; and if they escaped the raging furnace of the plague, they merely added to the troubles of the harassed commander. Meanwhile, with the royal army of 20,000 under the constable closing in on all sides, with huguenots as well as catholics rallying for the assault, with his ovens smashed by the enemy guns and his bakers mostly dead, Warwick found himself unable to feed his rapidly diminishing troops. On 15 July his effective strength was only 1,200; and five days later the position was so bad that Sir Thomas Smith, who had contrived to negotiate with the queen mother, sent an urgent message to Cecil advising the immediate conclusion of peace if England's right to Calais was to be preserved. But before diplomacy had got properly to work, the plague and the investing army had sealed the fate of Le Havre. On 24 July Warwick lost command of the harbour and saw his communications with England cut. At the same time contrary winds set in, preventing the approach of Admiral Clinton with new succours. On the 26th he decided to make the best terms he could with the constable, and intimated his willingness to parley. Two days later he hauled down his flag, and the remnant of the plague-stricken and war-weary garrison, with its honour intact, dragged itself and its baggage to the ships to begin the evacuation. France was again freed from the foreigner.

It was a terrible blow to Elizabeth's pride; and the humiliation was deepened by the severity of the terms imposed by the victorious French. Arguing that she had forfeited all claim to Calais by her entry into Le Havre and by making war on the kingdom, they steadily refused to accept any compromise; and against this resolute attitude Smith and Throckmorton (who had now joined him) bent their diplomatic skill in vain. Elizabeth might threaten to stop the import of French wines into England, or to 'impeach their great west and north fishing', or to deprive them of Newcastle coal, or to bombard their coasts with her navy: it was all to no good. Eventually, after hanging fire for nine months, an agreement was arrived at in April 1564, when the queen, relaxing her demands, allowed her ambassadors to accept 120,000 crowns in lieu of the hostages she had held since 1559,[1] and satisfied herself with the empty proviso that the rights of both princes in regard to Calais should be reserved. In this way the treaty of Troyes (11 April) to all intents and purposes annulled the obligations of the French under the treaty of Cateau-Cambrésis, and gave them effectual possession of Calais.

---

[1] As a matter of fact many of them had escaped.

III

MARY STUART AND THE SUCCESSION

IF Francis II had lived, it is more than doubtful whether Mary Stuart would ever have set foot on so uncongenial a soil as Scotland. Neither the land nor its inhabitants had any attraction to offer comparable to that of the rich inheritance she had enjoyed as queen consort of France. Turbulent and still heaving with the ground swell of revolution, Scotland must have seemed to its exiled sovereign little better than a semi-barbarous outpost on the frontier of civilization: a *triste pays*, stubborn and self-willed, whose people had rebelled against her mother's well-intentioned rule, and perhaps hastened her untimely end. But her career in France was finished. All she had to look forward to there was the flat, stale, and unprofitable existence of a dowager, and the still more disagreeable prospect of playing second fiddle to the jealous daughters of Catherine de Médicis, who monopolized the court. In Scotland she would at least be a queen; and who was to say that the journey to Edinburgh, circuitous as it might appear, would not be the shortest way to the goal on which her affections were set, the establishment of her right to the English succession? From conversations she had held with those of her subjects who visited her in France, she learnt that she might reckon on the support of a large and influential section of the nation. The return to her patrimonial kingdom was not, therefore, a blind leap in the dark, but an adventure, the end of which might be crowned with glorious success.

Mary was a very different type from her cousin the queen of England. Educated at the cultured and luxurious French court, she had experienced nothing of the trials that disciplined the character of Anne Boleyn's daughter. She was expansive, eager, and sanguine in disposition, much given to the gentle arts of dancing, masking, and 'joyusitie'[1]—in a word, objective and cheerful in her outlook on life. If it had suited the introspective Elizabeth to become a student, and to find compensation in books for the narrow and straightened circumstances of her outward estate as a princess, Mary, as a woman of the world, had never fallen a victim to the spell of the new learning, nor

---

[1] i.e. *joyeuseté* or merriment.

had she been driven to it by a craving for sympathy in a comfort-less universe. She was no scholar. Her knowledge of languages was limited to Latin and French. But she could indite a sonnet after the manner of Ronsard like other court poetasters, and her letters are unassuming, fluent compositions, beautifully written. Her education, in fact, sat lightly on her, and she did not take it seriously. For the rest, she was no delicate and cloistered heroine of romance, but a clever, masterful woman: self-reliant, bold, indomitable, with a masculine feeling for comradeship and high adventure. It was characteristic of her genius that she loved to hear of brave deeds, of perils overcome, and hardships endured, not with the passive wonder of a Desdemona, but with the eager longing that she herself were a man, to bear the buckler on her arm, the sword upon her thigh, and to share the hard life of the soldier in the tented field. Unlike Elizabeth she had a warm, passionate temperament, in which the primitive emotions of love and hate, hope and fear, found an unusually free expression. She was violent in her attachments, and equally violent in her hates; but in neither was there much permanence. More sensitive than the English queen to the changes of fortune, and probably finer in the grain, she was readily moved to take a rosy view of events, and just as easily projected into the blackest despair. Success brought jubilation: failure was attended by copious weeping. The buoyancy of her nature, however, was remarkable, and her recuperative power immense. Emotionally she was undoubtedly capable of greater heights than Elizabeth; but she lacked precisely those prudential qualities that make for solid achievement in the realm of practical affairs.

The moment she severed her connexion with France was not unpropitious. A crisis had arisen in Anglo-Scottish affairs, which brought about a revulsion of feeling among the Scots against Elizabeth. It had developed out of the English queen's second and final refusal to marry the Earl of Arran. This marriage proposal had been the rallying cry of the protestant party in Scotland ever since their war against the regent, and so strongly committed were they to the idea that when Elizabeth finally brushed it aside they jumped to the conclusion that she intended to snub the whole nation. Was not a Hamilton good enough for a Tudor, especially a Tudor whose descent was not above demur? *Amour-propre* was a force to be reckoned with

at all times north of the Tweed, but never more so than when
the two countries were beginning to draw towards a common
understanding and to cast off their medieval antagonisms. In
any case, the effect of the ill-starred marriage negotiations was
to galvanize into life the latent jealousy of the Scots for all
things English, and to carry the politicians of Edinburgh back
once more to their old stereotyped formula of national independence. Driven back on this primary instinct, so to speak,
they began to reconsider their bearings. They had fought for
the overthrow of French rule and the establishment of spiritual
liberty, and hitherto they had feared their queen as the ally
of the reactionary forces they wished to destroy. Now, by a
happy turn of fortune's wheel, they saw her a widow, and no
longer queen of France. She was young and a political cipher
so far as the direction of French policy was concerned. If she
would only be reasonable, there seemed to be no apparent
reason why she and her subjects should not pull together in the
common task of national effort. At all events they were prepared to give her the benefit of the doubt and receive her as
became good citizens.

Still, it must not be supposed that antipathy to Elizabeth
meant any warmth of feeling for their own queen. On the
contrary, the prevalent attitude among the people seems to
have been one of reserve. It could hardly be otherwise considering the trying circumstances through which the nation had
so lately passed, and the fact that the queen's character and
intentions at this time were largely unknown quantities. In
the prevailing ignorance, all sorts of rumours and exaggerations
took shape in men's minds. Keen protestants like Knox feared
lest her arrival might herald another revolution, involving the
loss of all they had fought for. They knew, for example, that
Mary was a staunch catholic, like her uncle the cardinal, and
to their excited imaginations it seemed not improbable that she
would take advantage of the still precarious condition of the
new religion to crush it once for all, and on its ruin to restore
the Mass, the hierarchy, and the burnings. If she did not venture
to make this clean sweep of the Reformation and all its works,
she might quite well endeavour to arrest its development; for
no catholic sovereign could rightly countenance the growth of
heresy in her dominions. There was also another disquieting
fact to be taken into account, namely the French upbringing

and sympathies of the queen. Presumably she was ignorant of the temper, tastes, outlook, and mentality of the Scots; and if she came as an apostle of French culture there would be a barrier from the very start between her and the nation. Scotland had had enough of the French for the time being: it desired above all things a thorough-going national existence. Would Mary fall in with the common will, or would she take a line of her own and thwart it?

For these reasons, therefore, Scotland was in no very confident mood on the eve of the queen's return. Randolph, the English ambassador, spoke of a 'mad world' as likely to supervene when she arrived. Maitland of Lethington, probably the acutest politician in Scotland, hinted that 'wonderful tragedies' might ensue. And the 'godly' brethren of the kirk, who had most to lose by an upsetting of the existing order, steeled themselves to resist, if need be, 'to the uttermost'.

Accordingly, when Mary landed at Leith in August 1561, she received a poor reception from the good people of Edinburgh. Hardly was she settled in Holyrood when she was treated to a protestant *auto da fé*, in which the effigies of the old Testament idolaters, Korah, Dathan, and Abiram, were committed to the flames. The meaning of this fiery tableau was too obvious to be missed.

Little did the pessimists know their sovereign! True, she was a catholic; and before leaving France she had the hardihood to tell the English ambassador, Throckmorton, who questioned her on the subject of religion, that she had no intention of altering her faith. 'Constancy becometh all folks well, but none better than princes,' she said; '. . . I am none of those that will change my religion every year.' Nevertheless, she had not come as a missionary or crusader to Scotland. She said quite sincerely that she harboured no designs against the religious freedom of her subjects. She was probably too fond of hunting and amusement to give much consideration to the graver questions of the hour. Her motto was liberty of worship for all, herself included: constraint of conscience was to be a thing of the past. In politics, still more clearly, the same note of friendly conciliation was sounded. Evidently, then, Scotland was not to be compressed, as so many had expected, into a French or 'papist' mould; the queen had made it plain that her policy would be, simply and solely, secular and national. By this wise moderation, which came as a great relief to all reasonable men, she soon gathered round her a strong following— every one, in fact, who desired to see the country freed from the tutelage of France, and dreamed, it may be, of one day seeing a Scottish dynasty seated on the English throne. Court life, which had virtually disappeared on the death of James V and the subsequent influx of French adventurers, was now resurrected at Holyrood, and the young nobles flocked eagerly to the capital to take their lawful place at the queen's side. In order still further to strengthen her position, as well as to bring all the resources of the country under her control, Mary formed her council of twelve lords, representing all the national interests, religious and territorial: including, on the one hand, the catholic earls Atholl and Huntly, and on the other the well-known pillar of the Reformation, the Lord James Stuart, and the laird of Lethington. So successful was she in winning the hearts of all that Knox complained bitterly of the waning enthusiasm of those who had formerly been so 'hot for the Gospel'. 'The holy water of the court', he remarks in his *History*, took the 'fire-edge' off their zeal. He alone of the leading men in Scotland refused to take the proffered hand of fellowship, and kept sounding alarms from time to time; but it was difficult even for him to preserve his 'fire-edge' when the whole people was fast drawing to the queen. Lethington strikes the true note when he writes, 'The queen behaves herself as reasonably as we can require; if any thing be amiss, the fault is rather in ourselves.' In short, a wonderful feeling of loyalty seems to have sprung up practically at once between the nation and its queen.

Thus did Mary successfully navigate the troubled waters of Scottish politics, disarm the critics, and attach to herself, by unmistakable tokens of good sense, a naturally warm-hearted people. Not only so: the whole country, so much disturbed of late by civil war and dissension, settled down, after her arrival, *in pacem incredibilem*.

Meanwhile she had turned to the question of her title to the English succession, the dominant interest of her life, and the source of all her subsequent troubles. In what mood and under what auspices did she approach this grave matter? Clearly in no spirit of bravado. The 'trial by battle', two years before, had sobered and chastened her mind: she was no longer under the control of her imperious uncles, whose sabre-rattling she had left behind her in France. What Mary now wanted was simply the right to be designated 'second in the kingdom', or heir presumptive to the English crown; and her quarrel with the treaty of Edinburgh was that, if she signed it as it stood, she would be deprived of this humbler satisfaction. Clause 6 of the treaty, to which she took objection, ran as follows: '. . . that the Most Christian King and Queen Mary and both of them shall at all times coming abstain from using the said arms and title of the kingdoms of England and Ireland. . . .' Undoubtedly these words could only be construed to mean one thing—total exclusion; and Elizabeth never ventured to suggest that they were susceptible of any other meaning. The point was not arguable: it was undeniable. The question was, would Elizabeth be willing to have the obnoxious clause revised? Lord James Stuart put it in this way: 'What if your title did remain untouched as well for yourself as the issue of your body? Inconvenient were it to provide that the queen my sovereign her own place were reserved in the succession to the crown of England, which your Majesty will pardon me if I take to be next by the law of all nations, as she that is next in lawful descent of the right line of king Henry VII, and in the meantime this isle to be united in a perpetual friendship?'

Accordingly, fortified by the moral support of Scotland, Mary dispatched Lethington, two weeks after her arrival in Edinburgh, to find out how the land lay at the English court and to discover whether Elizabeth would consider a revision of the treaty. Lethington, who enjoyed the sobriquet of the 'Scottish Cecil', handled his mission with a frankness that betokened complete conviction as to its reasonableness. He argued that the belief in his sovereign's right was widespread and incontestable, whereas Elizabeth herself was generally held to be a usurper. Consequently, if she insisted on ratification of the treaty as it was, such a settlement could not last; for if Mary were successfully forced into signing away her claim, she would seize the first opportunity to cancel her signature. On the other hand, if she were acknowledged 'second', there would be no further trouble. Elizabeth was equally frank in her reply. She admitted that she knew of no better title than Mary's, and solemnly stated that she intended no harm to her. Her objection was not to the justice of the claim, but to the consequences that might flow from a public declaration of it. To give Mary the position of heir presumptive would rouse controversy, not only in regard to the propriety of the step, but also on the subject of her own title—a contingency she contemplated with the liveliest apprehension. Moreover, she doubted whether, in the event of her acceding to the request, she could retain the loyalty of her own subjects; for 'more people worship the rising than the setting sun'. As 'second' in the kingdom, the queen of Scots would inevitably become, as she herself had become during her predecessor's reign, the centre of rebellion and plot against her rule. With a curious touch of pathos, she remarked: 'It is hard to bind princes when hope is offered of a kingdom. . . . Do you think I could love my own winding-sheet?' Here, then, was the core of the whole matter, so far as Elizabeth was concerned; in Lethington's words, she was resolved 'to be queen while she lived, and after her death, let that one succeed who had most right'. As for the equivocal clause in the treaty, round which the discussion moved, it was to stand, and each of them could take her own meaning out of it—*unusquisque in sensu suo abundet.*[1]

Although his mission was only a qualified success, Lethington was not discouraged, nor was his mistress; and when Sir Peter Mewtas appeared shortly afterwards in Edinburgh with a fresh demand from Elizabeth for ratification of the unrevised treaty, Mary took advantage of the occasion to reaffirm her desire for amity between the two kingdoms and her love for her 'good sister'; but Mewtas could get nothing positive from her in regard to the treaty, only a reiteration of her objections and a request for revision. In October the duke of Guise took a hand in the game, and stated the case for his niece with such cogency that he won over Throckmorton to his side, and caused him to write a letter to Cecil urging acceptance of the Scottish proposal. Lethington, too, concentrated his guns on the English secretary of state, pointing out to him that 'if this be overthrown . . . it may be judged that God wills that one nation shall ever be a plague to the other'. In January, after waiting many weeks for some answer to her message by Mewtas,

---

[1] Lethington, however, elicited from Elizabeth the verbal offer 'that if the queen, her sister, would require that commissioners were appointed to review the treaty, she would be content thereof, and by them come to a qualification'. (See *Scot. Hist. Soc.* xliii, App. 1.)

Mary took up her pen and wrote a personal letter to Elizabeth protesting her love and pleading for an interview, 'when the unfeigned nature of her good meaning would appear more clearly than in her writing'. Then, in May, Elizabeth yielded to the persuasion of Lethington and agreed to meet the queen of Scots in amicable conference. Mary had won her first success. Alone with Elizabeth all their difficulties which the politicians boggled over would vanish as by the stroke of an enchanter's wand. In a mood of girlish exultation, she wrote to her uncle, the duke of Guise: 'You can think', she said, 'how astonished others will be when they see us, the queen of England and me [agreeing] so well.'

Alas for her anticipations! No sooner had Elizabeth agreed to the conference proposal than Throckmorton's dispatches poured in from Paris with alarming reports of the imminence of civil war, the March *coup d'état* of Guise, and the perilous position of the huguenots. On 17 April he urged the queen seriously to consider the danger of leaving London when the situation abroad was so menacing. Three weeks later came the news of Sidney's failure to mediate a settlement between the government and the huguenots, and on 7 June Elizabeth decided to postpone the conference 'pending the troubles'. But Lethington, who had returned to London, succeeded in persuading her to change her mind; and she concluded, against the will of her council, that 'go she would' to the conference, unless she had further news from France 'that might cause her to stay'. On 18 June instructions were sent to the earls of Northumberland and Cumberland, the earl of Rutland, president of the council in the north, the archbishop of York, and many other dignitaries to prepare for the reception of the queen of Scots. Meanwhile Throckmorton, acting on Elizabeth's advice, put pressure on Guise to make an accord with Condé, by representing to him that unless there was a settlement in France the conference in England was doomed; and on 29 June he informed the queen that the troubles were successfully composed. On receipt of this report Elizabeth at once made arrangements for the court to move north to the rendezvous with Mary, which was now fixed to take place at Nottingham on 3 September; and Lethington left for Edinburgh with the articles for the conference in his pocket, in high spirits at the success of his mission. Then, like a bolt from the blue, came a dispatch from Throckmorton

intimating that he had spoken prematurely, and that the negotiations between the court and Condé had been broken off. On
the strength of this disconcerting news Elizabeth was reluctantly
compelled to postpone the conference. Thus, while Mary was
jubilantly preparing for her journey south, fortified by the
assurances of Lethington that all was well, Sir Henry Sidney
was dispatched to Edinburgh, on 15 July, with the melancholy
information that the whole business was off until the 'early
summer of the following year'.

With the failure of the plan for an amicable conference we
enter into a new phase of the queen of Scots' chequered career.
At first Mary was deeply affected by the destruction of her high
hopes, and kept to her bed all that day. But the acuteness of
depression soon wore off, and her naturally buoyant temper
reasserted itself. She was shrewd enough to see that if nothing
had been gained by the efforts to win Elizabeth's favour,
nothing, at all events, had been lost; she could begin again, if
need be, from the beginning. On the other hand, she was
convinced that some pressure must be brought to bear on the
English queen, if any further progress was to be made. The
question was what form the pressure should take. War was
further from her thoughts than ever, though there was an excellent opportunity for intervention in the Anglo-French struggle
of 1562–3. But she did not desire to irritate the English nor
to render a peaceful and diplomatic solution impossible: so she
preserved her neutrality. There was a much more feasible way,
namely to marry 'powerfully' and strengthen her feeble resources. If she could procure some rich and influential catholic
prince as a husband, it might be possible, she thought, to overawe Elizabeth and wring the desired concession by a kind of tacit
threat. At any rate there would be weight behind her representations, and she could carry on her attempt to persuade the
English queen in the confidence that if the worst came to the
worst, she had something to fall back on.

Mary's marriage was indeed a serious matter to Elizabeth,
for the queen of Scots was no mean attraction to suitors by
reason, it was said, of what she had title unto and 'is *in potentia
propinqua* to obtain'. To her ardent admirers she was the best
match in Europe. Elizabeth's main concern, of course, was to
see that whoever was selected to husband her rival should not
be a scion of the royal houses of Austria, Spain, or France: such

a person might easily become a real source of danger. Consequently if Mary moved in any of these directions, the only thing
to be done was to head her off by threats of enmity or by
diplomatic measures of a subtler sort, while at the same time
urging her in a friendly way to accept a husband of less august
status, preferably an English nobleman. On this point there
was bound to be trouble, for, as Lethington averred, his queen
would 'never yield to any marriage how fit or profitable soever
it be to her, unless she sees that her reputation [i.e. royal posi-
tion] shall not diminish by the match'. Nevertheless Elizabeth
pursued her plan with considerable success for the next two
years, albeit in the end the husband chosen by Mary did not
correspond to her wishes.

Of all the eligible princes the only one in whom Mary was
really interested, and who could be of any use to her, was
Don Carlos,[1] heir of King Philip and first catholic prince in
Europe. The Archduke Charles of Austria, whose claim was
favoured by the cardinal of Lorraine, was eventually ruled out
by his poverty; and Charles IX of France, whose name was
also mentioned, was also dropped. All things considered, the
heir of the Spanish empire appeared to be the man marked out
by destiny to husband the Scottish queen and re-establish her
fortunes. But it was no easy matter to carry through so grandiose
a project with the secrecy it demanded; and the politicians of
Edinburgh found they had undertaken the impossible. Elizabeth, of course, opposed the match tooth and nail. To Mary
she intimated bluntly that if she proceeded with it, she could
not avoid being her enemy, and therefore charged her to consider well her steps, for if, on the other hand, she married agree-
ably to her (i.e. Elizabeth) she would not fail to be her good
friend. Concurrently but independently Catherine de Médicis,
greatly fearing an alliance between Spain and Scotland and
profoundly suspicious of the Guises, strove to attract the prince
to a marriage with her own daughter Margaret. Mary's chance
of success was probably lessened by the death of the duke of
Guise in February 1563—an event that plunged her into deep
melancholy. But the event that finally shattered her hopes
was the grave illness of Don Carlos in the autumn of the same
year. Thereafter King Philip seems to have lost all interest

---

[1] Don Carlos is usually described as a cruel, sullen, gluttonous epileptic; but
Brantôme depicts him as 'a brave and gallant prince'.

in the matter. By the month of August 1564 it was as good as dead.

It is impossible to say what would have happened if Mary had been successful; but in all probability it was fortunate for her that she failed. She had not consulted her people in the matter, and she knew that she had no general support in the furtive policy she was pursuing. Moreover, the prince of Spain was both a foreigner and a catholic. If the protestants, in alarm for religion, poured out their vials of wrath on the secret negotiations, the rest of the community was almost certain to take offence at his being a Spaniard. And if Don Carlos had set foot in Edinburgh with a Spanish guard, the old cry that had rallied the nation against the French in 1560 would have been repeated against the Spaniard in 1563; and either Mary would have re-enacted the tragedy through which her mother had passed three years before, or she would have been precipitated prematurely into the humiliating position in which she found herself in 1567. It was well that the Spanish match of 1562–3 vanished into thin air like its more famous successor some sixty years later.

The elimination of Don Carlos from the field of Mary's matrimonial activities greatly circumscribed her horizon and at the same time simplified her choice. If a foreign match, suitable to her ambitions, was not to be compassed, might she not achieve her end equally well by marrying nearer home? The point was worth considering; and during the winter of 1563–4 Randolph, acting on Elizabeth's instructions, began to grope his way cautiously towards a proposal that she might with advantage take Lord Robert Dudley for her husband. In March 1564 the offer was definitely made, but without any accompanying pledge or assurance that, if accepted, it would advance Mary's interest in the slightest degree. Whether the anxiety of the queen of Scots blinded her in some measure to the burlesque character of the proposal, or whether, as is more likely, she was surprised by its effrontery, she would not commit herself at once; she must know more about the conditions attached to the offer; and she suggested a conference of ambassadors, committing the matter to the experienced hands of Murray[1] and Lethington. Elizabeth, whose purpose it was to 'drive time' and play out the farce for what it was worth,

---

[1] Lord James Stuart was created Earl of Murray in Feb. 1562.

willingly responded to the suggestion of a conference, and in October Bedford was joined with Randolph in commission to meet the Scots at Berwick.

When the discussions opened, and the Scottish commissioners insisted upon a clear statement on the subject of the succession, it was soon discovered that Bedford and Randolph had no authority to communicate anything of importance. Wrapping up their arguments in vague generalities and ambiguities, they tabled the following strange proposition: 'If this queen follow Elizabeth's advice and counsel in her marriage, she will thereon proceed to inquire into her title with all favour, and if it falls out on her behalf, then, on plain knowledge whom she will marry, she will deal with her as her natural sister or daughter.' Perplexed and baffled by such a farrago of 'obscure words and dark sentences', Murray and Lethington appealed to Cecil, the real stage-manager of the conference, to define more clearly the intentions of his sovereign. But Cecil was not to be trapped into anything in the nature of a contract; and replied, in effect, that if they wanted a parliamentary declaration of Mary's right to the succession as a condition of the marriage, he did not think it probable that it would be granted either in form or substance. Whereupon the Scots reported that 'lacking the chiefest argument that may work any persuasion', they could not conscientiously recommend their queen to marry an Englishman, or make herself a lifelong 'thrall' to Elizabeth. The suspicions of Murray and Lethington were corroborated by Sir James Melvil, who in the meanwhile had returned from his historic visit to the English court. His report to Mary on the circumstances of his mission was singularly discouraging: 'In my judgement', he remarked, 'there was neither plain dealing nor upright meaning; but great dissimulation, emulation, and fear lest her [i.e. Mary's] princely qualities would soon chase her [i.e. Elizabeth] from the kingdom.'

In this way the situation dragged wearily on through the winter of 1564–5: unsatisfactorily for the queen of Scots, but greatly to the advantage of the time-driving Elizabeth. Relief of a melancholy kind came in March 1565, when Randolph transmitted another communication from England. After prolonged meditation, Elizabeth now condescended to explain that, even if her good sister married Leicester, it would be impossible to examine into or publish her title, until she herself had either married or declared her intention never to marry. In other words, Mary was politely informed that discussion was, for the time being at all events, at an end; for nothing was surer than that Elizabeth intended to keep her own marriage an open question until it ceased to be a feasible proposition. She had frustrated the efforts of her rival to marry outside the realm: she had driven time with prodigious success for a year; and now, instinctively feeling that a critical point had again been reached, beyond which she could not go without involving herself in undesirable commitments, she resolved to bring the comedy to a close. Any step the queen of Scots in her desperation might be led to take could be awaited with tranquillity, and in the confident assurance that it could be met by adequate counter-measures as before.

But the game was far from over. Mary had played most of her cards; but there was still a strong one in her hand, which might yet turn the tables against her enemies and pluck victory from the very jaws of defeat. While the diplomatists were busy with the intricacies of the Leicester-*cum*-succession comedy, she had practically made up her mind to play it boldly. Randolph had an inkling of what was coming, but would not commit himself to more than that Mary would 'let fall her anchor between Dover and Berwick, though not in that port, haven, or road that you wish she should'. Gradually, however, the mystery cleared up. In the autumn of 1564 the earl of Lennox was invited to Edinburgh on the plea that he was to be restored to his family estates confiscated during the regency. His son, Lord Darnley, followed him to the Scottish capital in February 1565. How it happened that the latter obtained permission from Elizabeth to do so it is hard to say, for she must have known that some intrigue was afoot. Melvil was of the opinion that Cecil was responsible—'not that he was minded that any of the marriages should take effect, but with such shifts to hold the queen unmarried so long as he could'. The young nobleman would provide Mary with a fresh distraction, and thereby help to postpone the day of reckoning. On the other hand, should Darnley and the queen fall in love, it was always possible to recall him: he had estates in England, and his mother, the countess of Lennox, could be held as a hostage for his good behaviour.

For once, however, Cecil found himself hoist with his own petard. Darnley not only fell in love with the queen but remained in Scotland defiant of all summonses, threats, and penalties, 'lured on by the hope of a crown'; while Mary with unseemly haste promoted him to be lord of Ardmanach, earl of Ross, and duke of Rothesay—'honours never before given to any but the king's sons'. Parliamentary sanction for their marriage followed; and a papal dispensation was sent for, to remove 'the impediment in blood'.[1] Sensibly the atmosphere of the court and the behaviour of the queen underwent a complete change, from depression to gaiety, from caution to reckless abandon, culminating at Easter in a crescendo of festivity and fun that astonished all beholders.

'Organs', writes Randolph, 'were wont to be the common music', but the queen lacked 'neither trumpet, drum, nor fife, bagpipe, nor tabor. . . . On Monday, she and divers of her women apparelled themselves like bourgeois wives, went upon their feet up and down the town; of every man they met, they took some pledge of money to the banquet; and in the same lodging where I was accustomed to lodge, there was the dinner prepared, and great cheer made, at which she was herself, to the great wonder and gazing of man, woman, and child.'

With incurable optimism, the queen of Scots now believed that the goal after which she had so long and unsuccessfully striven was at last in sight, and the irritating tutelage of Elizabeth at an end. Throckmorton, whom the latter sent to remonstrate against the Darnley match, found her irrevocably committed to it, and in no mood to be 'fed' any longer 'with yea and nay'; and when he attempted to create a diversion by reviving the Leicester proposal with definite promises in regard to the succession, it was patent that his mission was utterly fruitless. In a haughty message 'wanting neither in eloquence, despite, anger, love, nor passion', Mary bade the English queen not to be offended at her marriage . . . and for the rest she would abide by such fortune as God would send her. Late in July she created Darnley duke of Albany, and on the 29th of the month, before the papal dispensation arrived, married him 'with great magnificence' in the royal chapel of Holyroodhouse. On the following day he was proclaimed king with fitting celebrations.

Although the marriage was viewed with dismay by the

---

[1] For Darnley's relationship with Mary and his position *vis-à-vis* the English succession see genealogical table II.

protestants, and with the greatest dislike by Murray, Châtel-
herault, Argyll, and others, who feared it would lead to their
political extinction as well as the overthrow of the kirk, there
was much to be said in its favour. Darnley's claim to the English
crown was almost as good as his wife's, his orthodoxy was un-
impeachable, and he was popular with the catholics south of
the border. Moreover, the many grave defects of his character,
though perceptible to the lynx eye of Randolph, were as yet
undeveloped: he was still the callow, pink-and-white lover of
the queen's day-dreams, passionately fond of music, horses,
and sport.

Nevertheless serious trouble began to brew in Scotland
immediately after the marriage. Murray continued his opposition and refused to be reconciled, sulked his way into rebellion,
and had to be 'put to the horn',[1] in August, for attempted
revolution. Elizabeth, whose interest it was to support the
protestant faction in Scotland, tried to intervene with an offer
of mediation; but her offer was curtly dismissed, and her envoy,
John Thomworth, returned to London with a virtual ultimatum
on the succession question. Meanwhile Mary and Darnley,
believing that the catholics of both kingdoms were behind them,
set their forces in motion to pursue the rebels to the death.
During the turmoil that followed—the so-called 'Chase-about
Raid'—Elizabeth was compelled, much against her will, to
preserve an attitude of neutrality; but she contrived to send
Murray £1,000; and when he and his party were driven south
to the border by the superior strength of the royal army, she
instructed Lord Scrope, warden of the West March, to cover
his retreat into England. Beyond this, however, she would not
go: partly because she dared not assist rebels openly, and
partly because the French ambassador, de Foix, was on the
alert for any sign of English intervention in Scotland. When,
therefore, Murray hastened from Newcastle to London to place
his case before her, in defiance of strict orders to the contrary,
she trounced him mercilessly in the presence of the privy
council and the French ambassador.

When the news of Murray's 'discomfiture' reached the Scottish court it was received with great jubilation; while Elizabeth's conduct was generally interpreted as a confession of fear.
In November the movable goods of all who had taken part in the

[1] i.e. outlawed.

rebellion, already declared forfeit to the Crown, were dispersed by public auction, being 'greedily taken up and sold for ready money at half their value'. Three months later Randolph was summarily dismissed from Edinburgh, on the ground that he had helped the rebels with money. In an age when ambassadors were often left to shift for themselves this did not necessarily mean a declaration of war; but it certainly indicated that Anglo-Scottish relations were rapidly drifting into the danger zone. For three years Randolph had worked whole-heartedly in the interests of peace and goodwill between the two realms— a highly respected and diligent frequenter of Holyrood, whose advice was much courted by the queen, and he himself, by his own admission, one of her most devoted admirers. But now the fateful marriage had ruined everything: replacing the amity he had striven for by a seemingly implacable enmity, and bringing the sovereigns, whose co-operation and conciliation had been the inspiration of his life, to the very brink of war. Anger mingled with disappointment. With a bitter and jaundiced mind he turned to rend the idol he had formerly worshipped: 'And in the whole world, if there be a more malicious heart towards the queen my sovereign than is she that here now reigneth, let me be hanged at my home-coming or counted a villain for ever.'

But the activities of the queen of Scots at this time had a much wider range than the facts just recounted would appear to show. Until the Darnley marriage she had acted on a very narrow basis, relying principally on her own resources and the skill of her able politicians in Edinburgh. She had consistently avoided any semblance of 'papistical' intentions; and nothing was farther from her thoughts than to stir up a religious war in Scotland. Indeed so strong was her desire to maintain the good-will of her subjects, protestant as well as catholic, that she resisted even the direct appeals of the Holy See itself, and thus gave all her catholic friends at home and abroad the impression that she had not the good of the church at heart. Sending no representative to the council of Trent, she had hustled Nicholas of Gouda, the nuncio, out of the country in 1562, pleading that the times were unpropitious for a religious revival.[1] Not unnaturally, therefore, the construction put upon

[1] J. H. Pollen, 'Papal Negotiations with Mary Queen of Scots, 1561–7' (*Scot. Hist. Soc.* xxxvii, 1901).

her actions was that she was cold in the cause of religion and a time-server. Likewise, too, in the political field, she had disappointed her friends by her neutrality and apathy. Now, however, all this was dramatically changed. Throwing aside the mask of forbearance, Mary made a strenuous appeal to both the supreme pontiff and the king of Spain. In the autumn of 1565, when the civil war was rampant in Scotland, the bishop of Dunblane was in Rome beseeching the pope for ten or twelve thousand men and a sum of money to enable her to overcome her enemies; and a certain Yaxley was accredited to the Spanish government of the Low Countries with the message 'that the queen of Scots, having reason to doubt the credit of her uncles in the court of France, was advised to address all her causes to the king of Spain, and would commit herself, her husband, and her country to his protection'. Philip was to understand that she would follow his advice without swerving so much as a hairsbreadth.

Great was the joy in Rome and Brussels at this news. Now at last, it seemed, there was ocular demonstration of the fact that Mary was sincere in her protestations of loyalty to the faith; many thought that the counter-reformation was begun in Britain. The amazing energy of 'this woman with a man's heart' stirred the enthusiasm of the aged pope to an unwonted degree. Despite the embarrassed state of his finances and the many other cares distracting the head of the church, he responded to the appeal for succour with alacrity. The aggressive Turk might conquer Malta, the anarchy in the papal states might continue, but the heroic queen must not be left alone in her war against heresy. If need be, he would even stint his table and sell his plate rather than see her beaten for lack of money. Had he been younger he would willingly have spent life itself in her service. Pius V was a born crusader, and, like a true knight, credited Mary with the same high intentions as he himself possessed. He offered her all she asked, 200,000 crowns, payable in monthly instalments of 40,000; and urged the Spanish king to send the necessary military aid.

Much was expected from this combination of forces; and for a brief space, to be reckoned in months, the queen enjoyed the satisfaction of riding the whirlwind. But the foreign element, which had lent so imposing and dangerous a character to the scheme, on which Mary principally calculated for its ultimate success, hardly came into action at all. The English catholics, whose active help the Darnley marriage was specially designed to enlist, remained unresponsive, preferring their 'silent compromise of conscience' to the dubious hazard of war; the promised Spanish troops never landed—were, in fact, never sent—because Philip, afraid of precipitating a war with England for which he was as yet unprepared, 'shamelessly deserted the young princes', leaving them 'naked to the assaults of heresy'; while, of the papal subsidy, only 40,000 crowns reached Edinburgh, the remainder being delayed at the pope's express order, until the queen's intentions became clearer. Thus prop after prop crumbled and fell away. But the mortal blow was delivered in Scotland by a confederacy of the malcontents.

The impending forfeiture of the rebel lands, fixed for 12 March 1566, gave the exiled nobles a vested interest in revolution; and, joined with the bitter personal quarrel between Darnley and Mary which had developed since the 'Chase-about Raid', provided the lever for the complete overthrow of the royal policy. The king had now shown himself in his true colours —in private life irascible, vicious, overbearing, impossible to live with: in public life futile and incapable, but possessed of a vanity as great as his incapacity, and desirous of parading both in the government of the country. Knowing his weaknesses, the queen studiously ignored his pretensions, and conducted the business of the State through her Italian secretary Rizzio: with the result that Darnley, feeling himself insulted, and profoundly suspicious of the secretary's relations with his wife, could only think of blind revenge. With a singular treachery and a callous indifference to the queen's health—for she was expecting the birth of a child—he conspired to restore the exiled lords, destroy by their aid the 'usurper Davy', and so make good his claim to the 'crown matrimonial'; while his coadjutors would regain their position and power. Thus the bargain was struck and the plot concocted, that ended in the brutal murder of Rizzio on 9 March (1566).

The details of the sordid conspiracy, together with the kaleidoscopic revolution to which it gave birth, belong to Scottish history, and are too well known to be recounted here. But several salient facts emerge from the confusion, with a direct bearing on Anglo-Scottish relations, and may therefore be emphasized. In the first place Elizabeth's ministers were in the secret of the plot, and cognizant of every step taken by the conspirators from the inception of the murder to its execution. In the second place, although Murray's part (if any) in the crime is obscure, his timely appearance in Edinburgh immediately after its perpetration, and the masterful way in which he handled the situation, suggest that the deed in itself was secondary and the *coup d'état* that followed the really important thing. In the third place, the effect of the upheaval was to throw all the queen's plans out of gear, and to place both her and the government once more under the influence of Murray. It would be too much to say that the revolution was engineered in England's interest—the facts do not warrant such a conclusion; but it cannot be denied that the outcome was eminently in England's favour. As for Mary, she accepted her fate with resignation, and did her best to pacify the country; but the only bright spot in her gloomy outlook was the birth of Prince James in June. The enthusiasm this happy event evoked showed that she still enjoyed the loyalty of her subjects. Undoubtedly, also, it afforded Mary a measure of personal triumph over the childless Elizabeth, who is reported to have said, when the news reached her, 'The queen of Scots is this day "leichter" of a fair son, and I am but a barren stock'.

So far we have considered the succession question mainly as affecting the two queens. But it had a wider import; and before we resume the thread of our history, a glance at the matter from a different angle is advisable.

Ever since 1559, when Elizabeth replied to her first parliament's recommendation to marry, with honeyed words about her care for the public weal and the advantages of virginity in a queen, who, as she said, had already given herself in marriage to her kingdom, the problem of dynastic security and the safety of the religious settlement, which depended upon it, had grown to be the dominant theme in English politics. Parliament did not know at first what to make of the queen's evident antipathy to marriage. An unmarried queen was an intriguing novelty; but to men of the world it was unnatural, and, in view of the prevailing uncertainty as to the future, distinctly disconcerting. After all, as the commons had hinted in their petition, princes are mortal and commonwealths immortal, and the kings of England had always shown great solicitude about their issue.

Therefore the queen must marry and provide an heir, her private preferences notwithstanding. The force of the argument was incontestable; and it received an accession of strength when, in October 1562, Elizabeth fell ill of the small-pox, and for several days seemed to hover between life and death. It was an anxious moment for the council and, indeed, the country at large; for it needed no flight of imagination to foresee the chaos that would supervene if she died prematurely. Protestants, confronted with the likelihood of an unconditional catholic succession, in the person of Mary Stuart, began at once to canvass the claims of Lady Catherine Grey,[1] the representative of the Suffolk line which, failing heirs by Henry VIII's children, had the 'remainder' in the succession by act of parliament. Others favoured the earl of Hertford or the earl of Huntingdon;[1] and doubtless the catholics and all partisans of the principle of strict hereditary succession secretly hoped to revive the Stuart claim, which Henry had barred. Consequently, when the second parliament met in January 1563, there was a general desire on the part of both houses to have the order of succession established by law. Petitions were presented to the queen by the lords and the commons. Let the queen marry, said the former, and designate a successor: 'thereby she shall strike a terror into her adversaries and replenish her subjects with immortal joy.' If she followed the example of Pyrrhus, and left her kingdom to the person with the sharpest sword, it would mean calamities innumerable, civil war, the destruction of religion, foreign invasion, and bondage. The commons argued on much the same lines, but laid more stress on the necessity for delimiting the succession than on marriage.

Elizabeth was perturbed; but again she had recourse to evasion and procrastination. Her own safety and theirs, she said in reply to the commons, was her paramount consideration, but she could not 'wade into' so 'deep a matter' without deliberation. To the lords she pointed out that it would cost England much more blood if she declared her successor: let them consider well what they were urging her to do. As for marriage, it was not impossible: they must not mistake the marks left on her face by small-pox for the wrinkles of age; and if she did marry, even although she were old, God would provide her with children as He did Saint Elizabeth. Once more the

---

[1] See genealogical tables II and III.

familiar assurances about the public weal, and hints that she might marry, but no pledges or commitments of a positive kind, and, above all, no indication that the question of the succession would be settled in the immediate future. Admittedly the queen played her cards well. She finessed boldly, employed all the cunning she knew, or ingenuity could suggest, and so mingled firmness and reproof with an evident desire to conciliate that she circumvented her difficulty with comparative ease. After parliament was prorogued, in April, she imprisoned John Hales, a chancery lawyer, for writing a pamphlet against the Stuart claim, and defending that of Lady Catherine; and Bacon, his patron, was exiled from court. So matters stood until the ominous events in Scotland following the Darnley marriage —the expulsion of the protestant lords, Mary's machinations with the pope and the king of Spain, and the birth of Prince James—exposed in startling light the peril in which England was placed by an unmarried queen and an unsettled succession.

Needless to say, when the third parliament assembled, in November 1566, a lively session was inevitable. Acrimonious discussions began at once on the all-important question of the hour. The queen, Cecil, and the royal physician were each attacked in turn—the queen because she was careless of her country and posterity, Cecil because he was a corrupt councillor, and the physician because he thought more of some alleged infirmity of the queen than of the nation's safety. In the upper house the earls of Leicester and Pembroke and the duke of Norfolk insisted that a husband should be forcibly imposed on Elizabeth, and the succession regulated by parliamentary statute. For this rank impertinence they were excluded from the presence-chamber, and had to sue for pardon on their knees. In the lower house excitement drove men to amazing licence of speech. It was contended by Bell and Monson, two lawyers, supported by Paul Wentworth, that monarchs are bound, in the interest of their peoples, to make known their successors. If the queen refused to do so, she was no mother to her country but a step-mother, nay, a parricide; for she would rather that England, 'which now breathed with her breath, should expire with her than survive her'. Not princes, but cowards and timorous women, they asserted, have ever stood in fear of their successors. The wrath of God and the alienation of her subjects' hearts would be her portion if she refrained from so obvious a duty.

This was plain speaking with a vengeance. But the queen, too, could speak plainly, and she had reason on her side. While the succession remained open, she was safe: both catholics and protestants must be her bondsmen, the former from hope, the latter from fear; and in her safety lay the peace and security of the state. History, moreover, had shown that there were dangers to successors as well as to those in possession, if a public pronouncement were made in the matter of the succession. She herself had been in peril at the time of the Wyatt rebellion; and the fates of Roger Mortimer, earl of March, and John de la Pole, earl of Lincoln, signalled still graver warnings from the more distant past. Is it surprising that Elizabeth felt justified in her resistance to parliament's demands? Their way was the way to chaos: hers was the way to peace. On the last day of the session she took a high hand with the assembled lords and commons: hectored them for their disingenuousness in making a 'vizard of liberty and succession' in order to triumph over her; but finished her harangue by assuring them, in characteristic manner, that 'a more loving [prince] towards you ye shall never have'. 'Thus by a woman's wisdom', says Camden, 'she suppressed these commotions, which Time so qualified, shining ever clearer and clearer, that very few, but such as were seditious or timorous, were troubled with care about a successor.'[1]

Meanwhile events in Scotland were moving with accelerated pace from tragedy to tragedy. Rizzio was hardly in his grave when the fates began to spin their toils about the luckless and wayward figure of Darnley. Hated and distrusted by the queen for his unforgivable treachery in the March conspiracy, disliked by Murray, Argyll, and others of the nobility, who chafed under his arrogant pretensions, and viewed with mortal enmity by his late accomplices, whom he had abandoned and denounced on the morrow of the crime, this petulant and feckless youth, who bore the name but not the dignity of king, now found himself the butt of all men's aversion, and farther off than ever from the 'crown matrimonial'. As the summer advanced, reports of his hopeless estrangement from his wife were coupled with sinister rumours of the queen's affection for the earl of Bothwell, the most turbulent, debonair, and un-

[1] For further details as to the constitutional crisis see chapter vi.

principled of the Scottish nobles. There was no mistaking the fact that trouble was brewing; for the queen made no secret of her desire to be rid of Darnley. 'I could wish I were dead,' she confided in Du Croc, the French ambassador; and Lethington, who also shared her secret, wrote to Archbishop Beaton in France that 'it is a heartbreak for her to think that he should be her husband, and how to be free of him she sees no "outgait"'. In November came the conference at Craigmillar castle, where the question of the 'outgait' was solemnly discussed by the queen and five of her council—Murray, Lethington, Argyll, Huntly, and Bothwell. Divorce was considered only to be dropped, as being dangerous to the status of the infant prince. Then Lethington suggested that the matter might be committed to the discretion of himself and his fellows. In all probability the 'murder band' was drawn up before the meeting dispersed—not, indeed, with the privity of the queen, or perhaps even of Murray, who was a master in the art of alibi—but without any effective protest from either, though both must have guessed, if they did not know, what was implied in Lethington's significant suggestion.

After the Craigmillar conference the movements of the chief actors in the drama become shrouded in a thick mist of conjecture. The main events alone are clear—the queen's pardon to Morton and others of the Rizzio exiles, at the instance of Murray and Bothwell: the retirement of Darnley, smitten with small-pox, to Glasgow, and his subsequent return at the queen's express wish, and in her company, to Edinburgh: the strange domicile chosen by Mary for her sick husband; and finally the murder on 10 February, when the house of Kirk o' Field was demolished by gunpowder, and Darnley's body was found strangled in the grounds.

The murder of the king is one of those Serbonian bogs of history 'where armies whole have sunk' in the endeavour to find bottom. Fortunately it is not necessary to establish the guilt or innocence of the queen, or to discuss the bearing of the foul deed on Scottish politics. The main fact, on which everything turned, is not legal but moral, and is connected with Mary's behaviour after the crisis, not with her doings at the time of the murder or before it. Hardly was the deed known when common fame pointed an accusing finger at the earl of Bothwell as the perpetrator and at the queen as his accomplice.

Yet, although the gravity of the situation must have been apparent to her, she never took any convincing steps to vindicate her reputation. Her indifference was not traceable to lack of good advice, for the whole world was shocked at the crime and told her so. In a letter of transparent sincerity Elizabeth implored her to bring the criminals to justice and do her duty as a noble princess and loyal wife. Mary took no notice of this friendly exhortation, nor did she reply to it. On the contrary she chose to hurry on the 'cleansing' of the murderer designate at a packed and terrorized assizes, and shortly afterwards created him duke of Orkney, thus raising a general presumption that, if she did not actually instigate the murder, she was accessory to it and sympathized with it. When she married him on 15 May, with protestant rites—Bothwell was a protestant— her depravity in the eyes of catholics was placed beyond dispute. For the purpose of the marriage Bothwell had been divorced from Huntly's sister twelve days before!

In the catholic world it was widely held that her downfall extinguished the last hope of better days for the old faith in Britain. 'The catholic religion,' said one, 'which in Scotland had no greater foundation than the good intention always evinced by the queen to support it, now is, according to general opinion, deprived of all hope of ever again raising its head, because she, without fear of God, or respect for the world, has allowed herself to be induced by sensuality or else by the persuasion of others to take one who cannot be her husband, and gives thereby a suspicion that she will go over by degrees to the new fashion.' The papal nuncio, writing from Paris to the pontiff, contented himself with the remark that 'one cannot as a rule expect much from those who are subject to their pleasures'. James Beaton confessed that 'it would have been better in this world that she had "lossit" life and all'; and the bishops of Ross and Dunblane pleaded witchcraft and destiny, respectively, as the source of the catastrophe. Pius V, after weighing the pros and cons, expressed himself in some doubt which of the two queens in Britain was the worse. It was mortifying that the plans of the Holy See for the spiritual regeneration of Britain should be so badly thrown out of gear by the caprice of a passionate woman.

The queen's position was, indeed, that of one who had just made the great renunciation of her life. Disowned by her own party and anathematized by her subjects, she figured in the popular imagination as some monstrous Clytemnestra, who, to satisfy an illicit passion, had imbrued her hands in her husband's blood. Yet such was the infatuation of this woman, in this moment of dire extremity, that the words of condemnation and execration sounding in her ears only served to goad her into more desperate courses. With Bothwell she would defy the world. Less than a month after her marriage the lords of the council summoned the lieges to their banner, proclaiming their intention of executing justice against the 'murderer of the king and the ravisher of the queen'; and at Carberry hill, on 15 June, the brief drama came to an end with Mary's surrender. Bothwell, under cover of his wife's self-sacrifice, made his way safely out of the kingdom. Days of anguish followed for the captive, while the fanatical mob of Edinburgh clamoured for her blood, and her captors sat in solemn conclave, as a committee of public safety, to decide her fate. As a precautionary measure they sent her a prisoner to Loch Leven castle.

At this point Elizabeth found it necessary to intervene with more than hortatory epistles. Mary's incarceration and the deliberations of the Scottish leaders cast a totally new light over the relations of the two queens: far-reaching political issues, involving the rights of crowned heads in general, were at stake. But the situation was a delicate one, calling for the greatest tact and care, for Scotland was touchy on the subject of its independence, and public opinion was strongly moved. The kirk was unanimous for the queen's condign punishment as a murderess and adulteress: the politicians were not so unanimous, some being in favour of her restoration, if only she would divorce Bothwell; but all were determined to have no interference from England. Thus when Throckmorton made his appearance in Edinburgh, early in July, with a proposal for mediation, and spoke of the majesty of crowned heads, he was informed that he could not be allowed to see her, and had to content himself with a message urging her to comply with the lords' demands and divorce Bothwell. Elizabeth was furious. She would teach these rebellious Scots a lesson in political science. Nay, if they did anything to deprive their queen of her royal estate, she would make herself 'a plain party against them, to the revenge of their sovereign and for example to all posterity'. But Mary destroyed her chances of lenient treatment by her stubborn refusal to be separated from Bothwell, and her deposition was inevitable. On 24 July, under threat of a public trial, exposure, defamation of character, and possibly death, she signed the articles of abdication. Three days later, Prince James was crowned king at Stirling.

Meanwhile the English intervention had produced the opposite effect from what was intended. 'Therefore take heed', said Lethington to Throckmorton, 'that the queen your sovereign do not lose altogether the goodwill of this company irrecuperably. For although there be some which would retain our prince, people, and amity to England's devotion, yet I can assure you if the queen's majesty deal not otherwise than she doth, you will lose all.' More pointedly still: 'For now we begin to hold all things suspected that cometh from you; and if you be overbusy with us, you will drive us faster to France than we desire to run.' And again: 'My lord ambassador . . . I assure you that if you should use this speech unto them [i.e. the nobles] which you do unto me, all the world could not save the queen's life three days to an end.' Still Elizabeth persisted with her adjurations and threats, to the despair of her ambassador and the dismay of Cecil and Leicester, who had never seen her so incensed as she now was against the Scots. She not only refused to acknowledge the *faits accomplis*, but tried to frustrate the establishment of the regency under Murray, and entered into negotiations with Lord Herries and the Hamiltons for the release of the captive by force. Eventually, when these efforts also proved fruitless, she withdrew Throckmorton without recognizing the legality of the new government in Edinburgh, and approached France with a proposal for joint action against Scotland by means of a blockade. But by this time Murray had the situation well in hand: all the strongholds of the country, except Dumbarton, were in his possession, and the revolution was virtually complete.

Fortune's wheel, however, had not yet come to rest. Before seven months were past it took another turn. On 2 May 1568 Mary escaped from her island prison, joined her partisans, the Hamiltons, and again the torch of civil war was brandished in Scotland. Elizabeth hastened to offer her congratulations, help, and mediation. But before the messenger had started on his

way the campaign came to a sudden and unexpected end by
the defeat of the Marian troops at Langside and the flight of
the queen to the border. In a tempest of baffled rage and
wounded pride, and with a blind passion for revenge, she
crossed the moors of Galloway and the Solway, placing herself,
as she fondly hoped, in the friendly protection of Elizabeth.
'I have endured injuries, calumnies, imprisonment, famine,
cold, heat,' she wrote to her uncle the cardinal, 'flight not
knowing whither, ninety-two miles across the country without
stopping or alighting, and then I have had to sleep upon the
ground, and drink sour milk, and eat oatmeal without bread,
and have been three nights like the owls, without a female in
this country.' Her thoughts were far from complex: her one
hope, when she touched English soil, was to procure Elizabeth's
aid for the destruction of her enemies. 'What she most thirsteth
after', wrote Sir Francis Knollys, who met her soon after her
arrival, 'is victory . . . and in respect of victory, wealth and all
things seemeth to her contemptuous and vile.'

Elizabeth was now in a quandary. When she offered help to
Mary she had not contemplated the possibility of being called
upon to restore her to the throne by force of arms. A war with
Scotland for such an object was, indeed, unthinkable. On the
other hand, to surrender her to Murray would be an act of
gross inhumanity and folly. Nor would it be politic to accord
her freedom of movement in England, because of her position
vis-à-vis the English catholics and the succession. Nor, again,
would it help matters to allow her to go abroad and seek the
aid of France, for French influence in Scotland was a contingency more to be feared than any other. From whatever point
the problem was approached, the difficulties seemed to be
insuperable, and yet some solution was imperative. In the
circumstances, Elizabeth decided that for the present her own
interests might best be served by detaining Mary under observation in England. She therefore instructed Lord Scrope and
Sir Francis Knollys, who had the supervision of the fugitive at
Carlisle, to inform her that it would be *malaisé* to receive her
at court while the stigma of her husband's murder still rested
upon her. This elicited from Mary the request that she should
be allowed to clear herself, by a personal interview, of the foul
aspersions cast upon her by her ungrateful subjects—a somewhat rash offer, since it gave birth in Elizabeth's mind to the

idea of getting both parties to plead their case before an English tribunal. Eventually, after considerable reluctance on Mary's part, and on the assumption that, whatever the outcome of the examination might be, her restoration to Scotland was assured, either by force of arms or by reconciliation with her subjects, the Scottish queen consented to the scheme. Murray was also persuaded to comply, on the understanding, apparently, that if he succeeded in justifying himself there would be no restoration. Having thus manœuvred herself into the position of arbiter, which she had always coveted, Elizabeth arranged the Conference of York (Oct. 1568).

But the York conference did not go according to plan. Much to Elizabeth's annoyance, it resolved itself into a series of intrigues, and threatened to end in a stalemate or even a reconciliation between the opposing Scottish factions. Norfolk, head of the English delegation, and Lethington, who were anxious, though for different reasons, to save the queen of Scots' future, put pressure on Murray to withhold the 'rigorous accusations' with which he had come armed. The former had already begun to dally with the project of himself marrying Mary: the latter feared he might suffer by the inevitable revelations if the Darnley murder were investigated. Murray, too, had his own hopes and fears regarding his position in Scotland, and refused either to abstain from defaming his queen or to go on with the defamation, except at a price, viz. the confirmation of his regency. It was all very disconcerting, mysterious, and confused. Meanwhile, however, Murray had exhibited the 'casket letters'[1] to the English commissioners, who took rough

---

[1] These letters, usually attributed to Mary, provide the main proof of her complicity in the murder of Darnley. It is impossible, here, to refer even briefly to the long drawn-out controversy regarding their genuineness or spuriousness. Suffice it to say that, in the opinion of the writer, the question stands pretty much where it stood twenty-three years ago, when Andrew Lang, the best informed of those who believed that the letters were forged or partially forged, capitulated to the arguments put forward by T. F. Henderson on behalf of their substantial authenticity. (For the last phase of the controversy see T. F. Henderson, *Mary Queen of Scots*, 1905, vol. ii, app. A, the *Scot. Hist. Rev.*, Oct. 1907 and Jan. 1908, and A. Lang, *The Mystery of Mary Stuart*, ed. 1912, prefatory note.) Since then an attempt has been made by R. H. Mahon in *Mary, Queen of Scots* (1924) and *The Tragedy of Kirk o' Field* (1930) to restate the pro-Marian case from a new angle; but the conclusions of this writer, although they are backed by a prolonged and laborious investigation, cannot be regarded as more than an unconfirmed hypothesis. An interesting critical discussion of the entire controversy, from the Marian standpoint, is given by R. Chauviré in the *Revue historique*, vols. 174 (1934) and 175 (1935).

copies of them and transmitted their verdict to London, thereby giving Elizabeth the opportunity to adjourn the conference to Westminster, where she could keep an eye on it, compel the enigmatic Murray to speak more openly, and at the same time submit the whole question to a more careful investigation. Once this was done, the result was a more or less foregone conclusion. Under the influence of threats, cajolery, and promises that he would not suffer by taking the bold course, Murray tabled his evidence against Mary in detail; and after a minute examination lasting for days, followed by a survey of the case by the privy council, the conclusion was arrived at that 'in view of these vehement allegations and presumptions', Elizabeth could not 'without manifest blemish of her own honour . . . agree to have the same queen come into her presence, until the said horrible crimes may be, by some just and reasonable answer avoided and removed from her'. Since Mary could not be induced to take up the refutation seriously—her representatives had been withdrawn from the conference—a presumption was created that she had no valid defence to offer. On 10 January (1569) Elizabeth summed up the situation by declaring that nothing had been brought forward to impair the honour and allegiance of Murray and his supporters, and that nothing had been 'sufficiently proven' against Mary. Not proven! It was a prudent resolution for the English queen, because it left the door open for future negotiation; but it was a damning one for the queen of Scots, whose character was now hopelessly smirched. Ten days later she was removed from Bolton to Tutbury to begin her long imprisonment in England, and at the end of the month Murray returned to Scotland, under a safe-conduct, to complete, if possible, the destruction of the Marian party.

For the moment Elizabeth had seemingly won a victory over her defenceless rival; but she was soon to realize that it was a Pyrrhic victory. By imprisoning Mary in England she all unwittingly lost her own freedom, and for the next nineteen years lived in ceaseless anxiety lest her prisoner might escape. While the two queens lived, there could be no quiet in the realm. Far better would it have been—such is the irony of history—had Elizabeth allowed her defeated and discredited enemy to go whither she pleased. The catholic world, scandalized by her recent behaviour, would have treated her with cold contempt: she would have sunk into comparative insignificance, suffered, it may be, complete political eclipse; and her name would have left no mark on history. But the incarceration saved her from this. As the victim of an unjust fate she became invested with a halo of martyrdom, recovered her position in the eyes of her co-religionists, and lived to plague the inventor of her misery with even greater misery than she herself endured.

IV

YEARS OF CRISIS: 1568–75

IN no department of state action is it more difficult to foresee the future course of events than in that of foreign policy. The enemies of yesterday become the friends of to-morrow, and, vice versa, the friends of yesterday become the enemies of to-morrow, according as interests coincide or clash. The reign began with Spain in a benevolent role and France as the national foe to be combated sword in hand. But before twelve years had elapsed a regrouping became necessary. The conflict with France had given way to a dawning appreciation of the fact that differences were not irremediable, whereas the early protestations of goodwill by Philip II were found to cover a wide range of opposing interests that threatened to become, and did become, insoluble except by the arbitrament of war. These differences were partly political, partly economic; they were in some measure due to the aggressiveness of England; but they were also caused by the refusal of the Spanish king to accept the religious settlement in England as final. Indeed, it would be no exaggeration to say that, in so far as any event in history may be described as inevitable, the great struggle between England and Spain, that convulsed the later years of the century, partook of this nature. Its shadow lay heavy across the path of Anglo-Spanish relations almost from the commencement of Elizabeth's reign.

Let us endeavour to see how the quarrel originated. To begin with, the queen's policy of establishing relations with the elements of disorder in the dominions of the continental monarchies operated no less effectively in the Netherlands than in France. The seventeen provinces which formed this outlying bastion of King Philip's empire had never taken kindly to his rule. On the contrary, their insistence upon old liberties, charters, and immunities, dating back to Burgundian times, ran counter to the royal conception of monarchical authority; while their love of religious liberty, fostered by an intense absorption in material pursuits, collided with the rigid Spanish belief that the catholic religion was indispensable to the maintenance of civil obedience. Grievances multiplied apace when it was seen that the king intended to treat them as possessions, to convert

them into a *point d'appui* for the Spanish power in the north of
Europe, and to tax them more heavily—because they were
wealthier—than any other part of the empire for the furtherance of purely Spanish interests. The fact that Philip was a
foreigner, unimaginative, slow of mind, maladroit, and altogether devoid of that higher statesmanship that takes into
account the idiosyncrasies of the people committed to its care,
probably increased the difficulties in the way of a mutual
understanding, and lessened the chances of fruitful co-operation. Nor did it mend matters—rather the reverse—when he
abandoned the country for the more congenial soil of Spain,
in the summer of 1559, leaving it to be governed by a viceroy
and a council of state, whose policy was provocatively Spanish,
and whose authority was sustained by an army of 3,000 Spanish
veterans. The simmering discontent ripened quickly into defiant outbursts of popular feeling against the cardinal de Granvelle, *le diable rouge*, president of the *consulta*, who, rightly or
wrongly, was singled out as the evil genius of the administration.[1] Attacked on the one side by the native nobility, who
refused to co-operate so long as they were excluded from the
inner counsels of the government, he was exposed, on the other,
to the odium of the masses, who regarded him as the instigator
of the obnoxious 'placards' against heresy, the activities of the
inquisition, and the other repressionist measures undertaken in
the interest of public order. The situation, in fact, was precisely the same as that which prevailed in Scotland during the
regency of Mary of Lorraine. A nationalist revolt was clearly
presaged as soon as the political and religious discontent joined
hands.

On this fertile soil fell the seeds of English propaganda. The
object aimed at was not to challenge Philip's sovereign rights
over the Netherlands, still less to deprive him of his inheritance:
it was simply to increase the difficulties of the Spanish authorities in Brussels, to stir up, if possible, a 'Flemish Amboise', and
thus prevent the complete absorption of the provinces into the
Spanish political system. To this end, therefore, the campaign
against Granvelle was fomented by pamphlets written in England and circulated by English agents among the Flemish

---

[1] Granvelle was a native of Franche-Comté: 'sa vraie patrie était la Cour et sa
fidélité au prince lui tenait lieu de patriotisme'. (Pirenne, *Histoire de la Belgique*, iii.
400.)

cities; and at the same time agitators in the pay of England mingled in the calvinist preachings, with a view to fostering religious discontent. The extent of the propaganda cannot be fully known, because it was carried out with great secrecy and relied for its success upon nameless individuals, often renegade Spaniards as well as Flemings, and high-placed officials in the government, who chafed at the autocratic rule of the cardinal.[1] That the king knew of it is certain from the dispatches of his ambassadors as well as from the correspondence of Granvelle himself. But it was impossible for him to check it, owing to the close commercial intercourse of England and the Low Countries at this time, and the comparative ease with which the disturbed areas could be reached. So serious did the situation become that not only was chaos produced in the administration of the Netherlands itself, but the European policy of Spain was also profoundly affected. The dismissal of Granvelle by the king in 1564 appeased, but did not end, the turmoil; for the nationalist movement was only just beginning to take definite shape, and the protestant agitation, stimulated by huguenot immigrants from the south, received a fresh access of strength. A climax came in 1566, when the fanatical populace, fired by the preachers, began a widespread attack on the churches, creating general anarchy. With this direct challenge to social order and the royal power, Elizabeth was careful not to identify herself. Her purpose had been served: she could afford now to condemn the rebels who had hitherto been her tools, and profess to the Spanish ambassador in London that her sympathies were entirely with his master. Meanwhile thousands of fugitive Flemings—there were from eighteen to twenty thousand of them in London and Sandwich in 1563, and thirty thousand in 1566—escaping from the troubled country, flocked over to England to be warmly received by the queen, who gave them permission to settle, practise their trades, and worship in their own calvinist conventicles. Thus England was doubly the gainer by the convulsed state of the Netherlands—Spain was weakened and held in check, and the immigrant foreigners were a valuable asset to the economic and industrial life of the English nation.

But if Elizabeth was treading on dangerous ground with

---

[1] e.g. Simon Renard, formerly Spanish ambassador in England, and Erasso, secretary to the holy inquisition, both of whom were enemies of Granvelle.

respect to Spain's authority in the Netherlands, Philip was far
from passive in his relations with the queen's catholic subjects
in England. Although he remained outwardly friendly, and
the instructions he gave to his ambassadors were uniformly
prudent, it soon became common knowledge that the Spanish
embassy in London dispensed largesse to needy catholics,
opened its doors to all who desired to practise the forbidden
ritual of the Mass in secret, and compromised itself with
notorious rebels like Shane O'Neill. The consequence was that
a careful watch was kept on its activities—so strict, indeed, that
de Quadra grumbled of the 'distrust that is publicly shown
to all who associate with me'. 'Not a prisoner is arrested for
state reasons', he remarked, 'without his being asked whether
he had any conversation with me.' But worse was to follow. In
the summer of 1562 Cecil bribed the secretary to the embassy,
Borghese Venturini, to turn queen's evidence against his master, and de Quadra found it difficult to reply to the formidable
dossier of unfriendly acts which the council drew up. A few
months later (Jan. 1563) he was charged with sheltering a
criminal who attempted the life of the vidame de Chartres, an
envoy of the huguenots, and deprived of the keys of his house,
a government custodian being placed in charge. At the same
time de Quadra was solemnly warned that if he overstepped
his duties as ambassador he would be tried by the laws of the
land, his extra-territorial privileges notwithstanding. In February his privacy was invaded by the marshal of the court
with a company of halberdiers in search of certain English
catholics, who were said to be at Mass in the embassy chapel;
and a number of Flemings, Italians, and Spaniards were carried off to prison in defiance of his remonstrances. Although
de Quadra complained bitterly to the king, the only comfort
he received was an instruction to bear his trials with fortitude.
Matters improved under his successor, de Silva, who was caution personified; but the potential danger from the embassy
continued, as may be seen from the fact that, during the year
1565-6, it became the principal channel of communication
between Mary Stuart and Spain. All this, however, was merely
preliminary to the period of turmoil that set in when de Silva
was replaced by Don Guerau de Spes, in 1568. To him we
shall return presently.

Political friction was bad enough, but not so dangerous to

the maintenance of good relations as the economic rivalry which followed as a necessary consequence upon England's expanding trade. Hitherto, as a result of the opening up of the world in the fifteenth century, the most delectable parts of the earth had fallen to Spain and Portugal. West Africa from Senegal to the Congo, a region of incalculable resources, was in the possession of the Portuguese, who claimed and exercised a prescriptive right to political jurisdiction and an exclusive trade monopoly over the entire area. A still more extensive and equally rigid monopoly was maintained by Spain in Central America, both on the mainland, from the Orinoco to the cape of Florida, and in the islands of the Caribbean sea. Although both monopolies had some legal basis, they cannot be said to have commanded general respect in a world ruled by force and not by right. The Portuguese sphere of influence had been invaded frequently by English traders before Elizabeth's accession, in defiance of protests from Lisbon, and the activities of the Portuguese naval squadron based on El Mina; and it was only a matter of time when the lure of tropical America would entice enterprising traders to the more remote and less familiar Spanish preserves on the other side of the Atlantic.

In 1561 Cecil scandalized de Quadra by remarking that the pope had no right to partition the earth and bestow kingdoms on whom he pleased—a significant indication that England at least did not recognize existing monopolies. A year later the validity of the Spanish monopoly was put to the test by John Hawkins of Plymouth, whose father, William, had trafficked with Guinea, and probably supplied his son with the initial incentive for his enterprise. Hawkins's idea was to effect an entry into the lucrative trade in African negroes which now formed an important constituent element in the internal economy of Spanish America. Relying on the urgent needs of the Spanish planters for slave labour to make good the wastage of the aboriginal population, and on the help of friendly Spaniards in the Canaries, who were prepared to furnish him with pilots for the journey across the Atlantic, he easily procured a cargo of negroes, partly by raiding native villages on the African coast south of Cape Verde and partly by purchase from Portuguese slavers, and set sail for Hispaniola in the autumn of 1562. As Hawkins was in no sense a pirate but a peaceful trader, reasonable in his charges and professing to be a friend of King Philip, the success of the venture surpassed expectations. He returned to Europe with a valuable freight of gold, silver, pearls, hides, and sugar, and with the consciousness that, in the Indies at least, the Spanish monopoly was not seriously regarded. Moreover, he had taken the precaution of obtaining certificates from the authorities as to his good behaviour and honourable intentions. But his moderation had little weight with the government at Madrid; for when he sent part of his cargo to be sold in Seville, as a further proof that his actions were above board, it was seized and confiscated by the port officials. In spite of this set-back, however, his ardour for the enterprise was in no way damped; and two years later he planned a second expedition on the same lines as the first but more elaborate. On this occasion he had the patronage of the queen, who loaned him one of her own ships, the earls of Leicester and Pembroke, and Cecil—all of whom took shares in the undertaking. Meantime, in view of the attitude of the Spanish government, which had issued strict injunctions to the administration in the Indies not to allow trade with the English, Hawkins had armed his flotilla. Fortunately there was no need as yet to abandon the pretence of peaceful trade; and although the Spaniards were reluctant to do business with him, Hawkins managed by cajolery and threats to dispose of his cargo of negroes at Borburata and Rio de la Hacha. He returned to Europe again with a substantial profit.

Spain was now thoroughly on the alert. Precautions were redoubled both in Europe and in America and at the Canaries, and every move of Hawkins was carefully watched by the Spanish ambassador in England, who reported his doings to Madrid. The consequence was that when he prepared his third expedition the circumstances from the start were far from favourable, and Hawkins must have known that his appearance in America might lead to trouble. The secrecy with which the details of the enterprise were screened from the prying eyes of de Silva seems to be a proof of this. Nevertheless in the autumn of 1567 he set out once more for America, supported on this occasion by no less than seven well-armed ships—the *Jesus, Minion, Judith, William and John, Swallow, Angel, Gratia Dei*—with Francis Drake in command of the *Judith*.

At every stage of the journey difficulties and troubles assailed him. An attempt was made at Plymouth to prevent his departure by the Flemish admiral, de Waachen. On the African coast he had to engage in inter-tribal war in order to obtain his negro cargo. At Teneriffe the Spanish governor tried to lure him under the guns of the fort. And when he arrived at his base of operations in America no one would have anything to do with him. But Hawkins was not to be denied. At Rio de la Hacha he opened fire on the treasurer's house, blockaded the port, and fought a pitched battle with the Spaniards on land; and it was only when he had partially burnt the town and seized its treasure that the inhabitants consented to take part of his wares. Cartagena, his next port of call, proved to be too strong for intimidation; and Hawkins was compelled to put into the harbour of San Juan de Ulloa to refit his ships and refresh his men (20 Sept. 1568). Three days later, and some time before it was expected, the Spanish fleet appeared in the offing, with the viceroy of Mexico, Don Martin Enriquez, on board. Hawkins was now in a position of some danger, for the Spanish ships were in the majority and he could not refuse to admit them to the roadstead. He therefore made terms with them, and the two flotillas anchored side by side. The viceroy, however, was bent on treachery from the first; and on a preconcerted signal, aided by troops secretly brought from Vera Cruz, the Spaniards attacked Hawkins's ships. A battle ensued, in the course of which several English ships were lost, including the *Jesus*, together with part of the treasure it contained. In January and February the remnant of the expedition, consisting of the *William and John*, the *Minion*, and the *Judith* arrived in England, much dilapidated, with over 120 men missing, dead, or prisoners of the Spaniards or Indians.

It was impossible any longer to disguise the fact that although England and Spain were nominally at peace in Europe, a state of war prevailed 'beyond the line'. The bad blood engendered by the affray had an incalculable effect in embittering the growing enmity between the two peoples, however much their political leaders might disguise the fact by diplomacy.

Meanwhile another field of economic rivalry was opening up in Europe. The main export trade of England at this time was conducted with the Spanish Netherlands, where the cloth from English looms found a ready vent in the busy market of

Antwerp. There was, of course, no monopoly in this traffic; nor was there any necessary opposition from Flemish merchants to the entrance of English goods: on the contrary both countries profited by the economic interchange. But government action introduced an element of unfairness into the traffic, which acted disastrously on the maintenance of good relations. Guided by the prevailing mercantilist ideas of the age, the Elizabethan government endeavoured to establish an advantage in favour of England against the Netherlands by restricting the privileges of Flemish merchants in England. Customs dues and other charges on many articles imported into England from Flanders were raised, and others were prohibited altogether, while the preference in the carrying trade was given to English ships. At the same time the Flemish merchants were placed under the obligation of spending the proceeds of their sales on the purchase of English goods before leaving the country. To this undoubtedly severe attack on the prosperity of the Netherlands the Spanish government at Brussels retorted with corresponding restrictions on English traders in Flanders, and on English goods imported into Flanders by Englishmen. Likewise the shipment of goods from Flemish ports in English ships was interdicted. Thus arose the economic *impasse* of 1563-4, when many English merchants temporarily abandoned Antwerp for the German port of Emden. The aggressive commercial thrust of England was undoubtedly an additional cause of friction.

In the third place there was the question of piracy on the high seas. The main sea-borne traffic of northern Europe passed along the commercial gut of the English Channel, and a large part of it was conducted between Spain and Antwerp, either in Flemish or Spanish ships. For the greater part of its route this traffic was subject to piratical attack, for the Channel swarmed with freebooters. Some were French, others English; but generally their lairs were situated on the English rather than on the French side of the Channel, because of the greater convenience of the numerous bays and creeks of the south and south-east coast of England. English shipping suffered little by the disorder at first, except by reprisals; but it was obviously to the interest of all countries abutting on the Channel to cooperate in its suppression. Spain naturally looked to England to give a lead in the matter, and numerous complaints were addressed to Elizabeth from Madrid as well as Brussels urging drastic action. The difficulty, however, of giving effect to measures against the pirates was great; and restitution was exceedingly hard to obtain. De Silva, writing in August 1564, was inclined to lay the blame not on the lack of energy of the English government, which indeed had behaved well, but on the remissness of admiralty judges, and the fact that the rogues, when captured, were usually unable to pay for their depredations. The desire of the government to combat the evil is apparent from the repeated efforts of the Privy Council in 1564, 1565, and 1566 to devise a satisfactory plan for tracking down the pirates, keeping a careful watch on their accomplices on land and their bases and sources of supply. Yet, in spite of all preventive measures, and the barbaric severities meted out to pirates taken red-handed, the pillage continued. Matters were made even worse when the huguenots and the Dutch *gueux de mer*, or sea beggars,[1] turned piracy into a weapon against the political aims of the French and Spanish monarchies. Although the Elizabethan government cannot be held responsible for the losses incurred by Spanish shipping in the Channel, the fact that the pirates were frequently English undoubtedly raised a presumption in the mind of Philip that Elizabeth was hand in glove with the nefarious traffic; and in this he was encouraged by de Spes, who, unlike his predecessor, refused to credit the queen with a single friendly act toward his master. Piracy, therefore, may be added to the causes producing trouble between England and Spain.

There were, of course, other matters, notably the treatment of English sailors in Spanish ports by officials of the inquisition, the arbitrary action of port authorities in both England and Spain, and the patronage by Spain of the polemical catholic literature introduced into England from the continental presses. But the prime source of antagonism was the clash of economic interests, and the political opposition induced by the religious changes in England.

By 1568 it must have been apparent that Anglo-Spanish relations were fast approaching a crisis. Everywhere the interests

---

[1] '*Gueux* signifie autant que *vagabond* ou *ribaud*' (M. Van Vaernewyck, *Troubles religieux en Flandres . . . au XVI° siècle*): in English *gueux de mer* is usually rendered as 'sea beggars'. For the incident out of which the term *gueux*, as applied to the rebels, arose, see Motley, *The Rise of the Dutch Republic*, i, chap. vi.

of the two countries were colliding. In the Netherlands the duke of Alva, who had taken over control from Margaret of Parma in the previous autumn, was engaged in extinguishing the last embers of the nationalist revolt, preparatory to a thorough-going 'hispanicization' of the provinces. His 'council of blood' with its proscriptions, judicial murders, and confiscations was as efficient an engine of repression as the revolutionary tribunal in Paris during the Terror. The whole country was soon cowed into abject submission; and Alva was now in a position to proceed with the second and more important part of his programme, viz. the conversion of the Netherlands into a *citadelle d'acier*, from which the king of Spain could lay down the law to the rest of Europe.[1] All pretence of constitutional government was abandoned in favour of a purely military dictatorship wielded by the duke and the little group of Spaniards who formed his immediate entourage—Vargas, del Rio, and Albornoz. The worst feature of the tyranny was not so much its ruthlessness as its complete disregard of the delicate machinery of industry and commerce, which was brought to a standstill, with disastrous repercussions on England's principal overseas market. Nor was this all. It was hardly possible to mistake the ultimate significance of Alva's measures, for it was observed that while he sedulously crushed the spirit of the Netherlanders, he expressly favoured the bands of English catholics who had taken refuge in the country for conscience' sake, and encouraged them to enlist in his armies. They, for their part, looked to him as the champion of the catholic cause to crown his career by reducing England to the Roman obedience. Thus there was every incentive to Elizabeth to be on her guard against a possible hostile move from the Low Countries, and to take such precautionary steps as would prevent it.

There were two obvious methods of doing this, without, at the same time, throwing down the gage openly to Spain. She might, for example, permit secret assistance to be sent to the patriot cause in the Netherlands from the Flemish and Dutch communities in England; and, on the other hand, she might recognize the *gueux de mer*, who, acting under commissions from William of Orange, were engaged in harassing Spanish commerce in the Channel. Neither action could be regarded as friendly to Spain, and either might lead to compli-

[1] Pirenne, op. cit., iv. 16.

cations; but the risk had to be taken, for the need was urgent. Consequently a good deal of assistance, both in the shape of money collected among the refugees in England and volunteers recruited in the same quarters, was dispatched from English ports to the Low Countries; and at the same time pirate commanders like La Marck, who flew the flag of Orange, were permitted to load ordnance and munitions in England. Of a similar nature, only more dangerous in its possible consequences, was the seizure of Spanish treasure in December 1568. It is difficult to believe that this action was not related in some way to the disaster at San Juan earlier in the year, for news of the Mexican occurrence had reached England by 3 December. In a letter to Cecil, written on this date, William Hawkins urged the secretary to 'advertise the queen's majesty thereof, to the end there might be some stay made of king Philip's treasure here in these parts till there be sufficient recompense made for the great wrong offered, and also other wrongs done long before this'. In any case, when the Spanish bullion ships, carrying the considerable sum of 800,000 ducats, the property of Italian bankers, who had loaned it to the king for the payment of Alva's troops in the Netherlands, put into Plymouth and Southampton to escape the clutches of the Channel pirates, a convenient pretext was easily discovered for detaining it. Obviously the queen could not be expected to have it convoyed to its destination in her own ships, and to let it proceed on its way unconvoyed was tantamount to making a present of it to the sharks lying in wait outside. For a few days she played the honest broker; and then, on learning that the money was not technically the king's, according to the contract, until it was delivered in the Netherlands, she decided to borrow it for her own use. Instantly, and without waiting to learn all the facts of the case, de Spes wrote to Alva urging immediate reprisals, and in the last days of December an embargo was laid on all property belonging to Englishmen in the Low Countries. News of the duke's action reached London on 3 January: on the 7th, orders were issued by the council for corresponding arrests of Spanish property in England; and on the following day de Spes was placed under arrest in his house, for acting without the king's orders to the prejudice of the ancient alliance and commerce of the two kingdoms.

Ostensibly de Spes had been sent to England, in July

1568, as a peace-maker. His instructions were to satisfy the
queen about the alleged ill-treatment of her ambassador, Dr.
John Man, whom the king had forbidden his presence because
he had publicly shown contempt for the catholic religion—a
serious offence in Spain—and had uttered opprobrious words
about the pope. De Spes seems to have imagined, however,
that his mission was in the nature of a crusade. Even before
he set foot on English soil his fiery zeal for the catholic faith
led him to overstep his instructions. When passing through
France he informed Mary Stuart's ambassador, the bishop of
Glasgow, that he had orders to do something for the royal
captive as soon as he reached his destination. A few weeks
later, when he had established contact with Mary's partisans
in England, he expressed the opinion that it would not be difficult to release the queen of Scots and 'even raise a revolt against
this queen'. On 9 November he thought the propitious hour
for action was approaching.

To de Spes the crisis provoked by the seizure of the treasure,
the embargoes, and the trade stoppage was nothing less than a
heaven-sent opportunity: it would provide him with the necessary lever for overthrowing Elizabeth's government, restoring
catholicism, and seating Mary Stuart on the throne. Early in
January Mary conveyed to him the message: 'Tell the ambassador that if his master will help me, I shall be queen of England
in three months, and Mass shall be said all over the country.'
In communicating this news to the king, de Spes stressed the
necessity of developing and extending the economic blockade.
'If they cannot work', he wrote, 'or there is any obstacle to the
disposal of their goods, they usually take up arms.' Let the king
check the vital imports—oil, alum, sugar, spices, and iron from
Biscay—as well as the exports, and he would have the fate of
the queen in the hollow of his hand. Philip was not impressed
by his ambassador's assurance; and, on 18 February, he consulted Alva as to the feasibility of the scheme.

In the meantime the indefatigable de Spes had established
connexion with a conspiracy among the older nobility for the
displacement of Cecil, that had been brewing for some time.
The causes of the antagonism were many and varied—discontent with the policy pursued in regard to Mary Stuart and the
succession: irritation at the growing estrangement from Spain:
personal pique at the power enjoyed by one who did not belong

to the ranks of the aristocracy; and, in some cases, a profound antipathy to the ecclesiastical innovations and the severance of the kingdom from Rome. The conspirators thought that if they could unseat Cecil and destroy his influence over the queen it would be possible to settle the succession in favour of Mary, set her free, and restore her to Scotland, with guarantees safeguarding Elizabeth during her lifetime. At the same time Anglo-Spanish relations might be rectified by restoring the treasure; and the peril created by the breach with Rome might be averted by a return to catholicism. In fact what the malcontents aimed at was a complete reversal of national policy. The plot arose in the minds of the duke of Norfolk and the earl of Arundel; but it was soon participated in by the northern earls, Northumberland and Westmorland, the earls of Pembroke, Leicester, and Derby, and Lords Montague, Dacre, Morley, and Lumley. On 29 February de Spes notified the king that he had been approached by a certain Roberto Ridolfi, a Florentine, in the name of Norfolk and Arundel, who had sent him to say that they were in favour of a settlement with Spain, that although they were not yet strong enough to challenge Cecil, they were gathering friends and hoped to turn out the present government, replacing it with a catholic one. They also thought that they could bring the queen to consent thereto. Behind them stood the serried ranks of a phantom catholic host who, de Spes fondly believed, were preparing to move at the command of the duke and his supporters.

Undoubtedly the situation was becoming serious—all the more so since relations with France were far from satisfactory. After several years of uneasy peace, that country had again plunged into the throes of civil war, and public opinion in England expressed itself strongly on the side of the distressed protestants. Although the queen professed her neutrality in the struggle, she did not prohibit her subjects from sending munitions and money to the huguenot stronghold of La Rochelle, and bands of enthusiastic volunteers were flocking to Coligny's help under Sir Henry Champernoun, Walter Raleigh, and other west-country gentlemen. In the Channel, too, English privateers co-operated with the huguenots in the destruction of 'Guisan' commerce. So intolerable did the state of affairs become that on 15 January 1569, at the very moment when the crisis with Spain began, Charles IX ordered the arrest of

English goods at Rouen. It was a retaliatory step, he said, undertaken in order to reimburse his merchants for the losses they had sustained by piracy. The tension lasted for some months, and de Spes was in high hopes of bringing France into the plot for Elizabeth's overthrow. Fortunately, however, although the French tried to make some capital out of the queen's embarrassment, they refused to play the Spanish game, being afraid that, owing to the intimate economic ties between England and the Netherlands, Alva and the queen might at any time make up their differences and leave France in the lurch. Nevertheless Anglo-French relations were strained almost to the breaking-point, in March, when La Mothe Fénelon, Charles IX's ambassador in London, presented a formidable list of grievances, and required Elizabeth to declare within fifteen days whether it was to be peace or war.[1] If the rupture was avoided, it was only by the exercise of considerable tact on the queen's part.

Meanwhile Alva's statesmanlike mind was assailed by grave doubts about the practicability of de Spes's plan for the reduction of England. On 10 March he wrote to Philip discountenancing war, and pointing out that he lacked both ships and money for the enterprise. The ambassador, however, was not to be thwarted. Norfolk and Arundel, he said, only wanted Spain to stand firm in regard to the trade stoppage, and the people, who were already beginning to move, would rise. It would not be difficult to bring the 'barefaced thieves' to account: he knew that the queen and Cecil were now 'suspicious even of the birds of the air', and that the former was 'abandoned by many and hardly any one really liked her'. If only Alva could seize the English cloth fleet on its way to Hamburg, it would precipitate the revolution. But the duke was not to be stampeded from reality by the heated imaginings of a fanatic. On 4 April he wrote to the king: 'Notwithstanding what Don Guerau writes, I am not yet convinced that they are not deceiving him.' Philip, too, was veering round to the view that it would be better to come to an agreement with Elizabeth, for his own subjects were petitioning him to bring about an early restoration of trade and their confiscated property. About

---

[1] 'Tout ce royaume est en suspens de la guerre, craignant de l'avoir tout à la fois avec la France et l'Espagne, ou separément avec l'une ou l'autre.' (La Mothe to Charles IX, 8 March 1569: *Correspondance diplomatique de La Mothe Fénelon*.)

the middle of May two letters crossed one another on the way to Madrid and Brussels, respectively. The first, from Alva to the king, stated that it was imperative that the restitution of the goods should precede the breach with England; the second, from the king to Alva, explained that the Spanish council of state had decided that war was undesirable, since it would not help the major matter, the recovery of the lost property. In other words, the plot was at an end so far as Spain was concerned, and the 'climb down' had begun.

In England the conspiracy was also at a standstill; for the leaders were afraid to move, and were waiting in vain for a sign from Alva and the expected popular demonstrations against the government. Nor was the blockade working according to plan. In spite of the trade stoppage, English goods were being passed into Europe through the friendly ports of La Rochelle and Hamburg, and imports were coming into England in Venetian and Ragusan bottoms. De Spes became more and more frantic in his efforts to stop the leakages. Spain, he said, must enlist the pope's help in arresting English trade with the seignories of Venice, Genoa, Lucca, and the dukedom of Florence, get the emperor to seize Hamburg and the mouth of the Elbe, and keep the English out of Poland, Muscovy, and the Easterling country. It would be advisable, too, to harass English ships in the Channel and neighbouring waters. In short, he had discovered that before England could be brought to her knees every port in Europe would have to be closed against her and a Spanish naval supremacy established on the high seas!

While de Spes was vapouring about blockades, the Norfolk-Arundel plan for seizing Cecil and overthrowing the government collapsed: partly through its betrayal by Leicester, who had been admitted into the plot in order to strengthen the attack: partly by the astuteness of Cecil, who had got wind of it from Leicester, and talked over Norfolk; and partly by a dispute between Norfolk and Dacre over the Dacre inheritance. Alva, too, had been busy separating the property-restitution question from the political revolution which de Spes was trying to promote; and on 12 June he had written to Philip stating that in his opinion war with England, which would assuredly follow any attempt to help Mary Stuart with men, was clearly impossible, and that money and advice was all that could safely be sent. He therefore proposed to sound the English government as to the possibility of a restitution by sending over a trustworthy Genoese merchant, Thomas Fiesco, to broach the matter unofficially.

In view, however, of de Spes's ardent advocacy of intervention by Spain in the internal affairs of England, it was necessary for Alva, if his new plan was to succeed, to stop the activities of the ambassador. On 2 July he wrote ordering him not to entertain any approaches made to him against the queen and her councillors, but to exercise the greatest reticence and neutrality in the internal affairs of the country. On 14 July he again wrote: 'I must again press this upon you and tell you that I am informed from France to-day that the queen of Scotland is being utterly ruined by the plotting of her servants with you, as they never enter your house without being watched. This might cost the queen's life, and I am not sure that yours would be safe.' On 8 August he complained to the king that in spite of all warnings de Spes would not do as he was told. De Spes, in fact, was still in full correspondence with the malcontent group of nobles. They had abandoned their plan of a *coup d'état* against Cecil, but they were still pressing forward their general plan for the declaration of Mary's title to the succession, and her liberation, coupled with a scheme for her marriage to Norfolk. In this they had the support of the whole council including Leicester, but the queen was in the dark and Cecil doubtful. Help from France was also invoked.[1]

The plotters, however, were soon to be disillusioned. About the middle of September Elizabeth got wind of what was on foot, and peremptorily forbade the marriage. Before the end of the month the back of the conspiracy was broken. Norfolk had been severely warned, Leicester had sought and obtained the royal pardon, de Spes was being carefully watched and his dispatches opened. Even La Mothe's correspondence was inspected, showing that the government was leaving nothing to chance. Moreover, throughout the country preparations were made to meet any emergency. Commissions were sent down from London to see to the musters: every family was ordered to possess itself of at least one serviceable weapon and practise

---

[1] Early in September Norfolk negotiated with Charles IX for the dispatch of a French force of 5,000 or 6,000 to Dumbarton; and the king agreed (20 Sept.) to give him all the help he could—'de gens et d'argent que Dieu m'a donné'. (*Correspondance diplomatique de La Mothe Fénelon*.)

its use: the coast defences from Yarmouth to the Lizard were placed in a state of preparedness: the fleet was augmented and provisioned, and lay at the mouth of the Thames: warships convoyed the cloth fleet to and from Hamburg, and the wine fleet to and from La Rochelle. Mary Stuart at Tutbury felt the rigour also, for in September she wrote to La Mothe: 'Si je demeure un temps ici, je ne perdrai seulement mon royaume mais la vie.'

While the tension was still at its height, the Northern Rebellion broke like a thunderclap over the country. In Tudor times the valley of the Trent marked, as it had done for centuries before, the great divide between northern and southern England. Beyond Trent, stretching from the Peakland to the Cheviots, lay a country whose aspect grew more and more forbidding as the great *massif* of the Pennines dominated the landscape and broke the plains into a tumbled mass of isolated dales and pasturelands, rocky uplands, bleak moors, forests, bogs, and wastes. To the southerner this distant and inaccessible region, with its sunless and tempestuous winters, its comparative lack of roads and bridges, its sparse and rude inhabitants, and its general suggestion of niggardliness, must have seemed little better than the unredeemed wilderness. The agrarian and industrial changes that had transformed the face of southern England, and were fast destroying the last vestiges of the old medieval economy, here found themselves impeded by natural obstacles and by a people whose spirit was opposed to change and slow to adapt itself to new conditions. Agriculture was largely confined to the vale of York and the coastal plains, and the woollen industry had made only a rudimentary beginning in the West Riding of Yorkshire, eastern Lancashire, and the palatinate of Durham. It was inevitable that a region so circumstanced should give rise to serious administrative problems, for its lack of an ordered life based upon economic pursuits perpetuated the turbulence of its inhabitants and made them a drag on the wheels of progress. For the young there was no choice of a career apart from service in the households of the nobility and gentry. In fact feudalism was a necessary element in the northern economy: it provided both a means of livelihood and a rallying point for the general life of the community. But it was a different kind of feudalism from that which prevailed in the south—less affluent, but deeper and stronger of root, more intimate, more militant. The great families of Percy, Neville, and Dacre, whose honours, baronies, tenements, and dependencies covered the bulk of the land between Yorkshire and the Border, were more like petty sove-reigns than subjects, for the feudal tie was so strong that it was commonly said of the Northumbrians that they 'knew no prince but a Percy'. Even the authority of the Crown, which made itself felt only intermittently when subsidies were levied or statutes proclaimed, or when justice was to be executed, was interpreted in terms of feudalism: the great magnates being the intermediaries through whom the commands of the queen were transmitted to the people. As lords lieutenant, justices of the peace, and dispensers of patronage they were at one and the same time Crown officials and the natural leaders of society.

But feudalism was not the only bond of union in the north, nor, perhaps, was it the most powerful. Lord and tenant were drawn closer together by a common interest in the defence of the old faith than they were by ties of a material kind; for the religious innovations of the queen and the reconstruction of the church which followed seemed to portend a complete overthrow of the main bulwark of the old order north of Trent. This was a matter on which all classes could combine, for it affected all. Hence as soon as the new Elizabethan bishops essayed to enforce the prayer book and church attendance they found themselves thwarted at every turn by a recalcitrant laity, backed by disloyalty among the Crown officials and stimu-lated by the obvious shortage of a protestant clergy. Recusancy began almost at once; and if in some parts the people con-formed so far as to go to church, the 'lees' of the old faith still lay 'at the bottom of men's hearts' ready to come to the surface if 'the vessel was ever so little stirred'. It was not the policy of the queen, however, to stir the vessel too much during the first ten years of her reign, and the despairing cries of the bishops were not heeded. The council in the north, which had done great service since its re-establishment by Henry VIII in removing many social grievances, was not disposed to interfere in the matter of religion; for its presidents were, with one excep-tion, complacent men, moderate in their opinions, and friendly to catholicism. The consequence was that England beyond

Trent remained an overwhelmingly catholic country—so much so that it almost seemed as if the government were content to recognize the territorial principle, *cuius regio eius religio*, as applicable to these parts.

There were dangers, of course, in such a situation; for the cleavage in religion might easily become the basis of a political antagonism, which, in its turn, might lead to another conflagration of the 'inly working north', similar in kind to that which shook the throne of Henry VIII. And if there was plenty of combustible material, there was no lack of flying sparks to set it aflame. Ever since the beginning of the reign Elizabeth had shown a disposition to weaken her over-mighty subjects by transferring important offices from their hands to new men, on whose reliability she could depend. Custodianships of castles, stewardships of royal manors in the north, posts of responsibility like the wardenship of the marches were taken out of their control. Similarly, when the youthful heir to the Dacre inheritance at Naworth died accidentally, in May 1569, the estates were conferred upon the Duke of Norfolk, to the exclusion of Leonard, uncle of the deceased, who by ancient usage should have taken both the title and the lands. Moreover, it was part of the royal policy to insist on the northern earls residing at court—no doubt because the queen believed they would behave themselves better under her eye than if they were left to brood in their native fastnesses. Residence at court entailed expense, and expense was a serious consideration to men whose revenues, like the countryside from which they were derived, were smitten by the blight of poverty. Finally to these private rancours were added public grievances—discontent with the queen's stubborn refusal to settle the succession question in favour of Mary Stuart, her peremptory veto of the Norfolk marriage proposal, and her imprisonment of the Scottish queen. Over all flickered the hostility to Cecil and his policy.

Strictly speaking, the rebellion should have begun when Norfolk, disappointed of his purpose, left the court in high dudgeon in September, and betook himself to his country house at Kenninghall. De Spes thought he had gone to raise the standard of revolt, and Mary counselled him to be bold. But the duke was a coward. Fearing the queen's resentment, and believing that his imprisonment was imminent, he craved the royal pardon in a letter of 24 September, and wrote to his fellow conspirators to dissuade them from action. He probably knew that Pembroke and Arundel were already under arrest, and that all who were privy to the plot were being watched. Eventually, trusting to Cecil's assurance that he would be leniently treated if he made his submission, he consented to set out for the court, and was intercepted at Burnham by Sir Francis Knollys and Sir Henry Neville, who conducted him to the Tower. Meanwhile the two northern earls, hearing of Norfolk's imprisonment, on 10 October, met the following day in Northumberland's house at Topcliffe in Yorkshire, to decide what course to take. Northumberland, who was loath to take extreme measures because of the danger in which Mary Stuart might be placed, and also because he had no stomach for rebellion, was overborne by the more fiery Westmorland, and by his 'company' who 'threatened him with daggs', saying, 'If ye will not cast yourself away, ye shall not cast us.'

About the same time the earl of Sussex, lord president of the council at York, had received instructions from the queen to warn all justices of the peace in his shire and throughout the north to keep good watch and ward in all boroughs, market towns, and places where people congregated, and to advertise any unlawful action attempted. Sons of northern families at the universities were detained in residence until the situation cleared. The lord president, whose information appears to have been defective, was inclined at first to make light of the rumours of rebellion, but in accordance with orders took preliminary measures for the defence of York, Hull, Pomfret, and Knaresborough, and obtained assurances from Northumberland and Westmorland that the 'bruits' were false. On 13 October he reported all quiet to the queen, and trusted 'the time of the year would cool hot humours'. But Elizabeth was troubled by the messages reaching her from other sources, and instructed Sussex to order the earls to repair to court forthwith (24 Oct.). The lord president did so on 30 October, but still maintained that he had heard of no conventions of multitudes at unlawful times and places. Sir Thomas Gargrave at Pomfret, however, had better information, and on 2 November reported to Cecil the names of the 'great doers in these matters' —Robert Tempest, John Swinburn, Thomas Markinfield, Francis Norton, Thomas Hussey, Christopher Danby, and Highingham of Yorkshire—but was also inclined to think there would be no rebellion. On 4 November Sussex, replying to the earls, who excused themselves from obeying his command, issued a fresh and more peremptory summons ordering them to be at York by the following morning. Before his messenger left Top-cliffe on the return journey, he heard the church bells sounding the tocsin for rebellion. But Sussex repeated his attempt to keep the earls to their obedience—to be informed on this occasion, by the countess of Northumberland, that her husband was gone, having received alarming messages from his friends at court. On the 12th it transpired that the earls were at Brancepeth, Westmorland's castle, with their followers in armour. He therefore decided to order a general mobilization of the armed forces of the north in Richmondshire and the bishopric; while at the same time he issued a proclamation denouncing the treason of the earls. Nevertheless, in spite of this, he made a fifth and last attempt to recall the misguided noblemen to their duty: 'If you have slipped', he wrote on 13 November, 'your friends will be suitors for you to the queen, who never shows herself extreme and has always borne you affection.' Two days later news arrived that the rebels had entered Dur-ham, celebrated Mass in the cathedral, and trodden the prayer book under foot. The die was cast. Westmorland, who now took the lead, informed Sussex that he proposed to remain at Brancepeth and fortify himself against his enemies who sought his destruction. No further parley was possible, and Sussex prepared himself for the worst.

Meanwhile the poor response to his call to arms was giving him grave concern: 'Except it be a few protestants and some well affected to me', he wrote, 'every man seeks to bring as small a force as he can of horsemen, and the footmen find fault with the weather and besides speak very broadly.' It was obvious that the north could not provide for its own defence; so Sussex had to be content with the protection of York itself, and dispatched an urgent appeal for help to the government.

The proclamation published by the earls after the occupation of Durham made it appear that their hostility was directed not against the queen, but against 'divers evil disposed persons about the queen', who by their 'crafty dealing have overthrown the catholic religion, abused the queen, dishonoured the realm, and seek the destruction of the nobility'. Force, they said, must be resisted by force. The ancient customs and liberties must be restored 'lest if we do it not ourselves strangers will do it to the great hazard of the estate of our country'.

On 20 November Sussex, having received more ample commission from the queen, proclaimed the earls rebels, offering pardon to all who repaired to their houses before the 22nd. On the same date he explained to Elizabeth that he had only 400 horsemen, ill horsed and badly furnished, and a great number with only bows and arrows: that the rebels had interrupted his mobilization, spoiling the men of their weapons and money: that the justices of the peace and gentlemen were mostly holding aloof; and that the rebels were strong in horse, having the reivers of Tynedale and Redesdale with them. Lord Hunsdon, who had been sent from the court to take command at Berwick, found his progress barred beyond Doncaster, and was compelled to go round by Hull; but he had heard from Lord Darcy's men, who were holding Doncaster for the queen, that the rebels intended to attempt the rescue of Mary Stuart at Tutbury. Sadler, who followed Hunsdon, also went round by Hull in order to reach York. Both men in their reports to the queen dwelt on the seriousness of the position and justified Sussex's inactivity in the face of heavy odds.

The march of the rebel host southwards indicates that the intention of its leaders was, as Hunsdon learned at the bridge of Don, to spring a surprise on Mary Stuart's guardians at Tutbury. Passing York, which would, in any case, have defied a force lacking artillery, they followed the line—Richmond, Ripon, Wetherby, Knaresborough, Tadcaster, Cawood, Selby, arriving at this last and southernmost limit of their march about 24 November. Whether a swift raid was actually made from here on Tutbury—a distance of over 50 miles—by a band of horsemen under Northumberland is uncertain but highly improbable. The fact that Mary was removed to Coventry on the 25th and that the whole expedition was back again in Richmond on the 28th, and at Brancepeth by the 30th—a rapid retreat—suggests, if it does not prove, that the retirement was not due to the failure of the projected *coup* but to other causes. In all probability the earls had heard of the great concentration of royal troops taking place on the south side of Trent, in Lincolnshire, Leicestershire, and Warwickshire, and determined to extricate themselves as speedily as possible from what might easily prove to be a death-trap.

Lord Clinton and the earl of Warwick, who were in charge of this southern army, were now completing the mobilization and were in touch with Sussex, although it was not until 10 December that they were able to begin effective co-operation. Meanwhile the retreating rebels, taking advantage of Sussex's immobility at York, captured Barnard Castle from Sir George Bowes—a useless piece of bravado which delayed them several days and very nearly jeopardized their safety. The royal forces had now begun to move. On 11 December Sussex, reinforced by some of Clinton's men, set out in pursuit of the rebels, leaving the main army to follow him a day's march behind. The fleet was also in occupation of Hartlepool, which the rebels had captured in November, in order to keep a port open for the reception of expected succours from Alva; and far away on the Scottish border Sir Henry Percy with garrison troops from Berwick was on his way southward to Newcastle.

The earls now realized that the game was up. At Durham, which they reached on 15 December, they decided on flight; and disbanding their weary and tattered footmen they themselves, with a handful of horse, made for Hexham and thence across the Pennines to Naworth and the Dacre country. Sir John Forster with the horsemen of the East and Middle Marches now took up the pursuit; but on the night of 20 December, with the help of Edward Dacre, they slipped past Scrope at Carlisle and reached the shelter of Liddesdale. Here they were safe for the moment, for the Scottish border families, the Scotts, Kers, Maxwells, Humes, and Hepburns gave them protection, esteeming it 'a liberty incident to all nations to succour banished men'.

The crisis was now past, but the danger was by no means over; and the queen's order for demobilization, issued on 16 January 1570, was somewhat premature. Elizabeth probably calculated on the kind services of the Scottish regent to secure the surrender of the fugitives; for Murray had been kept informed of their movements by Sussex. This, however, was a task beyond the regent, who, although he captured the earl of Northumberland by the treachery of a Liddesdale reiver, could not deliver him to the English because of the hostile attitude of the Scots. Nor could he do anything to effect the arrest of the others. Northumberland was therefore taken a prisoner to Edinburgh.

In the meantime further troubles were brewing on the
frontier. The Scottish borderers who had given hospitality to
the fugitives from England, being staunch catholics, now made
common cause with their guests; and rumours were rife that
Leonard Dacre was in league with them, albeit he had helped
Scrope to round up the stragglers of the rebel cavalcade in the
West March. Old wrongs were rankling in Dacre's heart, and
he probably suspected that the queen would attribute the
escape of her rebels to him as well as to his brother. As a
matter of fact Elizabeth had already ordered Sussex and Sadler
to effect his arrest, early in January, and they were in correspondence with Forster, Hunsdon, and Scrope with a view to
an attack on Naworth. But the castle was strongly fortified,
and the lack of artillery delayed operations. Eventually, however, the assassination of Murray on 23 January, and the aggressiveness of the Scots, made the immediate arrest of Dacre
imperative; and on 8 February a peremptory message from the
queen forced Hunsdon and Forster to attempt the task without
waiting for siege artillery.

Leaving Berwick with the comparatively slender force of
1,500 men on 15 February, Hunsdon reached Hexham on the
19th, where he was joined by Forster. Then followed a night
march to Naworth through the Dacre country, where 'the
beacons burned all night and every hill was full of horse and
foot crying and shouting as if they had been mad'. A surprise
attack was obviously out of the question; and Hunsdon decided
to continue his march past Naworth, it being still dark, in the
hope of reaching Scrope at Carlisle. The move was undoubtedly
wise, as Scrope had news of a considerable army of the Scots
hastening to Dacre's assistance. Happily, however, Dacre attempted to hold up Hunsdon's march at the crossing of the
Gelt, four miles west of the castle, and a general engagement
ensued. After 'the proudest charge upon my shot that ever I
saw', the rebel troops broke themselves against Hunsdon's
arquebuses, and their leader fled, leaving 300 dead and over
200 prisoners in the hands of his opponents. Hunsdon then
moved on to Carlisle, descrying in the distance, as he approached his objective, the Scottish reinforcements on which
Dacre had counted. 'If we had tarried until Wednesday', he
remarked in a letter to the queen, 'as Dacre thought I should,
he had been past dealing with, for he would have had four or

five thousand men more out of Scotland, besides increasing his own power'. It was clearly a narrow thing; and Hunsdon confided in Cecil that if he had 'taken the repulse, from Trent hitherward had been in great danger, and Carlisle would have been gone'. On 26 February the queen lavished her praise and thanks on her victorious general: 'I doubt much, my Harry, whether that the victory were given me more joyed me, or that you were by God appointed the instrument of my glory; and I assure you for my country's good, the first might suffice, but for my heart's contentation, the second more pleased me.'

While these decisive events were being transacted on the border, the main 'southern army' under Clinton and Warwick began an irregular spoliation of the north from Doncaster to Newcastle, driving cattle before them, seizing lands, goods, and leases, putting the miserable people to ransom 'as if all be rebels from Doncaster northwards'. Sussex tried to stop the reign of terror by a stern proclamation against the indiscriminate plunder, forbidding under penalty any one to take or spoil men's goods except the sheriffs or their officers. But Gargrave was afraid that the queen would 'lose most of the forfeitures', and could not 'get knowledge of the twentieth part of the goods bruited to be carried out of the country'. On 18 January he wrote to Cecil: 'I think a difference should be made between a rebellion in the queen's own dominions and a foreign realm.' And Hunsdon, who was likewise irritated by the behaviour of the southerners, commented with pardonable sarcasm: 'If the earl of Sussex had not been where he was, neither York nor Yorkshire had been at her [i.e. the queen's] discretion, and then the lusty southern army would not have returned laden with such spoils, nor put their noses over Doncaster bridge; but others beat the bush and they have had the birds.' The wastage of live stock was appalling. 'Cattle and sheep have come to my hands by seizure', wrote Gargrave, 'and I have no meat for them; if I buy it, they will soon eat up their value. Some have died by driving and lack of meat. The queen orders me to keep the goods without diminishing, but it would be better husbandry to sell some than keep all.' It was a cruel visitation upon an already poverty-stricken land.

By 4 February 500 of the rebels had suffered execution by martial law; they were of the poorer sort, and were put to

death to overawe the people. Another 129 were indicted of
treason, conspiracy, and rebellion, respectively, at York, Durham, and Carlisle. And in March the queen sent instructions
to Sussex and others associated with him to wind up the work
by compounding with such offenders as were deemed 'meet for
mercy and pardon', subject to their making their submission
and taking an oath of obedience to the queen. Thus persons
in possession of lands worth £5 annually or under might
redeem them by paying a reasonable sum: retainers of rebels
might be pardoned if their relatives were wealthy enough to
pay their fines for them: those who held office under the Crown
were to be discharged from their offices; and all payments were
to be in ready money or goods to be sold for the queen's use.
The lands of the important rebels, however, were reserved to
be attainted and resumed by the Crown, their valuation being
entrusted to royal surveyors to be certified to the court of
king's bench. By July the fine rolls and the valuation were
complete, and Sir Thomas Gargrave placed his accounts with
the balance of the money in the hands of Valentine Browne,
treasurer of Berwick.

Altogether there suffered for the rebellion some 800 persons;
but the chief loss was in goods and property. If the north had
been poor, it was now reduced to the verge of starvation. In
September 1570 Gargrave wrote to Cecil: 'I have not heard
the complaint so general of poverty as it now is. They have
been much touched with the late troubles, payments for armour,
assessment for repair of above a dozen bridges overthrown
last winter, payment of fines, enhancement of rents, lack of
traffic with Flanders, commissioners for concealed lands and
goods, and for the sale of wines, outlawries, &c.'

The collapse of the northern rebellion undoubtedly cleared
away Elizabeth's most pressing danger at home; but, inter-
nationally considered, her position was far from reassuring.
On the contrary, the political sky was everywhere clouded,
nowhere more so than in Scotland. Here, the assassination of
Murray had withdrawn the main pillar of English influence
north of the Tweed; and the Marian party, leagued with
Elizabeth's rebels, raised again the standard of the queen of
Scots. France, too, was eager to take advantage of the situation in order to increase England's predicament. Already, in

February, Sir Henry Norreys found it necessary to warn his government from Paris of the naval preparations in Brittany for a descent on Dumbarton in aid of the Marians: 'all men's mouths' being 'full of the invasion of England'. But Elizabeth was in no mood at first to pay much attention to these rumours of French aggression. 'It were a great folly and danger and against common sense', she said, 'to restore the queen of Scots.' One policy alone seemed feasible, to lend immediate aid to the party led by Morton and Mar, on whose shoulders had fallen the mantle of Murray, and at the same time to secure redress for the unfriendly action of the borderers during the late rebellion. In pursuance of this plan, Sussex was authorized to carry out a great punitive raid into the border country, which he did in April, in conjunction with Scrope and Forster, burning and destroying the castles of 'all such as helped the queen's rebels' as far as Kelso, Jedburgh, and Hawick. A few weeks later Sir William Drury was dispatched from Berwick with an English contingent to strengthen Morton's hands at Edinburgh; and very soon he was engaged along with the king's troops in the siege of Dumbarton castle. La Mothe protested vigorously in the name of France against this violation of Scottish territory, but was told that it would cease as soon as the English rebels were expelled from Scotland.

If Cecil and Bacon had had the sole direction of affairs at this time, the forward policy would have been followed by a thorough-going subjection of Scotland to England's will, defiance of France, and the utter ruin of Mary Stuart's cause. But other voices were now whispering caution in Elizabeth's ear; and with that sensitiveness to danger that always characterized her movements, she began to doubt the wisdom of becoming further embroiled in Scottish affairs. The publication of the papal bull in England (May 1570),[1] and the uncertainty of her relations with Spain, coupled with the menacing attitude of France, lent weight to the suggestions of Leicester and Arundel that she ought to pacify both France and Scotland by seeking a diplomatic restoration of Mary Stuart. Towards the end of May, therefore, Elizabeth suddenly reversed her Scottish policy, recalled Drury to Berwick, and sent Randolph to intimate her changed intentions to the king's party at Stirling. It was a sad blow to the hopes of Morton and Mar; but on being

[1] See p. 135.

assured that ample guarantees would be exacted for their security and that of the prince, they acquiesced in Elizabeth's will, and elected the earl of Lennox, her nominee, as Murray's successor in the regency (July). The next step was to bring about an armistice between the rival parties and begin negotiations for a treaty. Sussex took the matter in hand, and in September announced that a suspension of arms was arranged for six months. The way was now open for a settlement by diplomacy.

While events were moving towards peace in Scotland, great anxiety prevailed in England as to the activities of Alva in the Netherlands, where elaborate preparations were being made during the summer for the reception and escort of Anne of Austria, Philip II's bride elect, on her way to Madrid. The royal flotilla was to be accompanied from Antwerp by the Spanish fleet; but fears were expressed in London that under cover of this apparently legitimate display of Spanish courtesy some blow was intended against England. France was also thought to be in the plot, and it might even be that both powers were preparing to carry out the papal decree. The consequence was that during the month of July England was in the grip of an acute scare. 'They are so alarmed here', wrote Antonio de Guaras, 'that they fear their own shadows.' Throughout the country sheriffs were busy enrolling men for home defence: hackbut practice was instituted on every village green: mobilization of the fleet was ordered at Rochester under Clinton: Lord Charles Howard was sent with ten great ships to guard the narrow seas; and signalling beacons were set up in various parts to notify any attempted landing of the Spaniards. Howard was instructed to be punctilious about exacting the salute when the Spanish ships passed through the strait of Dover.[1] In August the whole coast from the Thames to Berwick, from Dover to Cornwall, was guarded as if the enemy were in sight, and scouting pinnaces were patrolling the French side of the Channel, while Hawkins was on the look-out at Plymouth, and watch was kept night and day on the hill near Margate. This state of tension lasted all summer until the Spaniards had completed the double journey. It was the month of October before the government felt justified in ordering a general relaxation of precautions.

[1] The strait of Dover was 'her Majesty's stream'.

Eventually, after the excitement over the movements of the Spanish fleet had subsided, the terms on which Elizabeth was prepared to restore the queen of Scots to her kingdom were communicated to the royal captive at Chatsworth by Cecil and Sir Walter Mildmay. They were necessarily severe—involving, among other things, the surrender of her son as a hostage for his mother's good behaviour, the conclusion of a defensive alliance with England, an undertaking not to contract any marriage without Elizabeth's consent, and the retention by the English of Hume castle until restitution was made for the spoils committed by the borderers in England. Objection was raised by the French ambassador to the surrender of James and the defensive alliance with England; but the real obstacle to a convention arose not from the French, nor from the queen of Scots, who was willing to purchase her liberty at almost any price, and apparently satisfied Cecil as to her substantial acceptance of all the articles, but from the king's party in Scotland, who had become increasingly afraid of their position in the event of Mary's return. After a long delay of more than three months, Morton and his fellow-commissioners ventured to explain to Elizabeth, in a personal interview, that they could not enter into any treaty that invalidated the abdication signed by Mary at Loch Leven and legalized by the Scottish estates, or involved any abridgement of the king's sovereignty, or his removal to England. If, therefore, Mary returned to Scotland, it must be as a private person. Elizabeth was visibly annoyed with this recalcitrant reply, and threatened Morton with her hostility if he persisted in his attitude; and La Mothe also exerted pressure on him 'par prières et même par menaces'. But the Scots refused to modify their position, and returned to Edinburgh in March 1571, on the plea that they must consult parliament, but really resolved to fight out their quarrel with the Marians to the bitter end. The restoration of the queen of Scots was now an impossibility, for Elizabeth could not coerce the party on whom alone she must rely for the maintenance of English influence in Scotland. Meanwhile, the armistice having come to an end, the rival parties plunged again into war for the mastery of the kingdom; and with the return of anarchy, all hope of achieving an Anglo-French agreement with regard to Scotland temporarily vanished.

The principal conclusion to be drawn from the failure was

that the solution of the Scottish question must wait until more substantial grounds of co-operation with France could be discovered. Nor was it difficult to see that public opinion on the other side of the Channel was also moving towards the same conclusion, albeit from a different angle. As early as 1568, the house of Châtillon had mooted the idea of an Anglo-French alliance, based upon the marriage of the duke of Anjou and Queen Elizabeth: hoping thereby to get rid of the fanatical and superstitious heir apparent to the Crown, rescue their country from the recurring chaos of civil war, and restore foreign policy to its traditional anti-Habsburg groove. The idea, though creditable to the prescience of its promoters, was somewhat premature and found less support than it was entitled to, at the time it was first discussed; but two years later, when the pacification of St. Germain freed the hands of Charles IX for a more ambitious role in European affairs, it had much to commend it. Like his father and grandfather before him, Charles aspired to strike a blow at the Habsburgs; and like them, too, he determined to build up a system of alliances between France and the protestants abroad. The first step was the treaty of St. Germain, which established liberty of worship for the huguenots and gave them legal security: the next was the summons of Coligny to the council, a token that co-operation was to be the watchword of the future; and on this solid foundation could be established alliances with the protestants of Germany and the Netherlands. The union of the houses of Bourbon and Valois by the marriage of Henry of Navarre and the king's sister Margaret would complete the protective circle which Charles, magician-wise, hoped to draw round himself; and if he could marry his brother Anjou to the English queen, he might even draw England into his schemes for the dismemberment of the Spanish empire.

The isolation of England, owing to the rupture with Spain and the hostility of the pope, and the troubles in Scotland, had likewise created the need for an ally; and, as we have seen, there was a strong pro-French party in the council, headed by Leicester ('plus français que nul autre de ce royaume'), who were largely responsible for the new Scottish policy announced by the queen in May 1570. Cecil, too, began to turn over in his mind the 'commodities' that might result from a marriage between Elizabeth and Anjou. The consequence was that unofficial feelers were put out during the winter of 1570–1 with a view to discovering whether France was sufficiently interested in the matter for it to be committed to diplomacy. Then, in February 1571, Lord Buckhurst went to Paris, ostensibly to congratulate Charles on his marriage, but really to announce in Elizabeth's name that England was ready to entertain official *pourparlers* on the subject of the Anjou match. Thereafter the negotiations passed into the capable hands of Sir Francis Walsingham, the resident English ambassador in Paris, and became the main substance of Anglo-French diplomatic correspondence until January 1572.

If the failure of her negotiations for the pacification of Scotland drove Elizabeth into seeking a *rapprochement* with France, it found the queen of Scots deeply involved in the dangerous mazes of the Ridolfi conspiracy. This pretentious but thoroughly impracticable plot took shape in England at the very moment when Lord Buckhurst initiated the Anjou conversations in Paris. Its notoriety lies chiefly in the irreparable damage it did to Mary Stuart's cause and its tragic consequences for the duke of Norfolk. Essentially a Hispano-papal plot, it was the work of the cosmopolitan financier Roberto Ridolfi, who had already graduated in conspiracy by taking a hand in the intrigues of 1569. Being an Italian, versed in the swift and sudden revolutions characteristic of his own country, he conceived it possible, by combining the elements of unrest in England with the queen's enemies abroad, to stage a spectacular *coup* against the English state, overthrow Elizabeth, and seat Mary Stuart on the throne of a catholic England. He was nothing if not methodical; and he conducted his conspiracy as he conducted his business transactions, in the best counting-house manner, procuring signatures, letters of recommendation, statistical proofs of his assertions, and carefully noting the obligations of everybody. But he was too masterful, too confident of his own ability, and too much given to make light of difficulties he could not understand and still less foresee. His ideas on international politics were bound by the horizons of an ordinary Italian *cittadino*, in whose eyes the decisive power in European politics lay with the catholic powers under the hegemony of the pope.[1] Of the intricate, involved,

---

[1] J. H. Pollen, *The English Catholics in the Reign of Queen Elizabeth*, p. 132.

and complex relationships of the powers to each other he knew nothing; nor was he really conversant, in spite of his elaborate array of figures, with the feelings and aspirations of the English catholics. His greatest weakness, however, was his gullibility, and his confidence that a conspiracy entailing the co-ordination of centres so far apart as London, Rome, Madrid, and Brussels could be controlled and directed by one man, or could long escape the eyes and ears of the English government and of other governments friendly to Elizabeth. It is difficult to see how the Ridolfi plot could have achieved success even in the most favourable circumstances.

The focus of the conspiracy was of necessity England, where the decisive events would occur. Here, the cause of Mary Stuart would provide the battle-cry: the papal bull, news of which spread through the country during the year 1570, would supply the incentive to action; and the duke of Norfolk with his associates among the nobility, whom Ridolfi numbered at over sixty, would give the plot its natural leaders. Behind the nobility, massed in their tens of thousands, would rally the discontented catholics. Then, on a preconcerted signal to be given by the landing of a Spanish expeditionary force composed of veterans from the Netherlands and Spain, on the coast of Norfolk, the queen would be seized or assassinated, the Tower and the fleet captured, parliament overawed or broken up, and Mary Stuart freed from prison. Such was the substance of this grotesque conspiracy.

The first to be drawn into Ridolfi's net was the queen of Scots, who readily gave her assent to it, with the pope's approval, in February 1571, and furnished the arch-conspirator with letters of credence for Alva, the pope, and the king of Spain. Norfolk was harder to win; for he disliked Ridolfi's 'Italian devices', doubted the possibility of keeping the matter secret, and insisted upon knowing what help might be expected from Spain. Eventually, however, he allowed himself to be overborne by Ridolfi and the persuasive arguments of the bishop of Ross, his chief coadjutor, who showed him that, as Morton had returned to Scotland, nothing further could be done towards a Marian restoration by negotiation; and about 10 March he gave by word of mouth his assent to do what he could for the queen of Scots and the advancement of the catholic religion. Probably also, though the point is not verifiable, he acknowledged himself a catholic.

Having thus laid the mine to his own satisfaction in England, Ridolfi then left for the Continent on 25 March with his
copious notes and documents to concert measures with Alva,
the pope, and Philip II. He was in Brussels in April; in May
he was in Rome; and late in June he reached Madrid. His
reception varied in accordance with the political acumen of the
men with whom he dealt. Rome welcomed his scheme with
open arms: in Madrid it was submitted to a long and careful
scrutiny before it was accepted; but Alva was from the first
sceptical, critical, irresponsive. The astute governor of the
Netherlands, who desired above everything else the recovery of
his confiscated treasure and the property lost in England by the
embargoes, poured cold water on Ridolfi as he had already
done on de Spes in 1569, and let it be known that he thought
as little of the ability of the one as he did of the other. He
described Ridolfi as a *gran parlaquina*: a great chatterbox. Nor
would he abandon his stubborn resistance to the plot unless
and until he had proof positive that the English were prepared
to carry out the essential part of the business themselves. In
several letters to his master he analysed with merciless cogency
the dangers and difficulties in the way of success, and the fatal
consequences of failure: the disastrous effects a war with England would have on Spanish policy both in northern Europe
and in the Mediterranean, and the impossibility of sending
an army from the Netherlands at the present juncture of international affairs. At the back of Alva's mind lay the fear that
England would seize the occasion to accelerate the alliance
with France which was already beginning to loom over the
horizon, and the complementary fear of French aggression in
the Low Countries, which was also coming daily nearer. So
far as England was concerned, failure of the plot would
probably mean ruin of the catholic cause and the forfeiture
of Mary Stuart's life. In the face of this opposition from
Alva, it was impossible for Philip to move, for he had complete trust in the judgement of his lieutenant. Thus by the
end of August the bottom had fallen out of Ridolfi's elaborate
scheme.

Meanwhile the activities of the plotters were gradually coming within the purview of the English government. The first
authentic evidence of Mary's relations with Alva came from
Scotland, where it fell into the hands of Regent Lennox after

the capture of Dumbarton castle on the night of 2–3 April, and was remitted to Cecil. News of Ridolfi's plans was sent to Elizabeth by the grand duke of Tuscany, with whom he had stayed during his journey from Rome to Madrid. From the queen of Navarre came information extracted from the dispatches of a Spanish courier intercepted in the Pyrenees. In the month of May a certain Charles Bailly was arrested at Dover with a packet of letters and books consigned by Ridolfi to the bishop of Ross, and this led to the discovery of Ross's implication in the conspiracy, and his arrest. 'Trente' and 'Quarante' (Norfolk and Lumley), the addressees of the letters, to whom Ross had distributed them before his arrest, were not yet identifiable. But shortly afterwards the arrest of Higford, the duke of Norfolk's secretary, and his agents, Barker and Bannister, in connexion with the transmission of a sum of money by Norfolk to Mary Stuart's supporters in Scotland, brought guilt home to the duke, and he was arrested and taken to the Tower on 7 September. Finally, in October, Ross revealed the whole plot, under threat of torture, and placed the government in possession of a damning case against Norfolk. Further revelations led to the implication of Arundel, Southampton, Cobham, and others of Norfolk's associates, and they were likewise arrested. As the ramifications of the conspiracy were discovered, excitement grew. The catholic nobles were carefully watched, the port guards were changed, the watch was reinforced in London and other cities by day and night, and all roads leading to and from the capital were picketed.

In January 1572 a process for treason was begun against Norfolk, and on the 16th he was sentenced to death by his peers. Six months later, after prolonged hesitation on the part of the queen, he was executed. Thus perished the greatest noble in England—the last aristocratic champion of reaction—a victim to his own inordinate ambition and crooked ways.

While Ridolfi was posting across Europe endeavouring to weave together the various strands of his complicated plot, Walsingham was busy in Paris wrestling with the exasperating problem of the Anjou marriage. From the first his task was practically hopeless; for as soon as the articles were reduced to writing it was seen that no compromise was possible on the question of religion. The French insisted upon freedom of worship for the duke, on the plea that he could not be expected to abandon the faith in which he was nurtured. Elizabeth, for political reasons no less cogent, could not yield on the point that he must conform to the religious observances established by law in England. Nor could the wit of man devise a formula that would meet the objections on either side. Walsingham's efforts to cajole or browbeat the duke into acceptance of Elizabeth's demand were countered at every step by the combined influence of the papal nuncio, the cardinal of Lorraine, and the ambassadors of Spain and Portugal, who played on the fears and ambitions of Anjou with consummate skill. He stoutly refused to do violence to his conscience even with the hope of a kingdom dangled before him. By the month of July something very like an *impasse* had been reached in the negotiations, and the promoters of the match on both sides were reduced to despair. Charles IX, who longed to see his brother safely out of the kingdom, was angered at his intransigence; and the queen mother 'never wept so much since her husband died'. Lord Burghley (Cecil) and Leicester were no less distraught before the obstinacy of the queen, who flatly refused to sanction the use of the Mass in England 'how secret soever', and seemed wilfully blind to the gravity of the issues involved. 'This amity', lamented Burghley, 'were needful to us; but God hath determined to plague us, the hour is at hand, His will be done with mercy!' 'God protect and defend us,' wrote Leicester; 'thus we are with our neighbours, in all places without friendship!' A last desperate effort was made, in August, by a mission of Paul de Foix and the Marshal de Montmorency, to procure some concession from the queen; but it only served to prove the utter impossibility of solving the problem by diplomacy. About a month later Leicester, who had means of knowing the queen's mind better perhaps than any one, concluded that there would be no marriage. 'Surely', he remarked in a letter to Walsingham, on 20 September, 'I am persuaded that her Majesty's heart is nothing inclined to marry at all, for the matter was ever brought to as many points as we could devise, and always she was bent to hold with the difficultest.'

This avowal doubtless came as a great relief to Walsingham, for his interest and sympathies were elsewhere than in the interminable marriage negotiations. Deep in the confidence of the huguenot leader Coligny and his brother in arms Lewis of

Nassau[1]—kindred spirits with whom he could converse on terms of perfect equality and frankness—he had become convinced, like them, that the need of the hour was a combined onslaught on 'the proud Spaniard'. In August he adumbrated to Burghley a grandiose scheme, emanating from the fertile brain of Lewis of Nassau, whereby the Spaniards could be ejected from the Netherlands, and their inheritance divided between France, England, and Germany. France would take Artois and Flanders: Brabant, Guelderland, and Luxemburg would fall to the Empire; and England would annex Zeeland. It was a brilliant conception, satisfying at one and the same time the crusading impulse in Walsingham, the territorial ambitions of France, and the aspirations of the Flemish and Dutch patriots. Spain would be thoroughly humiliated, and England, at a small cost of financial assistance and the loan of a few ships, would see all her dangers vanish in the smoke of a European conflagration. In urging Burghley to recommend the project to the queen, Walsingham strongly emphasized the latter point: if the scheme prospered, 'the fire that is now kindling may grow to a flame and we take comfort at the heat thereof'. But although Leicester 'never found cause . . . that moveth me more to further it', the cautious Burghley, whose insular mind and self-regarding ideas of foreign policy cut sharp across the imaginative schemes of dreamers and firebrands like Coligny and Nassau, refused to be tempted. 'I fear the offers of so great amity', he wrote, 'will diminish or divert the former intention of the marriage without which the French amity shall serve to small purpose but to make us ministers of their appetites, and those fulfilled, to cast us off.' A French occupation of the Netherlands was in his opinion a contingency more to be feared than all the threats of Spain—a subject on which he would have more to say when the storm broke over the Low Countries early in the following summer. So Walsingham was gently but firmly shepherded back again to the main purpose of his mission. He protested, it is true, but acquiesced, and the 'great enterprise of Flanders' was shelved so far as England was concerned.

Meanwhile the inexorable logic of events was steadily driving home the conviction, both in France and in England, that an immediate settlement, marriage or no marriage, was imperative

[1] Brother of William of Orange.

in the interest of national defence. The continued anarchy in Scotland, leading to the *coup d'état* of the Marians at Stirling (September), in which Lennox was killed, together with the discovery of Spanish machinations in the Ridolfi plot, accelerated the pace on the English side; while the French were likewise stimulated by the victory of the Spaniards over the Turks at the battle of Lepanto (October), and the rumours of a treaty between Alva and the Marian party in Scotland for the invasion of that country and the removal of James to Spain. Accordingly, when Sir Thomas Smith and Sir Henry Killigrew went over to Paris at the end of December, the Anjou match was quietly dropped, and diplomacy addressed itself in earnest to the elaboration of an Anglo-French defensive league based upon the real needs of both countries. Difficulties were encountered, of course, owing to the reluctance of the French to abandon the cause of Mary Stuart, and the insistence by the English on the inclusion of an *etiamsi religionis causâ* clause in the *casus foederis*.[1] Eventually, however, the French agreed to withdraw their opposition in regard to Scotland, and Elizabeth contented herself with a 'letter missive' from Charles IX granting the substance of what she asked, although the matter was omitted from the text of the treaty. On 21 April 1572 the document giving effect to the settlement arrived at was signed at Blois; and the diplomatic revolution, as it has been called, was complete.

In actual fact the Treaty of Blois was more of a diplomatic convenience than a diplomatic 'revolution'. It did not abrogate the Anglo-Burgundian alliance on the one hand, nor did it impose a veto on a reconstruction of a Franco-Scottish on the other. Briefly, it committed England and France to mutual military and naval assistance if attacked by a third power, arranged for the establishment of a 'staple' in France to compensate England for the loss of her Flemish trade, and provided for a joint effort to pacify Scotland. The importance sometimes attached to the treaty is probably excessive. The promised staple in France was never set up, because the settlement between England and Spain in 1573 rendered it unnecessary. Nor were the military and naval clauses ever put into operation. In fact the only substantial result was the triumph of Elizabeth's

---

[1] England desired the treaty to be operative even in the event of invasion for religious reasons.

policy in Scotland, and the abandonment of Mary Stuart by the French. From this standpoint England rather than France was the real gainer by the treaty.

Some three weeks before the Treaty of Blois was concluded, a revolution broke out in the Netherlands, which upset all the calculations of the diplomatists, plunged France into one of the gravest crises in her history, and shook the Anglo-French alliance to its foundations. The immediate cause of this sudden and unexpected sequence of events was the expulsion of the *gueux de mer* from English waters by a decree of Queen Elizabeth on 1 March. The measure was undertaken primarily to protect English commerce, and that of countries friendly to England, against the indiscriminate plunder indulged in by the disorderly crews of La Marck, the accredited admiral of the prince of Orange. Previously the queen had recognized the prince's right to issue naval commissions, and La Marck was granted the freedom of the port of Dover for victualling and munitioning purposes. But the exercise of the privilege led to gross abuses, to the slander and disgrace of the realm, and the time had come to teach the offenders a salutary lesson. Elizabeth, however, was not responsible, either as accomplice or accessory, for the action of the sea rover after his expulsion: that was due solely to the urgent need of a fresh base from which to continue his depredations on Spanish shipping in the Channel. On 1 April La Marck made a surprise descent on the Dutch coast, captured Brill from the Spaniards, and proceeded to fortify it with a view to permanent occupation. It was the first foothold of the patriot cause on the soil of the Netherlands, and was taken as a signal for a widespread revolt against Spanish authority. Within a very short time the inhabitants of Flushing, Rotterdam, Schiedam, Gouda, and many other towns followed the lead given by Brill, expelled their garrisons, and placed themselves under the flag of Orange, who was proclaimed stadholder of Holland, Zeeland, Friesland, and Utrecht.

Great was the excitement when news of the rebellion reached London. 'The least thing they shout on 'change and in the street', wrote Guaras, the Spanish *chargé d'affaires*, to Alva, 'is that the states are utterly lost to us and that your Excellency and the Spaniards will have to leave.' The Flemish refugees organized expeditions in support, sending money and munitions, and presently a steady stream of adventurers, English as well as huguenot, began to join the movement. Although the enterprise had been achieved without the knowledge or sanction of Orange and Nassau, they realized quickly that the critical moment was at hand for launching their long-projected offensive in the Netherlands. In May Nassau, accompanied by the huguenot La Noue, led an army of French levies across the frontier. Valenciennes and Mons fell before them, to the cry of 'France et Liberté'; and with Orange on the move from Germany with powerful reinforcements, it must have seemed as if the 'great enterprise' was fairly on the way to success.

Clearly the time was ripe for Elizabeth to make some pronouncement on English policy; for the French were daily strengthening their hold on the country, especially in the island of Walcheren, the key to the entire Netherlands from the seaward side. In June Burghley analysed the situation for the benefit of the council, pointing out that the encroachment of the French in the maritime parts constituted a direct menace to England, to be resisted at all costs, even if it meant giving help to the duke of Alva. Shortly afterwards (11 July) Sir Humphrey Gilbert landed with a strong company of volunteers at Flushing, and was instructed by the queen—in the spirit of Burghley's memorandum—not to embark on any inland adventures, but to prevent the occupation of Flushing by the French. Not a word was said about the defence of Dutch and Flemish liberty: the eyes of the English government were directed exclusively to the danger from France, whose designs on the Low Countries were well known through Walsingham's dispatches of the previous year.

The drama unrolled itself with surprising swiftness, for Alva left no stone unturned in his anxiety to drive the French back again across the frontier. He retook Valenciennes, invested Mons, and cut to pieces a relieving army under the huguenot Genlis, who had been sent by Coligny with Charles IX's approval to assist Nassau. Mons then surrendered by capitulation, and the whole undertaking collapsed in ruin. Orange was compelled to beat a retreat into France with a rapidly dwindling army. Not only so: the capture of Genlis led to the discovery of information showing the French king's complicity in the affair. Charles IX was now confronted with the probability of an imminent war, single-handed, against Spain—a contingency he had made no provision for, and Coligny alone accepted with complacency. The queen mother, backed by the catholics in the council, protested that war could only lead to the triumph of the huguenots and the subjection of the king to their will; and hot words passed between her and the admiral. In order to escape from his dilemma, and at the same time avert a possible catastrophe, Charles decided to denounce the 'enterprise'. It was now Coligny's turn to fall into the pit he had unwittingly dug for his master. Regarded with suspicion by the king as the author of the disaster to Genlis, he found himself exposed to the concentrated animosity of all who resented his power in the state. The chronology of events places it beyond doubt that his removal by assassination was now the uppermost thought in the minds of Catherine de Médicis and the Guise faction.

News of Genlis's overthrow reached Paris when the huguenot leaders from all over France were assembling for the celebration of Navarre's marriage to the king's sister. On 19 August the wedding was solemnized with festivities and revelries; and on the 22nd, before the dispersal began, Coligny was shot at and wounded in the streets of the city. Charles IX, unaware of the part his mother had played in the deed, and thinking it was simply another instance of the interminable private feud between the houses of Guise and Châtillon, vowed vengeance on the culprits, and ordered an inquiry. But the suspicions of the huguenots were roused, and accusing fingers were pointed at the queen mother and the duke of Anjou: with the result that Catherine determined to destroy the leading men of the party, including Coligny, before the inevitable revelations came out. She took Tavannes, Nevers, Guise, and others of the admiral's enemies into her confidence, concocted a bogus plot of the huguenots to kill the king and all the members of the royal family, and induced her son to sanction the murder of the ringleaders. Navarre and Condé, being of the blood royal, were to be spared. Thus on the early morning of 24 August, St. Bartholomew's day, the 'caged birds' were brutally done to death; and the fanatical mob converted an act of private and essentially political vengeance into a signal for a wholesale massacre.[1]

Catholic Europe did not hide its satisfaction at the atrocity.

[1] For details of the massacre, its causes and consequences, see Lord Acton, *History of Freedom and Other Essays.*

'Canticles of holy joy' resounded in the churches of France. At Rome, Pope Gregory XIII avowed himself better pleased than with fifty Lepantos. He ordered the *Te Deum* to be sung at St. Peter's and *feux de joie* to be lit at the Castle of St. Angelo, and summoned Vasari from Florence to decorate the hall of kings at the Vatican with paintings of the massacre. Profane minds everywhere gave a cynical approval to the skill with which the trap had been set; and the pious acknowledged the genuine religious spirit of the French court. In Italy the 'stratagem of Charles IX' became a byword. Philip II, relieved of his anxiety about Flanders, mirthfully referred to the 'long dissimulation' of his most christian brother, while Alva dismissed the bogy of the Anglo-French alliance as a thing of the past. The general impression was that France had vindicated her catholicity and thrown in her lot irrevocably with the cause of the counter-reformation. Protestant countries, however, were struck dumb with amazement and horror at the cruelty and barbarity of the deed. So far as England was concerned it came like a bolt out of a rapidly clearing heaven. The summer had begun with rejoicings and festivities over the ratification of the treaty, and Walsingham in Paris had no glimmering of what was brewing round him, the only thing that troubled him being the parlous position of William of Orange, and the seeming indifference of Elizabeth to his fate. The queen was palavering, now hot, now cold, with Catherine de Médicis's new proposal that she should marry her younger son Alençon, if she could not see her way to accommodate Anjou. On 6 June the execution of Norfolk had closed the episode of the Ridolfi plot, and Northumberland, who had been surrendered by the Scots, paid the same penalty on 22 August for his share in the rising of 1569. Scottish affairs were also proceeding towards a settlement under the joint guidance of England and France. Against this brightening landscape was suddenly projected the grim spectre of a France dyed red in the blood of the huguenots. Crowds of panic-stricken fugitives from Dieppe began to arrive at Rye on 27 August with harrowing details of their sufferings, and the news spread like wildfire through England. 'Il n'est pas à croire', wrote La Mothe, 'combien cette nouvelle émeut grandement tout ce royaume.'

In the absence of authentic tidings—for Walsingham's dispatches were held up, and the French king was slow to send his ambassador the official account of the occurrence—the worst construction was put upon the motives and intentions of the French court. The interview between Catherine de Médicis and Alva at Bayonne (June 1565) was recalled as a proof that the massacre had been planned long before.[1] Nor did the horror and excitement abate when Charles IX first threw the blame on Guise and then retracted this specious excuse in favour of the huguenot murder-plot theory advanced by the queen mother. When Elizabeth heard this revised version of the responsibility for the massacre she commented caustically on the barbarity of the French method of administering justice, and urged an early publication of the details of the alleged conspiracy. The council was more outspoken in its criticism. La Mothe was told that France had been guilty of the most heinous crime since the days of Jesus Christ. The general impression, so far as the ambassador was able to gauge public opinion, was that the word of a Frenchman was no longer to be trusted. 'Leur défiance est si grande', he wrote, 'qu'ils croient que tout que je leur dis de votre part est pour les surprendre et tromper.'

The difficulty of believing the official assurances of the French court that the matter was now finished and done with was greatly increased by the spread of the atrocities to Rouen and other centres of population, and by the pessimistic reports received from Walsingham. It was impossible to disguise the fact that the mortal enemies of the huguenots were now directing the government, and that the edict of pacification had been abandoned. Everything pointed to the enthronement of the principles of the counter-reformation as the presiding genius of French domestic policy. Navarre and Condé, who had escaped death, were offered the alternative of the Mass or the Bastille, and the papal legate, the Cardinal Orsini, was on his way to Paris. To make matters worse, hostility to England was beginning to show itself in suspicious movements of the French fleet under Strozzi, and the pillaging of English merchants at Rouen

---

[1] The Conference of Bayonne discussed the question of a marriage between Margaret of Valois and Don Carlos, and the enforcement in France of the decrees of the Council of Trent. It is certain that Catherine undertook to have the decrees of the council examined by a body of prelates and to expel the protestant clergy from France on condition that the marriage took place. There is no record, however, of any talk about a massacre of the huguenot leaders. In any case the conference produced no tangible result. (Lavisse et Rambaud, *Histoire Générale*, v. 132.) For full details see E. Marcks, *Die Zusammenkunft von Bayonne*, esp. pp. 234-8.

and on the high seas. Not all the eloquence of Cicero or Demosthenes, said La Mothe, could obliterate the bad impression created in England by the notorious evidence of the facts. 'I think it less peril', wrote Walsingham, summing up the situation, 'to live with them as enemies than as friends.'

Elizabeth's attitude during the crisis was a curious blend of dignity, restraint, fear, and dissimulation. She could not penetrate the ultimate designs of the French court, for Walsingham found it increasingly difficult, after the massacre, to learn what went on in the council. On the all-important matter of France's relations with the pope and Spain conjecture alone was possible, and, in the circumstances, the worst was apprehended. On the other hand, the conciliatory tone of the king's communications, and his evident desire to retain the goodwill of England, together with the queen mother's hasty revival of the Alençon marriage proposal, seemed to indicate that Walsingham was probably over-severe in his judgements. All things considered, Elizabeth decided to move cautiously in her dealings with France. Distrust must not be allowed to endanger the continuance of amicable relations, nor must the Treaty of Blois be regarded as a dead letter; but at the same time every necessary precaution must be taken to protect England's interests and to render France incapable of doing any harm. The first object was served by retaining Walsingham at his post in Paris, and by the dispatch of the earl of Worcester, in January, to represent Elizabeth at the christening of Charles IX's infant daughter, to whom she had consented to act as godmother. The renewal of the Alençon negotiations, however, was coldly received, it being pointed out by Burghley to La Mothe that no good could come of discussing the question so long as the French government maintained its unsatisfactory attitude to the huguenots.

From Elizabeth's point of view the dangerous position of the huguenots was the crux of the situation. The shock of the massacre had driven the province of Gascony into armed rebellion, and the terrified sectaries were now concentrated in La Rochelle, with nothing to shelter them from their enemies but the walls of the fortress. If they succumbed, England would be deprived of the one effective means she possessed of influencing French policy. To aid them overtly was, of course, impossible if the Treaty of Blois was to be rigidly observed; but to refuse assistance might be more dangerous to England than a breach of the treaty. It was imperative that the French government should be kept busy within its own borders. Elizabeth therefore compromised. Publicly she proclaimed her neutrality: privately she connived at assistance being sent by her subjects; and when La Mothe protested against the activities of the count de Montgommery, the huguenot agent, on English soil, she disclaimed all responsibility and denied that any ship, man, or gun of hers was engaged in the enterprise.

With regard to Scotland, the only really weak point in her armour, where France, if evilly disposed, could do her serious injury, she proceeded more vigorously to work. Here the crucial matter was the strong position of the Marian party in Edinburgh castle and their confident hope that France would rally to their assistance as soon as the huguenots were disposed of. Edinburgh castle was, in effect, England's La Rochelle. For a brief space (Sept.–Oct.) Elizabeth toyed with the idea of destroying the Marian cause altogether by handing over Mary Stuart to Regent Mar, to be tried by process of justice and executed by the Scots themselves. But Mar died in October, and the plan was dropped. Killigrew, who was entrusted by Elizabeth with the negotiation, now applied himself to the task of procuring a settlement by agreement; and in February 1573 he succeeded in inducing Huntly and the Hamiltons, under guarantee of English protection, to make their peace with the new regent, Morton, at Perth. A similar proposal was made to Grange and Lethington at the castle, but it met with a summary refusal, and Elizabeth was reluctantly compelled to authorize the reduction of the fortress by English artillery. On 17 April Drury once more crossed the border at Berwick, his siege-guns being conveyed by ship to Leith. A month later the bombardment began; and on 28 May, disappointed by the failure of France to lend any help, overcome by the intensive artillery fire, and cut off from their only water-supply, the garrison capitulated. Of the gallant defenders, Lethington only survived the surrender by eleven days, and Grange was put to death some months later by order of the regent; but Elizabeth's recommendation to mercy saved the others from reprisals. Thus ended the contest that had raged in Scotland since 1570; and the 'postern gate' that had worried Walsingham was temporarily closed. It was an unexpected result of the turmoil in France, but a happy one for England.

The triumph of Elizabeth's Scottish policy was followed by

the failure of Charles IX to subdue the Rochellois. The great huguenot fortress, defended by a desperate population in receipt of continuous help from England and the Netherlands, defied the concentrated might of catholic France; and in July peace was restored by the grant of freedom of conscience to the protestants, with full liberty of worship in the specified *enclaves* of La Rochelle, Montauban, Nîmes, and Sancerre. Now was the king's opportunity to attempt the restoration of Anglo-French relations to their former cordiality, and with this in view he reopened the question of the Alençon marriage. Elizabeth, who made it clear that the oppression of the huguenots was the only serious obstacle in the way, received his advances favourably; with the result that an arrangement was arrived at, in March 1574, whereby Alençon should cross the Channel secretly and meet the queen near Dover.

But the duke had other schemes on hand which wrecked the project. Desirous of playing the leading role in French politics that his brother Anjou had played before him, he had become associated with Montmorency, Navarre, Condé, and Montgommery in a combined movement of catholics and huguenots to force 'politique' doctrines on the court. As a result he involved himself in conspiracy and rebellion and was thrown into the prison of St. Germain along with Montmorency and Navarre. Shortly afterwards Charles IX died (30 May), and the succession passed to Anjou, the 'tyrant from Poland'.[1] As the new king was well known for his fanatical catholicism and his subservience to the clergy and the Guises, it was confidently expected that his accession would initiate a fresh outburst of religious strife, the abandonment of the league with England, and the championship by France of the counter-reformation in its most aggressive form. The danger was fully realized in England and Germany, and diplomatic pressure was immediately brought to bear on Henry III by Elizabeth and the protestant princes to avert the threatened catastrophe. For nine months the weak-willed king 'floated between the storm and the rock', tormented by his natural desire to follow the counsel of the Guises and the queen mother, but afraid to take a step that would rally protestant Europe to the support of the huguenots. The situation was so menacing that in February 1575

---

[1] Anjou had been elected king of Poland in February 1574: his reign lasted only a few months.

Elizabeth entered into communication with the count palatine about dispatching assistance to Condé. Eventually, however, Henry III ended the period of tension by sending an envoy to England with the news that he was ready to confirm the Treaty of Blois. On St. George's day he was elected to the order of the Garter at a chapter held at Greenwich, and within a month the league was formally confirmed by both governments. Elizabeth could now rest assured that she had nothing further to fear from France, for the treaty was a guarantee that the French government would not indulge in excesses against the huguenots, and the strength of the huguenots was a guarantee that the treaty would not be overthrown.

Meanwhile the shock of the reaction against France consequent upon the massacre of St. Bartholomew had generated a strong feeling in England in favour of a settlement with Spain. For three years Alva had struggled in vain, interrupted by the foolish plotting of de Spes and Ridolfi with the English catholics, to end the deadlock created by the seizures of 1568–9; but every ambassador he had sent to England—D'Assonleville, Chapin Vitelli, Fiesco, Sweveghem—had returned with the same tale: Elizabeth refused to negotiate unless Spain would agree to bring all the outstanding questions at issue between the two governments within the scope of the settlement. The first indication that a change was coming was given by Burghley in a conversation with Antonio de Guaras, at the time when the league with France was nearing completion. Burghley, whose interest in the Anglo-French alliance was never very keen, stated that a concord between England and Spain was eminently desirable, and might be brought about on the basis of 'the most just restitution of the confiscated property', the reopening of trade, the mutual banishment of rebels, and the confirmation of all the old treaties. It was not, however, until the catastrophic events of the summer opened every one's eyes to the shifty, incalculable, and untrustworthy tactics of the French court that the question entered the field of practical politics. The ill treatment of English merchants in France during the troubles, the insecurity and inadequacy of Rouen as a market in lieu of Antwerp, and the shrinkage of trade converted many others of the privy council, including Leicester, to Burghley's standpoint; and negotiations were begun through Guaras with Madrid and Brussels.

In April 1573 an agreement was reached between Alva and

Elizabeth, in virtue of which traffic was restored for a period of two years in order to allow diplomacy to get to work on the economic and political questions involved in the projected settlement. A year later (July 1574) Bernardino de Mendoza arrived in London as representative of the king of Spain, and was given a cordial reception. On 28 August of the same year there followed the treaty of Bristol, which wound up the matter of the confiscated property by assessing the losses on both sides and striking the balance. The Spanish claim was placed at the net figure of £89,076 and the English at £68,076; thus leaving Spain England's creditor to the extent of £21,000. In the meantime Requesens, who had succeeded Alva during the previous year as governor of the Netherlands, had busied himself with the problems arising out of the navigation of the Scheldt and the banishment of Elizabeth's rebels from the Spanish dominions. Both questions raised difficulties; for Orange and the Dutch, who were at war with the king of Spain and commanded the mouth of the Scheldt, insisted upon controlling the traffic on the river, and the English catholic refugees were anxious about their future and the pensions they received from Spain.

A satisfactory settlement, however, was reached in March 1575. Requesens agreed to banish all English refugees from the Netherlands and to allow English merchants to trade with Antwerp provided they gave their bond not to have any dealings with the Flushingers. He even granted them freedom to practise their religion so long as they did nothing scandalous or derogatory to the established religion of the country. In return, a month later, Elizabeth forbade the prince of Orange, his aiders and abettors, to frequent English ports or her subjects to give them any succour until they returned to the obedience they owed to their natural sovereign. The good relations now established between England and Spain seemed to presage a happier future. They are reflected in the friendly reception accorded to the Spanish fleet which put into Portsmouth, Dartmouth, and the Isle of Wight in October—a marked contrast to the hostile preparations that were made only a few years before, when a similar armament passed along the Channel on its way to the Netherlands.

V

CATHOLIC AND PURITAN

WE have seen that the bulk of the laity found no great difficulty in transferring its allegiance from the old church to the new. The queen, for her part, had made the establishment as comprehensive as possible, taking her stand on the principle that all members of the state were *ipso facto* members of the state church. Consequently, since the existence of dissent was not admitted, there was no persecution; and the first decade of the reign passed away without any serious challenge to the established order. This does not mean, of course, that acquiescence was complete, or that those who clung to the Mass in defiance of the law escaped hardship. Whenever the political heaven became overclouded, by reason of diplomatic tension with Spain, or uncertainty as to the attitude of Rome, the peace of the London catholics was disturbed by domiciliary visits and the rounding up of those who were suspected of celebrating Mass in secret. But these petty vexations, harassing enough to the sufferers, perhaps, cannot be described as persecution, at all events in the sense in which the word was then understood. It was intermittent, variable, and not due to any fixed or deliberate government policy, but rather the result of fluctuations of the political barometer. After ten years had elapsed, however, a number of new and far-reaching influences began to affect the course of events; and the government, confronted with a possible catholic break away, was compelled to modify greatly its attitude to church attendance and the exercise of the old religion.

In the first place must be reckoned the effect of the papal bull of 1570. Pius V, who succeeded to the tiara in the last days of 1565, is generally regarded by his co-religionists as the greatest pope of the counter-reformation period. Unlike his predecessor, Pius IV, he was not a politician, nor did he set much store by the many secular and worldly factors entering into the organization of religion; but he had the gift of leadership in a marked degree, a remorselessly logical and sanguine temperament, and a very clear conception of his duty as head of the catholic church.[1] Originally a Dominican friar trained in the service of

[1] Ranke describes him thus: 'A strange medley of singleness of purpose, loftiness

the Inquisition, of which he became in due course the supreme officer, he was essentially a medievalist in his mode of life and thought; and when he was promoted to the Holy See, he resolved to cut the gordian knot of the English question in a manner that would have commended itself to Innocent III. The actual circumstances leading to his fateful pronouncement against the queen are not clear, but there can be no doubt that he was strongly influenced by the optimistic reports reaching him from his agents in England and elsewhere as to the rising temper of the northern catholics in the autumn of 1569. He determined to strike while the iron was hot. Without communicating his plan to any secular prince, he opened a process for heresy against Elizabeth at Rome on 5 February 1570. Eight days later sentence was passed, and on the 25th a bull[1] excommunicating and deposing the queen was published by the papal chancery.

So slowly did news travel in those days that Rome was still without authentic tidings, at the moment the bull was released, of the outbreak of the northern rebellion in England three months before. And what is even more striking, the date of the publication of the bull practically coincided with Hunsdon's dramatic defeat of Dacre's men on the banks of the Gelt—a defeat that completed the subjugation of the north to the queen's government. Thus from the standpoint of mere timeliness the pope's action failed to assist the cause it was intended to promote. Another three months were to elapse before a copy of it was smuggled into England and affixed by one John Felton to the door of the bishop of London's palace in Paul's churchyard; and by this time the terroristic measures employed by the government in the disaffected districts had crushed the spirit of the northern catholics.

Historically, a special interest attaches to Pius V's pronouncement. It was the last effort of the papacy to check the progress of the reformation by medieval methods, and its spectacular character turned it into something of a *cause célèbre*. Writers have examined not only the grounds on which the pope acted, but also the form and substance of the bull itself, with a minuteness which is probably excessive and unnecessary. It has been

of soul, personal austerity, and entire devotion to religion with grim bigotry, rancorous hatred, and sanguinary zeal for persecution.' (*History of the Popes*, bk. iii, § 8, pp. 361–83.)   [1] Printed *in extenso* in Camden.

shown, for example, that Elizabeth is styled the 'pretended queen of England' (*pretensa Angliae Regina*), as if her right to the crown had never been acknowledged at Rome: whereas both Paul IV and Pius IV had recognized her as queen without qualification. Again, part of the charge on which she was condemned rested on the assumption that she had claimed the supreme headship of the church—a title which, as we have seen, she definitely declined. Furthermore, one of the iniquities imputed to her was the adoption of calvinism: although she cordially disliked, and continued to dislike, this form of church government. Finally, attention has been drawn to the fact that she was allowed no time for recantation—the excommunication and the sentence of deposition being simultaneous, which was a departure from canonical procedure in regard to the treatment of heretical princes. These and other criticisms seem to point to a certain confusion at Rome as to the real situation in England, and also to undue haste.[1] But when all is said and done, the errors and defects of the papal enactment are of trifling importance compared with the effect it was intended to produce as a sovereign remedy for the English question, and the consequences it drew in its train.

Briefly stated, the bull was directed to the achievement of a twofold result. In the first place, by declaring the queen excommunicate and deposed, it aimed at destroying the allegiance of her subjects. In the second place, by coupling all who continued to obey her laws and mandates in the same anathema,[2] it not only legalized rebellion but, by implication, positively commanded it. Thus to every one who in his heart respected the authority of the Holy See, the bull must have come as an ultimatum, ordering him to choose between his conscience and his political obligation. The dual obedience and the tacit compromise of conscience, on which the vast majority of catholics in England had hitherto acted, was for ever destroyed, and in its place was restored the duty of unqualified allegiance to the church of Rome. Theoretically, also, the sentence of excommunication involved the severance of all diplomatic and commercial relations with the doomed country, whose ruin would be completed by the invading army of any catholic

---

[1] For a full discussion of the subject see Meyer, *England and the Catholic Church under Queen Elizabeth*, 78–83.

[2] 'Qui secus egerint, eos simili anathematis sententia innodamus.'

power that felt itself strong enough to act as executor of the papal will.

Normally, then, the bull ought to have stricken the queen and her protestant subjects with dismay. And if the world had moved at the bidding of Pius V as it had done at Innocent III's, the result might have been disastrous enough. Caught between the fire of rebellion at home and the advancing tide of invasion from abroad the Elizabethan government would have been in a parlous position. But in reality there was little likelihood of either event taking place. The only disaffected catholics in England had already shot their bolt in the winter of 1569-70, and the rest remained outwardly at least quiescent. On the other hand, the interested catholic powers on the Continent betrayed no inclination to shoulder the unenviable task of making war on England. On the contrary, both the king of Spain and the emperor Maximilian II were astonished and angry at the pope for not consulting them before he proceeded to extreme measures, and they branded his action as a political blunder, protesting to the queen that they had no hand in it. Maximilian, in fact, made an unavailing effort to get the bull rescinded. In England the reception of the news that the queen was excommunicated and deposed varied greatly among the different individuals and classes affected by it. Elizabeth, it was said, displayed her indifference by inditing some Latin verses scoffing at the papal authority, and saying that no ship of Peter would enter her ports. In certain protestant circles, however, the position was regarded as dangerous in the extreme, and the pulpits resounded with appeals to patriotism and adjurations to both catholics and protestants to rally to the queen's support 'lest we all go together and row in the galleys of Spain'. Cecil, according to the Spanish ambassador, began to think of flight! Bishop Jewel, on the other hand, attacked the pope as a foolish Balaam, whose curses could not harm the people of God. 'O vain man!' exclaimed the worthy prelate: 'as if the coasts and ends of the earth were in his hands, or as if no prince in the world might rule without his sufferance.' The coarse jest and the vulgar lampoon supplied an outlet for the pent-up feelings of others; and pamphlets appeared ridiculing the 'bull of Bashan', 'the monster bull that roared at my lord bishop's gate', &c. In short, nothing was too scurrilous to say about the pope's folly.

Nevertheless the government resolved to leave nothing to

chance. A vigilant watch was kept on the ports for papal
emissaries and the importation of seditious books: the movements of the Spanish fleet were shadowed; and all catholics in
England were regarded with suspicion, as people in whom the
principles of sedition were, so to speak, in suspension and ready
to be deposited whenever a favourable moment arrived. In
1571 a statute was passed (13 Eliz., cap. 1) by which it was
declared to be high treason to affirm that Elizabeth was not, or
ought not to be, queen: that any other person ought to be
queen during her lifetime; or that she was a heretic, schismatic,
tyrant, infidel, or usurper of the Crown. Since the main question at issue, from the government standpoint, was one of jurisdiction rather than of religion—namely, the maintenance of
national sovereignty as against the 'false, usurped, and alien
authority of Rome'—stress was laid on the fact that to reconcile
the queen's subjects to Rome was to withdraw them from their
natural and due allegiance, to create sedition, and to connive
at the overthrow of the state. Hence it was made a treasonable
offence to introduce and 'put in ure' papal bulls and instruments of absolution and reconciliation, or to act in accordance
with them (13 Eliz., cap. 2). By the same act the importation
of *Agnus Dei*'s, crosses, beads, pictures, and other gear from
beyond the sea was brought under the pains and penalties of
the statute of *Praemunire*. By a further enactment (13 Eliz.,
cap. 3) confiscation of property was decreed against all who
had fled abroad without permission since the beginning of the
reign and refused to return within a year.

For the moment, however, the only effect of the bull so far
as the catholics in England were concerned was a certain
'frowardness' and 'stubbornness' in the matter of church attendance. Recusancy, or deliberate abstention from the Anglican
service, made its appearance as a phenomenon to be reckoned
with in the national life.

'I wis, I wis', cried Dr. Overton of East Grinstead, 'there are
many cursed calves of Basan abroad, which, since they have sucked
the bull that came from Rome have given over all obedience and
allegiance to God and the queen; for before that time they could
be content to come to the church and hear sermons and to receive
the sacraments, and to use the common prayer with the rest of the
congregation of Christ, and so forth. They were conformable in all
respects, and content to do anything that beseemed good christians

to do; but since they sucked that mad bull, they are become even as brainsick calves, froward, stubborn, disobedient in word and deed, not to be led nor ordered by any reason.'[1]

Something more, however, was needed to awaken the English catholics from their lethargy and to drive them out of the establishment in considerable numbers; and this was supplied by the missionary efforts of priests from the Continent. Shortly before the publication of the bull, an English college was set up at Douai (1568) by William Allen—himself a refugee—to provide a rallying-point for the scattered bands of catholics who had fled from their native land for conscience' sake, together with educational facilities for the young. The success of the scheme and the enthusiasm it evoked led its promoter to the conviction that something might be done by its agency towards the revival of the old faith in England. Thus, from being merely an asylum for refugees, Douai rapidly became the centre of a great missionary effort to wrest England from the grasp of heresy.

Accepting only youths of from fourteen to twenty-five years of age, of respectable family,[2] the college supplied them with an excellent education based on the curriculum of the jesuit schools. The key-note to the system of instruction was given in the vow taken by the novice shortly after entrance: 'I swear to Almighty God that I am ready and shall always be ready to receive holy orders, in His own good time, and I shall return to England for the salvation of souls, whenever it shall seem good to the superior of this college to order me to do so.' With this self-dedicatory motto for their inspiration, the youths who entered the seminary bent their energies to a seven years' study of theology, philosophy, dialectic, and the history of the English church, under the severest bodily and mental discipline then known. But at the same time as they were trained to co-operate in the struggle for the dogmatic mastery of protestantism, they were also nurtured in the idea that the apotheosis of the religious life was martyrdom, to die, if need be, for England. The walls of their rooms were hung with pictures of torture-chambers and the glory of Christ's sacrifice: prayer, meditation, and seclusion were carefully provided for in the daily routine; and everything was done to concentrate the mind of the neophyte on that

---

[1] J. O. W. Haweis, *Sketches of the Reformation . . . taken from the Contemporary Pulpit* (1844), pp. 193–4.

[2] 'Non ex faece hominum . . . sed nobilibus plerumque orti familiis.'

'glorious conquest of human nature' which was deemed necessary for the success of his mission. Self was extinguished by the
overpowering belief that God had saved them from the disasters
overtaking their country in order to make them instrumental in
saving others. Without a doubt no body of missionaries ever set
about their task with a deeper conviction of the majesty of the
cause for which they contended. So popular did the calling
become, in spite of its dangers, or perhaps in virtue of them, that
the English government grew alarmed at the numbers fleeing
overseas, and instituted heavy penalties to check the stream.
But it was in vain: the movement spread from Douai to other
centres, like Rome (1579), Valladolid (1589), Seville (1592);
and the colleges could hardly cope with the influx of students.

There is no reason to suppose—indeed the evidence is conclusively against such a supposition—that political aims entered
into Allen's scheme, or that the priests who enlisted under his
banner were other than they professed to be, crusaders for
the catholic faith. The *via dolorosa* that led from Douai to
Tyburn could not have been trod by men who were not profoundly imbued with the spiritual character of their work. But
the fact that the pope and the Spanish king were the chief
patrons of the colleges afforded a strong presumption against
the alleged innocence of the missionaries. It was difficult to dis-
connect the seminary movement from the avowed policy of
Pius V, or to avoid the conclusion that under the guise of saving
souls the priests were really acting as executors of the bull.
Moreover, even if the mission was devoted to exclusively religious ends, the law, as we have seen, made it treason to reconcile
the queen's subjects to Rome. Consequently, when they began
operations in England, in 1574, the unfortunate priests walked
straight into the trap which circumstances had prepared for
them. The very word 'seminarist' came to mean in common
parlance 'conspirator'. Nevertheless their ministrations gave a
great impetus to the growth of recusancy. By 1580 there were
a hundred seminary priests working in England.

The third influence affecting the attitude of the Romanists
came into play after Gregory XIII succeeded Pius V in the
chair of St. Peter (1572). The latter part of the period covered
by Gregory's pontificate (1579–85) is the most dramatic, if not
the most critical, in the conflict between the English government and the forces of the counter-reformation. The new pope

was as great a believer in coercive methods as his predecessor. He took his stand unflinchingly on the bull of 1570, professed supreme contempt for policies that truckled to worldly considerations, and gave himself to the advancement of the campaign against the 'Jezebel' of England with all the vigour of his fiery nature. His confidence in the strength of the catholic cause was unbounded. Daily he saw the sword of the counter-reformation driven deeper and deeper into Europe, and it seemed to him that a few more resolute blows shrewdly administered would give him the mastery of England. The destruction of Elizabeth and all her works was the master purpose of his life. The great Pope Gregory I had been the instrument of England's conversion from heathenism a thousand years before: the rescue of England from heresy would be the crowning glory of his sixteenth-century namesake.

As a man of action Gregory had many admirable qualities—optimism, initiative, concentration of thought, unquenchable zeal; but he was strangely ignorant of the complexity of international affairs, and hopelessly failed to combine his actions in one coherent plan. His schemes for the overthrow of Elizabeth were confused, impulsive, self-contradictory, and lacking in that steady purpose which is the essence of statesmanship. Every weapon that ingenuity could devise he pressed into his service without scruple, and with an almost criminal disregard of consequences. Nor was he helped but rather hindered by his secretary of state, the cardinal of Como, whose rashness in the field of action was even more pronounced than his master's.[1]

Gregory's main contribution to the solution of the English question was his patronage of the jesuit mission, which materialized at Rome during the winter of 1579–80, and began operations in England the following summer. Substantially there was little difference between this mission and that of the seminarists. The instructions with which the leaders were provided (April 1580) make it clear that the object aimed at was the 'preservation and augmentation of the faith of the catholics in England': they were forbidden to speak, or to allow others in their presence to speak, against the queen, or to correspond with Rome on political matters. But a certain discretionary

---

[1] J. H. Pollen (*The English Catholics in the Reign of Elizabeth*) describes him as a shallow, imprudent man, with little of that traditional caution for which papal diplomacy is generally distinguished.

power, to be used with the utmost caution, was granted them in regard to political conversations with 'those whose fidelity has been long and steadfast'. It is impossible to say how far, if at all, this permissive clause in the instructions, which ran counter to the whole tenor of the mission, was used or abused during the first year of its existence. But the good faith of Rome need not be questioned; for the clause was withdrawn in 1581, and the conditional prohibition of politics was made an absolute prohibition. On the other hand, the fact that the original instructions fell into the hands of Burghley enabled him to brand the enterprise as a disingenuous attempt to destroy the queen's government under the guise of saving souls.

But this was not all. The mission was dogged from the first moment of its appearance by unfortunate accompaniments. In his desire to meet the grievances of the English catholics, who had long been complaining of the intolerable position thrust upon them by Pius V's action in 1570, Gregory issued an *Explanatio* (April 1580), which he entrusted to the jesuits to publish in England. It was to the effect that, while the policy of the Holy See remained in principle unchanged, and the bull must be regarded as still operative against Elizabeth and the heretics, it did not bind catholics except when public execution of the said bull should become possible. The pope's intention was obviously not to establish peace with the English government, for that could only be done by a complete withdrawal of his predecessor's enactment—an impossible supposition in any case —but to proclaim a truce for an indefinite period, thereby saving the catholics in England from the formal imputation of treason, unless, of course, an overt act could be proved against them. But as a matter of fact the *Explanatio* produced quite the reverse effect. It increased the suspicion with which the jesuits were regarded by the government, and afforded Burghley stronger grounds for the accusation he brought against them.

A further blow was struck at the mission by the pope himself on the very eve of its departure from Rome. For many years Gregory had been urging the need for an expedition against England, either in the shape of a direct attack from Spanish Flanders or indirectly through Ireland. His efforts to give this *empresa* (enterprise), as it was called in the papal documents of the time, a concrete embodiment in men, money, and ships were warmly supported by his cardinal secretary, by papal nuncios like Ormanetto in Madrid, by the leading English catholic exiles, and by swashbucklers like Thomas Stukeley,[1] a renegade Englishman who, having fought under Don John at Lepanto, had blustered his way into high favour at Rome. But always Gregory had been thwarted by the superior claims of other and more urgent affairs, by the difficulty of finding a commander, and by the calculated indifference of the Spanish king, whose sounder knowledge and military experience rendered him dubious, critical, and even contemptuous of the scheme advanced by the pope. Eventually, however, in the summer of 1577, the cardinal of Como decided that the time had come to act, whether the Spaniard was ready to co-operate or not; and a plan was evolved, with the pope's approval, for sending Stukeley to Ireland in command of papal troops, so that he might harass and wear out the English queen 'as Orange has worn us out'. When he arrived in Ireland Stukeley was to join forces with the rebels in Munster, and rouse the whole country against Elizabeth. In February 1578 he sailed from Civita Vecchia with 600 Italian filibusters and a shipful of arms for the use of the native Irish.

With the subsequent history of this grotesque and ill-starred expedition—the desertion of Stukeley at Lisbon, the reorganization of the enterprise by the papal nuncio in Spain, the rebel Irishman James Fitzmaurice Fitzgerald, and Dr. Nicholas Sanders, and its final destruction at Smerwick by the English under Lord Grey—we need not at present concern ourselves.[2] The important fact to be grasped is its relation to, and reaction on, the jesuit mission in England. Thanks to Stukeley's bravados, which had been circulated in every port of call between Italy and Lisbon, the news that the pope was at war with England percolated to all European countries months before Campion and Persons left Rome on their journey to London. Is it to be wondered at that the two jesuits were much perplexed at the ill-timed action of the pope which exposed not only their own lives but also the whole future of their mission to imminent and dire peril? 'We plainly foresaw', wrote Persons on 31 May 1580, 'that this would be taken in England as though we had been

---

[1] There is no authoritative biography of Stukeley, but reference should be made to the *Dict. of Nat. Biog.*; to Z. N. Brooke's article in *Eng. Hist. Rev.* xxviii. 330–7; and to Pollen's in *The Month*, ci. 69–85; cf. also R. Simpson, *The School of Shakespeare* (1878).   [2] See chapter on Ireland.

privy or partakers thereof, as in very truth we were not, nor ever heard or suspected the same until this day.' Thus if the original instructions of 1580 compromised the jesuits deeply in the eyes of the English government, the coincidence of their arrival in London with the *empresa* against Ireland placed them in a still more sinister light, and gave colour to the belief entertained by most protestants that they were the advance guard of the main onslaught on England which, it was believed, the catholic powers might at any moment decide to launch.

Confusion could hardly be worse confounded when, in addition to sanctioning the 'Irish enterprise', the pope grasped at a still more doubtful weapon against Elizabeth, political assassination. The lawfulness of tyrannicide was much discussed at this time in Europe by both catholic and protestant writers; but the only example of its use by a responsible authority was in 1580, when the king of Spain outlawed William of Orange and put a price on his head. Unfortunately there can be no doubt that Gregory XIII was also a disciple of the doctrine, or that he advocated its use against the queen. Two plotters who were racked and executed in England in 1581 and 1584, Anthony Tyrrell and Dr. Parry, confessed that they had undertaken the murder of Elizabeth under promise of a plenary indulgence. But the decisive proof that they did not confess more than the truth is to be found in a letter written (Dec. 1580) by the cardinal secretary, Como, to the papal nuncio, Sega, at Madrid. The letter was in answer to a question addressed by the latter to the pope at the instigation of two English nobles, who, having entered into an agreement to take the queen's life, desired to know if they would thereby incur sin. They wanted to be assured on this point because the deed might cost them their lives. The reply was clear and decisive.

'Since that guilty woman of England', writes the papal secretary, 'rules over two such noble kingdoms of Christendom and is cause of so much injury to the Catholic faith, and loss of so many million souls, there is no doubt that whosoever sends her out of the world with the pious intention of doing God service, not only does not sin but gains merit, especially having regard to the sentence pronounced against her by Pius V of holy memory. And so if these English nobles decide actually to undertake so glorious a work, your lordship can assure them that they do not commit any sin.'[1]

---

[1] Meyer, op. cit., p. 271.

Thus, if Pius made assassination lawful by his bull of excommunication, Gregory positively encouraged it. Clearly he was not above his age, but rather deeply immersed in it. In his impetuous desire to make an end of heresy in England, he virtually cast aside the moral weapon—to the best catholics, the only legitimate weapon with which to combat heresy—and dragged the cause of religion in the mire. Fortunately the vast bulk of the catholic priests, engaged in the attempt to recover England for the faith, believed in moral heroism. But the rash action of the head of the church seriously compromised them in their single-minded work.

Under these unhappy auspices the jesuit campaign was begun. The two men selected by the general of the jesuits for command of the expedition, Edmund Campion[1] and Robert Persons,[2] were remarkably diverse in character, temperament, and genius. Englishmen both and Oxford bred, they were at one only in the enthusiasm and loyalty with which they embraced the perils of the journey to England. Campion was the *beau sabreur* of spiritual gladiators—a radiant figure, whose nonchalance in the face of danger won for him an imperishable renown among his followers, and a high place in the gallery of great Elizabethans. His saintliness, transparent sincerity, and glowing rhetoric were infectious, and it was largely through his ministrations that catholicism in England rose to the heights of heroism it reached during the 'eighties. Many a time he owed escape from death to the deep affection in which he was held by the populace; and to the very end he protested the non-political and essentially spiritual nature of his mission. 'We are dead men to the world,' he cried on the scaffold: 'we travelled only for souls; we touched neither state nor policy: we had no such commission.'

Of Persons, his coadjutor and immediate superior in the enterprise, it is impossible to speak with the same assurance and lack of reserve. His was a subtle and complex character. Gifted with a trenchant personality, he had all the qualities— resourcefulness, perseverance, foresight—that mark the man of action rather than the saint or the pastor. His mind never ran on small things; and from the very first it is more than likely

---

[1] For Campion's life see R. Simpson, *Edmund Campion* (Catholic Standard Library).

[2] There is no complete biography of Persons, but the *Dict. of Nat. Biog.* should be consulted, also Persons's 'Memoirs', ed. Pollen (*Cath. Rec. Soc. Misc.* ii).

that he felt himself irked by the limitations imposed upon him
as a missionary priest in the English field. His roving, adventurous, and ambitious nature is seen to more advantage when
he fled from England after Campion's execution (1581) and
became *par excellence* the politician and organizer of intrigue,
plot, and invasion against his native country. Despite the fact
that his aptitude for political work was seriously doubted by
responsible and wary diplomatists like Mendoza, the Spanish
ambassador, who remarked to his master (April 1582) that the
priest 'cannot be trusted with matters of state', Persons soon
found his way into the inner circles of Elizabeth's enemies and
became a kind of factotum, counsellor, and confidant of them
all. Every one consulted him—the pope, Philip II, the Guises—
fondly believing that he was invaluable to them owing to
his knowledge of England. His consuming ambition was to be
the instrument for the overthrow of Elizabeth; and to this end
he urged on the slow-moving Philip, and became the *deus ex
machina* of every important plot and intrigue of the age.

Campion spoke the simple truth about the mission when,
before his judges, he disclaimed all participation in politics, and
asserted that his charge was to preach the gospel, to minister
the sacraments, to instruct the simple, to reform sinners, to
confute errors, and to 'cry alarm spiritual against foul vice and
proud ignorance'. Yet the purely personal work of winning
souls, great as it was, was not sufficient in itself: literature was
necessary to help in upbuilding the converts, and supplying
the mental and spiritual needs of those who could not be reached
by a personal visit. Hence the idea of a secret printing-press
arose, and took shape in 1581. It was a matter of extreme
difficulty to arrange, because the government was fully alive
to the dangers of a catholic press, while the supervision of
literature was in the hands of the bishop of London, and the
stationers' company controlled the presses. Every week the various printing-houses in the capital were visited: every one was
known, and the type it used. Only Oxford and Cambridge,
outside London, were permitted the luxury of a press. Yet, in
spite of all difficulties, the new catholic press was set up at
East Ham in Essex. Workmen dressed up in gentlemen's clothes,
with ruffs, cocked hats, and swords, eluded for a time the
prying of the curious and the pursuivant, and drove a thriving
trade. Books were printed, marked 'Douai' (to mislead the

government), and carried by night to London, Oxford, and Cambridge, whence they were further colported among the catholic population. In the morning, so to speak, town and university awoke to find the new propaganda on the breakfast table. The experiment lasted about a year before the press was traced to its hiding-place by the government. The migrations of the illicit type and machinery would make a subject in itself: here it must suffice to say that the flight of Persons, who had taken a lion's share in the work, in 1581, and the subsequent dispersal of the company, ended the venture.

 'Book-running' from the Continent was easier to organize than printing under the eyes of the queen. And when Persons fled, he devoted much of his time to developing this branch of activity from the safer vantage-ground of Rouen. Thus the infiltration of 'seditious and papistical literature' into England received a new stimulus from the indefatigable energy of the fugitive jesuit. Boxes of books, printed at the seminaries and continental presses, were brought over under the guise of general cargo, and landed at remote places on the coast. Thence they were taken by an army of receivers and distributed to London and other centres.

One of the most daring of the band of professional 'book-runners' was a certain Ralph Emerson, who came over from France in the company of Father Weston, with a considerable package of books 'and other blest things of importance'. It was his last journey. They landed in 'broad daylight' on the 'open coast' of Norfolk, at a point 'between two ports'. Weston went on to London, leaving the intrepid Ralph to manage the transit of the baggage to the capital. After his departure, Emerson got his books conveyed up the river to Norwich, 'where the carriers take the goods and merchandise to London'. Unfortunately the package was delayed for inspection at an inn; and Ralph, fearing the worst, continued his journey without it. When he arrived at the address in London given him by the priest, without the invaluable gear, a hurried conference was held as to what to do. It was cowardice, they thought, to abandon the books, and yet 'to cling to them and redeem them was full of peril'. The upshot was that Ralph went back, was arrested, brought before a magistrate, and thrown into prison, where he remained for a year or more; and 'no one could tell where he had been taken'.

The career of Emerson is typical of many another, and shows the perils to which the traffic in illicit books was exposed.

Naturally the intensification of the catholic campaign by the appearance of the jesuits caused the government the liveliest anxiety; and in order to meet the increasing danger the penal laws were greatly amplified and strengthened. Recusancy, which hitherto was punished by a fine of 1s. per Sunday, in accordance with the provision of the Act of Uniformity, was now penalized by the infliction of the relatively enormous fine of £20 per month; and attendance at Mass cost the offender 100 marks[1] and one year's imprisonment (23 Eliz., cap. 1). Since, however, the collection of the recusancy fines now became a matter of great difficulty, the law had to be reinforced a few years later by a further enactment (28 &29 Eliz., cap. 6) which provided that if a recusant failed to meet the exactions laid upon him, the Crown might 'seize and enjoy' his goods and chattels, and two-thirds of his lands and tenements, leaving only one-third for the maintenance of the individual, his wife, and family. But if the lot of the recusant was hard, that of the proselytizing priest and his proselytes was harder still. By the act already alluded to in connexion with the recusants (23 Eliz., cap. 1), all priests who sought to withdraw the queen's subjects from their natural obedience, or to withdraw them 'with that intent' from the religion by law established, were adjudged traitors, and their proselytes with them. To *become* a catholic was thus a much more serious offence than to *be* a catholic, for while the latter offence was commutable for a monetary fine, the former could only result in the death of the offender. The severity of the government towards the priests reached a climax in the act (27 Eliz., cap. 2) which ordered all 'jesuits, seminary priests, and other priests' to leave the country within forty days on pain of death for high treason, their aiders and abettors being threatened with punishment as in cases of felony. By the same act all English students residing abroad in the seminaries were ordered to return under penalty for default of being declared traitors.

The application of these laws constitutes the bulk of the 'persecution' attributed to the Elizabethan government, and we must see what it amounted to. First, however, let it be understood that under whatever guise the issue appears, not only the

[1] The English 'mark' was worth 13s. 4d.

reformation settlement in England but also the safety of the state was at stake. It was a life-and-death struggle, in which both parties—the medieval catholic church and the English government—were endeavouring to crush each other to powder. That being so, it was natural that the apologists for the state should represent the conflict as political. Those who suffered, said Burghley, did so, not because they believed in transubstantiation or upheld the Mass, but because they were traitors and sedition-mongers. But the catholic priests, on whom fell the burden and heat of the day, were just as convinced that they were martyrs for their faith. Therein lies the profound pathos of this terrible time. Up to a point both were right; but when the whole sweep of the controversy is taken into account, it becomes clear that neither was wholly right. So long as loyalty to the state was interpreted in terms of loyalty to the anglican church, and loyalty to the Roman church involved acceptance of the papal bull, there could be no separation between politics and religion. Moreover, the confusion of religion and politics made it impossible for impartial justice to be administered. Juries were biased, judges were convinced that every priest was a traitor, and convictions were often obtained on evidence supplied by men of worthless character— renegades, spies, and informers—who throve on their nefarious trade. In fact, all the conditions incidental to a reign of terror prevailed. The innocent suffered with the guilty. Nor was it always a question of proving an overt act against an accused person: the opinion he held on such matters as the legality of the pope's power to excommunicate and depose the queen, or the validity of the lay supremacy over the church, or what side he would take in the event of an invasion by a catholic power, was regarded as sufficient warranty for his punishment. Doubtless the government had a right to be informed on these points, particularly the last; but in extracting the information, with the help of torture, and with the hideous penalty for treason lurking in the background, it introduced a system of judicial inquiry as bad as that practised by the Spanish inquisition, which it professed to abhor. Consciences were racked as well as bodies. The majority of priests who suffered death, died not because they were proved guilty of conspiracies—that was the exception rather than the rule—but because they held opinions that were considered dangerous to the existence of the state—

opinions, moreover, from which they could not dissociate themselves without ceasing to be catholics. At the same time, we
must remember the abnormal conditions prevailing in the
political world, when tyrannicide was sanctioned not only by
political thinkers but also by the head of the catholic church,
when the queen's life was believed to be in danger from fanatics,
and when England was in daily fear of a Spanish invasion.
The fairest criticism that can be passed is to say that both
parties to the conflict were the victims of a tragic dilemma, from
which there seems to have been no escape but by the shedding
of blood on the one side, and by self-sacrifice on the other.

The heroism of the priests cannot be denied. Strange as it
may seem, they were the real gainers by the ordeal through
which they passed. The moral stature of men like Campion
and Southwell has won them a place among the truly great.
On the eve of his execution, the latter wrote to a friend at Rome:
'We have sung the canticles of the Lord in a strange land, and
in this desert we have sucked honey from the rocks and oil from
the hard stone. . . . It seems to me that I see the beginning of a
religious life set on foot in England, of which we now sow the
seeds with tears, that others hereafter may with joy carry in the
sheaves to the heavenly granaries.' Campion's death produced
a whole cycle of poems, some of which rise to the level of literature:

> God knows it is not force nor might,
> Nor war, nor warlike band,
> Nor shield and spear, nor dint of sword,
> That must convert the land:
> It is the blood of martyrs shed,
> It is that noble train,
> That fights with word and not with sword,
> And Christ their capitaine.

A movement that produced poetry, and could reach such
heights of moral and spiritual sublimity, was representative of
many of the best elements in the nation. On the other hand,
those who plied the penal laws did not escape scatheless from
the ordeal to which they subjected their victims. From Walsingham to Topcliffe, the rack-master of the Tower, every one
engaged in the tracking down and condemnation of the priests
suffered a certain moral loss. Nor was this all. In the catholic
countries abroad, nothing was too bad to say about the inhu-

manity of the English government: indeed the whole nation was calumniated as a herd of ravening wolves:

> Et de loups ravissants l'isle en est toute pleine,
> Qui de patte et de dent, et de puante haleine,
> Font mourir à jamais et les corps et les âmes.[1]

Books like the *De Persecutione Anglicana*, published in Rome in 1582 and illustrated with gross pictures of arrests, tortures, and executions, created a foul impression of Englishmen that persisted for a century.

Yet the wonder is that so few suffered death under the Elizabethan penal laws. There were no holocausts of victims, such as blotted the contemporary history of catholic countries abroad. Only some 250 perished during twenty years, including those who died in prison—a notable fact, when we consider that in the five years of Mary's reign more than 300 protestants were put to death. The fact is the Elizabethan government preferred to work by fine, imprisonment, and even banishment rather than by putting into force the full rigour of the law; and few went to the scaffold compared with the number imprisoned in the various castles and strongholds throughout the country— Wisbech, Kimbolton, Wigmore, York, Durham, &c. There they lived under various forms of 'vexation', but at least their lives were spared, and their liberty, though restricted, was not greatly interfered with.

At the same time as the catholic revival threatened the Establishment from one side it was exposed on the other to the running fire of the 'precisians' or puritans, who desired to carry on the reformation to its logical conclusion in a church entirely emancipated from the Romanist tradition. It is probable that the queen regarded the puritan attack with even greater irritation than the catholic; for although the vast majority of puritans remained within the church, and professed to aim not at its overthrow but at its transformation, the political doctrines underlying their religious 'platform' were inimical to the whole structure of Tudor government. Had they triumphed, they would have subordinated the state to the will of a Genevan sanhedrim. Thus puritanism was branded equally with catho-

---

[1] G. Ascoli, *La Grande-Bretagne devant l'opinion française depuis la guerre de cent ans à la fin du XVIᵉ siècle* (1927), p. 154.

licism as a potential, if not actual, danger to the peace of the realm, and punished accordingly.

From the standpoint of its completed development, puritanism may be described as a comprehensive protest against the principle, the method, and the result of the settlement of 1559.

As we have seen, the court reformers had proceeded on the assumption that the church of Rome, despite its corruption in certain points of doctrine and discipline, was nevertheless a true church. They had therefore gone back for their inspiration and guidance to the fourth and fifth centuries, when the see of Rome was still without its universal sway, and by lopping off the 'errors' and 'superstitions' of the medieval church had contrived to preserve all that was essential to the maintenance of pure religion. To the puritan, however, this assumption was untenable. He refused to accord the church of Rome the title of a true church: to him it was simply the Antichrist of the Bible, and to have anything in common with it was to be contaminated. In order to recover the truth in matters of church polity it was necessary to go back, not merely to the fourth and fifth centuries, but to the apostolic age and the Scriptures. The will of God and not the works of man was the final court of appeal.

No less vigorous was the opposition of the puritan to the method by which the religious changes had been effected. A lay settlement of religion seemed to him, no less than it did to the catholic, a gross and unjustifiable encroachment by the state on a domain properly belonging to the clergy; or, as it was put in the second *Admonition to Parliament*, 'Church matters ought ordinarily to be handled by church officers.'

Lastly, the legislation of 1559 had resulted in the erection of a church that was an amalgam of protestant doctrine and catholic ceremonial, which was repellent to the conscience of the purist. To him there was no logical stopping-place between Rome and Geneva, no security for the individual who had left the Roman communion until he found a resting-place within the unassailable fortress of calvinism. To compromise with Rome in any shape or form was to endanger the future of protestantism. Moreover, to some of the more extreme puritans the sweeping of all and sundry into the established church 'at the blast of Queen Elizabeth's trumpet' seemed a crass betrayal of the first principles of a christian fellowship. To the

anglican contention that all members of the state were necessarily members of the state church, and that church and state
were different aspects of the same body politic, these puritans
replied that the church and state were separate and distinct
corporations: that the church was a voluntary association of
believers whose aims and ideals were entirely spiritual, whose
bond was the acceptance of a common creed, and whose only
head was Christ.

Puritanism, however, was more than a protest against religious
forms that, in the opinion of its votaries, had no basis in the
Scriptures. It was also a protest against the spirit of greed and
materialism that saturated society and corrupted the church.
In spite of Parker's effort, in the Canons of 1571, to check
pluralism by making the holding of more than two benefices
illegal,[1] many of the leading clerics indulged in the practice to
the top of their bent. Bishop Hughes, who was appointed to the
see of St. Asaph's in 1573, held *in commendam* an archdeaconry
and ten other livings; and added another six later, nine of the
total being sinecures. He also sold livings in his gift, and leased
episcopal manors for long terms to his wife, children, sisters,
and cousins. Sandys of Worcester was a man of similar stamp,
who systematically plundered his see. Even the celebrated
Bancroft held canonries in St. Patrick's Dublin, Westminster,
Canterbury, and St. Paul's, and, in addition, the rectories of
Feversham and St. Andrew's Holborn. Of the simoniacal
practices of several of the episcopate there is, unfortunately, no
doubt whatever. Scory of Hereford took his see on condition of
alienating seventeen manors to the queen; and he gave three
sinecure prebends in Bromwich collegiate church to his wife,
and another in his own cathedral church to his worthless son.
Harington says of him that he 'worshipped images (not saints
but angels)', i.e. the coin. When Young entered upon his tenure
of the see of York he pulled down the hall attached to the
cathedral 'for the greediness of the lead that covered it'. It
was indeed a common practice for a new bishop to enter an
action against his predecessor for the spoliation of the episcopal
lands and impoverishment of the revenues of the see. Nor was
there any respect shown by the queen and those about her
for the sanctity of church property. The see of Oxford was

---

[1] 'Non licebit cuiquam, cujuscunque sit gradus, plusquam duo ecclesiastica bene-
ficia obtinere eodem tempore', &c. (Prothero, *Statutes and Documents*, p. 202.)

kept vacant for a long time in order that she and her favourites might enjoy its revenues. Fletcher on becoming bishop of Bristol leased his manors to the queen's personal friends at a nominal rent. When he was promoted to London it cost him more than £2,000 in cash to gratify the queen's nominees. In fact, all the principal laymen battened on impropriations. Cecil surrounded his mansion at Burleigh with estates once belonging to the see of Peterborough, and Christopher Hatton compelled Cox of Ely to lease him his London house in Holborn for 21 years. When Coldwell was promoted to the see of Salisbury, he was compelled to alienate the manor of Sherborne, one of his best estates, to Sir Walter Raleigh. Wickham was similarly treated on his entry upon the see of Winchester, having to make over extensive leases to Sir Francis Carew. These are only a few instances of the worldliness that characterized the age and ate like a canker into the church. But they are sufficient to show that the puritans had something to grieve about when they contemplated the condition of religion in England.

Naturally the whole sweep of the puritan objection to the Establishment did not manifest itself at once: it was a slow development, stimulated by conflict and nurtured by persecution; but for all practical purposes it was substantially the detailed exposition of the differences just indicated.

The dispute began in 1559 over what appeared to be a relatively insignificant matter, viz. the vestments prescribed for the use of the clergy by the royal Injunctions. Many took offence at this enforcement of the 'livery of Antichrist', and protested that dress, being a 'thing indifferent', should be left to the discretion of the individual. 'Let the clergy be known', said one, 'by their demeanour and conversation and not by their dress.' 'We confess one faith of Jesus Christ,' said another, 'we preach one doctrine, we acknowledge one ruler over all things; shall we be so used for a surplice; shall brethren persecute brethren for a forkèd cap?' To which the anglican replied that if dress was 'indifferent', it might advisedly be left to the decision of the civil magistrate; but for decency's sake one form of dress was essential, and the vestments prescribed were better than any other because of their hallowed associations. Met on this point, the puritan had recourse to the argument of the 'tender conscience'. The vestments in which he was asked to officiate had been consecrated to Antichrist: they were in the same category as meat offered to idols. Weak brothers, seeing a protestant minister clad in the livery of Rome, would confound the two religions and lose their balance.

The question at issue appeared to be trifling, yet neither side could afford to yield—the puritans, because they had grounded their case on conscience: the queen, because she had merely made a legitimate use of her prerogative for the right ordering of the church service. Several continental reformers—Bullinger, Gualter, and Peter Martyr—tried to ease the deadlock by advising a temporary acquiescence, lest in the conflict of protestants the settlement might be jeopardized, preaching be brought to a standstill, and the Romanists again lift their heads. But they failed to moderate the rising temper of the puritans, and the agitation increased rather than diminished. The use of organs in churches, of the ring in the marriage ceremony, and of the sign of the cross in baptism, kneeling at communion, the excessive number of holy-days, and other 'dregs of popery' were added to the list of puritan *gravamina*. In 1563 the agitators petitioned convocation to make a clean sweep of these obnoxious practices, and to limit the use of vestments to the surplice only. So strong was the movement that the petition was rejected by only one vote, and that was a proxy vote.

Meanwhile great disorder prevailed in the church. In some instances the prayer book was rigidly adhered to, in others psalms in metre were added; in some churches the communion table stood in the middle of the chancel, in others 'altarwise a yard distant from the wall'; some ministers officiated at the sacraments in a surplice, others without it; some used a chalice in the communion service, others were content with a common cup; some favoured unleavened bread, others preferred leavened; some baptized in a font, others in a basin; some made the sign of the cross, others dispensed with it; and so forth. On 25 January 1563 the queen becoming alarmed instructed the archbishops of Canterbury and York 'to take effectual measures that an exact order and uniformity be maintained in all external rites and ceremonies . . .' and to see 'that none hereafter be admitted to any ecclesiastical preferment but who is well disposed to common order and shall formally promise to comply with it'. In virtue of this instruction Parker eventually issued his *Book of Advertisements* (1566) laying down fixed rules for the conduct of public service. He was careful to point out in the preface that his regulations were not prescribed 'as equivalent with the Word of God . . . but as temporal orders mere ecclesiastical'; but he soon found that as many as 37 per cent. of the clergy in the diocese of London alone refused to obey, and were ready to leave the church rather than 'make themselves hypocrites for the queen's sake'. Suspension from office was followed by deprivation, and the first serious split in the ranks of the protestants was consummated. Parker's problem, however, was by no means solved. Many puritans, less conscientious than the stalwarts who went into the wilderness, contrived to remain in the church, albeit openly contemptuous of the *Advertisements* and defiant of the bishops' authority. To cope with this nuisance the archbishop was compelled to give the screw another twist by commanding the clergy to obey their superiors as they obeyed the queen or the letters patent of the council. The only result was to stimulate the puritans to a more resolute and more comprehensive attack on the Establishment. Leaving the question of vestments for the moment, they proceeded to challenge the position of the bishops and the whole hierarchical organization of the church. 'Bishops must be unlorded' became the cry. 'All ministers must be equal.'

For this new development the bishops themselves were partly to blame. Many of them, including Jewel, Sandys, Grindal, Horne, Pilkington, and Parkhurst, had openly avowed their dislike of the vestments regulations, and professed sympathy with the puritans. There were many things, they acknowledged, 'barely tolerable' in the existing state of religion, which ought to be removed as soon as opportunity offered. 'I confess', wrote Pilkington of Durham, 'we suffer many things against our hearts; but we cannot take them away, though we were ever so much set upon it. We are under authority; and we can innovate nothing without the queen; nor can we alter the laws; the only thing left to our choice is whether we will bear these things or break the peace of the church.' Such men were clearly more of a hindrance than a help to the archbishop in his attempt to discipline the clergy: to engage in repression, as they were obliged to do if they remained in office, and at the same time sympathize with the party to be repressed, could only end in trouble. The more honourable course would have been to resign and let others wield the lash: weakness was fatal.

But there were other factors at work than episcopal indecision.

When Parker issued his *Book of Advertisements* it was noted by the puritans that it lacked the authority of the broad seal, and this gave rise to the conviction that the archbishop had acted illegally, in which case they were not legally bound to obey him. In actual fact the queen was just as desirous as Parker for uniformity in the church, but for political reasons she did not want the uniformity to be enforced by state action; she believed that the old visitatorial powers vested in the archbishop and the bishops from time immemorial, even if they were deprived of the authority they possessed in former times, were fully sufficient for present needs. Indeed her quarrel with Parker and his successor Grindal was that they did not make enough of the system, and she was inclined to lay all the troubles of the church at their door. But it is now clear that diocesan visitation, backed as it was by ecclesiastical commissions appointed by the Crown from time to time as occasion required, was quite inadequate to ensure that constant supervision which the church stood in need of in a period of emergency. Only at a later date, when the floating commissions were replaced by a permanent court of high commission, was the machinery of control and correction brought into correspondence with the actual requirements of the situation; but that was the work of Richard Bancroft after 1587. Lastly, we have to note that the puritans had many friends among the nobility and those who occupied important positions in the government. The influence of these men was often used to restrain or weaken the hands of the archbishop. The privy council itself was divided in its sympathies, and even Cecil was not altogether on Parker's side. In many ways, therefore, the task of keeping the puritans within bounds was beset by great difficulties.

Meanwhile puritanism had found a stalwart leader in Dr. Thomas Cartwright, the most outstanding figure in the early history of the movement, and a man of considerable learning. As a fellow of Trinity College, Cambridge, and Lady Margaret professor of divinity in the university, Cartwright had convinced himself through his studies in the New Testament that the whole fabric of anglican church government, no less than the Roman, was in flagrant contradiction with that of the primitive church. Diocesan episcopacy, he asserted, had no basis in Scripture, and ought to be abolished root and branch. His conception of the church entailed the removal of archbishops and archdeacons: the restoration of bishops and deacons to their proper apostolic functions—the former to preaching and teaching, the latter to the care of the poor: the substitution of presbyteries for episcopal chancellors and archdeacons in the disciplinary system of the church; and the election of ministers in the first instance by congregations. It will be apparent that Cartwright's views were simply undiluted calvinism. His originality did not lie in the reforms he advocated, but in the fact that he was the first puritan to take his stand boldly on Scripture as the court of appeal for discipline as well as doctrine, and to apply this principle with logical completeness. In the face of his sweeping affirmations, based on a seemingly unassailable foundation, the controversy about vestments sank into comparative insignificance; and the anglican appeal to history, which had hitherto been the main pillar in the defence of episcopacy, lost half its validity. Historical precedent was now confronted with the divine will revealed in Scripture.

There was no gainsaying the fact that Cartwright was the most dangerous man in the church. In 1570 he was deprived of his professorship: a year later he lost his fellowship at Trinity; and in 1574, in consequence of a summons issued by the ecclesiastical commissioners for his arrest, he was compelled to seek refuge abroad. But by this time he had set puritanism to march in a new direction, and supplied it with a battle-cry, a creed, and a clearly marked objective. In the year of his flight there appeared the *Ecclesiasticae Disciplinae . . . Explicatio*, by Walter Travers, translated by Cartwright, which became the canonical book of presbyterian nonconformity.

The advance made by puritanism under the impetus communicated by Cartwright may be seen in the *Admonition to Parliament* and the *Second Admonition*, published in 1572. Both distinguished from previous puritan documents by the comprehensiveness of the indictment they bring against the Establishment, although they differ greatly in tone and manner. The first was an emotional diatribe, strong in language and essentially destructive in character, addressed to the general body of the laity as well as to parliament: the second was a systematic exposition of the new discipline, less interesting from a literary point of view, but of more practical value because of its careful statement of the puritan position. After the publication of the second *Admonition*, there could be little doubt that presbyterianism had become the creed of the majority of puritans. Its popularity was so considerable that the bishops took alarm. 'The city will never be quiet', wrote Bishop Sandys of London to Burghley, 'till these authors of sedition, who are now esteemed as gods, as Field, Wilcox, Cartwright, and others be removed from the city; the people resort to them as in popery they were wont to run on pilgrimages.' Scambler of Peterborough was similarly troubled about the state of Northampton and Rutland.

But the strengthening of intellectual propaganda was only one aspect of the movement. The *Admonitions to Parliament* opened up a fresh line of attack; for it was arguable that just as the settlement of 1559 had been carried out by parliament, so it might be altered by parliament. Moreover, it was noticeable that the house of commons was becoming more and more puritanical in its sympathies, and deeply concerned about the religious state of the country. In 1566, for example, a petition was drafted drawing the queen's attention to the evils accruing from the lack of efficient ministers, pluralism and non-residence, the absence of a 'true discipline', and the insufficiency of preaching and instruction, 'whereby infinite numbers of your majesty's subjects are like to perish from lack of knowledge'. The parliamentary campaign did not begin in earnest, however, until 1571, when a bill for the reform of the prayer book was introduced into the commons by Walter Strickland; and the immediate result was a constitutional crisis beween the Crown and parliament on the question of rights and privileges. From the queen's point of view the intervention of parliament in religious matters was just as intolerable as that convocation should interfere in the secular government of the country. Through parliament she ruled the state: through convocation she ruled the church; and neither could be allowed to overstep its legitimate sphere without upsetting the balance of the constitution. In the same way as she had prevented convocation from taking an official part in the settlement of 1559, so now she refused to sanction the attempt of parliament to intrude into a department reserved for herself and convocation. In consequence of this attitude, Strickland was summoned before the privy council and temporarily forbidden the house 'for the exhibiting of a bill into the house against the prerogative of the queen'. A year later the queen found it necessary to inform the commons through the Speaker that 'henceforth no bills concerning religion shall be preferred or received into this house unless the same should be first considered and liked by the clergy'; and at the same time she impounded two bills 'touching rites and ceremonies'.[1] It was clear that there could be little hope of ecclesiastical reform through parliament; and the puritans turned to a more useful avenue of self-help.

For some little time there had grown up among the clergy in many of the south-eastern dioceses a practice of holding weekly or fortnightly meetings for the study of the Scriptures and the improvement of morals. To these 'prophesyings' or 'exercises', which were generally supported by the bishops as conducive to greater efficiency in the clergy, it was customary to invite the laity; but the conduct of the meetings, the exposition of scriptural texts, and the discussion that followed were understood to be matters for the clergy alone. In some cases, as in Scambler's diocese of Peterborough, a voluntary discipline was set up in conjunction with the 'prophesyings' whereby offenders against morality, both clerical and lay, could be brought to book and compelled by pressure of public opinion to reform their evil lives. In this way an attempt was made to supplement the canons of the church and to deepen and strengthen the religious life of the community in accordance with the puritan ideal.

Of the efficacy of such a movement there could be no doubt; but it had its dangers. Study of the Scriptures might easily lead to open or covert criticism of the existing rites, ceremonies, and government of the church; and the fact that the laity were present, even although they did not intervene in the debates, might cause the spread of discontent, schism, and rebellion among those who were least able to appreciate the significance of what was said. It was to meet these dangers that Archbishop Edmund Grindal[2] issued his regulations in 1576. His attitude to the 'prophesyings' was that of a sympathizer who believed that every effort of the clergy to better themselves should be recognized by their superiors and incorporated, if possible, in the organization of the church: provided always that ample safeguards were available against unconstitutional innovations. On this point, however, he was destined to be rudely disillusioned by the queen, who saw nothing in the 'prophesyings' but a veiled attempt to carry out a further reformation of the church

---

[1] See also pp. 182–4 for the constitutional crisis.
[2] Appointed to Canterbury on Parker's death (1575).

in defiance of the law and the royal prerogative. On the ground that they were causing disorder in the general body of the laity, she peremptorily ordered the archbishop, in 1577, to take steps immediately for their suppression, and to report the names of those who refused to desist to the privy council for exemplary punishment. When Grindal declined, with amazing audacity, to carry out these instructions, and pointed out that the *hoc volo sic jubeo* of the queen's letter was no better than the 'anti-christian voice of the pope', he was deprived of office and remained suspended from the exercise of his temporal jurisdiction for five years.

The full severity of the repression, however, did not really begin to break over the puritans until the appointment of John Whitgift as archbishop of Canterbury in 1583. The new prelate was a man after the queen's own heart—hard-working, consistent, and a resolute opponent of dissent. Unlike Grindal he had formed his views in conflict with puritanism, when, as master of Trinity College and Regius professor of divinity at Cambridge, he had been the chief defender of the church against the assault of Cartwright; and by the time he was promoted to Canterbury they had grown to be a kind of second nature to him.

'The question', he said, 'is not whether many things maintained in your platform of discipline [i.e. the puritan] were fitly used in the apostles' time or may now be well used in sundry reformed churches; but whether, when there is a settled order in doctrine and discipline established by law, it may stand with godly and christian wisdom to attempt so great an alteration as this platform must needs bring in, with disobedience to the queen and law and injunctions of the church, and offence to many consciences.'

As for the puritan demand for equality of ministers, he was prepared to recognize it *quoad ministerium*, but *quoad politiam et ordinem* there must be degrees of superior and inferior. In short the time for argument was past, and the time for drastic action had come.

Whitgift's panacea for the religious troubles of the time was to erect in England a church as absolute and infallible as that of Rome. His policy is contained in the six articles of 1583, by which the clergy were compelled to accept by unqualified subscription the following three points: (1) the ecclesiastical supremacy of the Crown: (2) the book of common prayer and the ordinal, as containing nothing contrary to Scripture; and (3) the whole of the Thirty-nine Articles as agreeable to the word of God. The immediate effect of this enactment was the suspension of over two hundred ministers in Lincoln, Norfolk, Suffolk, Essex, and Kent, and a great outcry against the inhumanity of the archbishop. There is no doubt that Whitgift's methods were those of a grand inquisitor: in tracking down puritan delinquents he followed, as Burghley pointed out to him in 1584,[1] the example of the Romish inquisition. His chief weapon of repression was the ecclesiastical commission, which, under his presidency, was drawn more and more into the disciplinary system of the church as an adjunct to the ordinary diocesan machinery whose inadequacy Parker and Grindal had already experienced. The advantage of having some such weapon to combat the spread of puritanism was great: it was swift and secret in action, could dispense with formalities of the law, and had behind it the whole coercive power of the state. Suspects and witnesses who were summoned before the commissioners were tendered an oath (the oath *ex officio*, as it was called) which obliged them to answer all interrogatories. If they refused to take it, they were handed over to the court of star chamber to be punished: if they accepted it, and through their subsequent admissions were convicted, they were sentenced to fines or imprisonment at discretion and without right of appeal. Needless to say the victims complained bitterly against the legality of such a régime. 'This corporal oath', said Cartwright, 'is to inquire of our private speeches and conferences with our dearest and nearest friends; yea, of the very secret thoughts and intents of our hearts, that so we may furnish both matter of accusation and evidence of proof against ourselves, which was not used to be done in causes of heresy or high treason.' By substituting inquiry by oath for indictment and trial by jury in open court the queen, it was said, was over-riding the common law, and setting up a system of legal procedure in flat contradiction with the basic principle of English law—*nemo tenetur seipsum accusare*. On the other hand, it is difficult to see how the integrity of the church could have been protected against puritan disruption otherwise than by some sort of inquisitorial process such as the high commission made possible. Statutory enactments for the repression of dissent were practically impossible when parliament

[1] Letter printed in Prothero, pp. 213–14.

itself was so strongly puritanical in its sympathies. Whitgift was fully convinced that he could not proceed effectively with the purification of the church unless the discretionary power of the commissioners was allowed; for in many cases parishioners were so friendly to puritan ministers that they would not 'present' them nor appear as witnesses against them.

As if to emphasize the need for drastic control, there had developed, shortly after Whitgift's appointment, the so-called 'classical' movement within the ranks of the puritans. The basis of the scheme was the introduction of a presbyterian discipline into episcopacy. The 'classis' was simply a secret synod after the Genevan pattern, and was composed of all the clergy in a district who were willing to follow a new book of discipline by Travers. The object aimed at was to keep within the letter of the law, but to seek ways and means of adapting the anglican service and polity to the presbyterian. Wherever possible, offending passages in the prayer book were to be omitted: puritan interpretations of the Scriptures were to be freely used in preaching: congregations were to elect their ministers and recommend them through the 'classis' to the bishop for consecration; and the names of elder and deacon were to be substituted for churchwarden and collector for the poor. Clearly such a movement, however law-abiding, was tendencious in the extreme, and might easily culminate in the overthrow of the Establishment if allowed to run its course unchecked. But owing to the secrecy in which it was wrapped it was difficult to obtain evidence that would secure conviction of any overt breach of the law.

The same could not be said of the other menace to the established religion which began to manifest itself about the same time—the separatist movement. Prior to Whitgift's tenure of the archbishopric of Canterbury, the presbyterian idea of reforming the church by constitutional means had maintained substantial unity of aim within the ranks of the puritans. The only serious attempt to break away from the Establishment and found a 'separatist' organization at Plumbers' Hall in 1567 had ended in failure and the imprisonment of the leaders. Thirteen years later, however, the effort to set up an independent church was resumed with increased vigour under more favourable auspices. By this time a group of men, led by Robert Browne and Henry Barrow, who belonged to a younger generation than Cartwright and cared nothing for the authority of Calvin or the sacro-sanctity of the Genevan system, had begun to preach a conception of the church that differed radically from both the puritan and the anglican. As against the inclusiveness of the latter and the cast-iron rigidity of the former, they held that voluntaryism or congregationalism was the only possible basis of a christian church. The root principle of their teaching was that the close connexion between church and state, symbolized by anglicanism and upheld by the presbyterians, was unscriptural and ought to be abolished. For the rest, they denied that a learned ministry was essential, laid supreme stress on inspiration, and exalted the idea of a non-hierarchic, non-ritualist form of public worship. In spite of inadequate organization, or possibly because of it, their spiritual enthusiasm gave an emotional touch to religion which neither anglicanism nor presbyterianism could supply, while their denunciations of Rome reached the fervour and intensity of the Hebrew prophets. In short the Brownists and Barrowists formed the spearhead of Elizabethan puritanism. Unfortunately their uncompromising opposition to the Establishment and their determination to effect a further reformation of religion 'without tarrying for any' brought them into conflict with the ecclesiastical supremacy of the Crown; and their democratic theory of church government ran counter to the spirit of the age.

Towards the close of the year 1587 yet another phase of puritan activity took shape in the Martin Marprelate attack on the bishops. It began with the issue of an anonymous pamphlet addressed to convocation, and 'compiled for the behoof and overthrow of the unpreaching parsons, fyckers, and curates that have learnt their catechism and are past grace'. Its ribald humour and bold sallies against the bishops, coupled with the anonymity of the author, at once caught the attention of the public; and very soon a second appeared still more outspoken, declaring war on the 'petty antichrists, proud prelates, intolerable withstanders of reformation, enemies of the gospel and covetous wretched priests'. 'Martin', it announced, would not only publish every one of their mistakes, but also put 'a young Martin in every parish . . . every one of them able to mar a prelate'. In all, seven famous tracts were published: 'The Epistle', 'The Epitome', 'Hay any Work for Cooper', 'Minerall and Metaphysical School-points', 'Theses Martinianae', 'The Just

Censure and Reproof', and 'The Protestation'. If the literary form adopted was novel, the attitude of the mysterious 'Martin' was even more novel: he swept away in a tide of unrestrained jocularity all the traditional reverence for the episcopate. For a time the martinists and anti-martinists bespattered each other to the amusement of the London public, while the government and the bishops conducted an unremitting search for the secret martinist press and the author of the tracts. The episode, how-ever, proved to be of short duration; for no secret press could exist for any length of time in the England of Elizabeth, as the jesuit experiment of 1580-1 had already shown. By Christmas 1589 it was over. Of the chief actors—John Penry and Job Throgmorton—the former escaped for a time to Scotland, but subsequently returned to England to be executed: the latter, for some reason which remains a mystery, was allowed to go free. The others who were associated with the movement in a minor capacity suffered fines and imprisonment, but no clear evidence is available even to-day as to who Martin actually was. Like the author of the letters of Junius his identity remains an un-solved problem of history.

Although the repressionist campaign conducted by Whitgift went far beyond the comparatively mild measures adopted by his predecessors, it was Richard Bancroft's appearance on the high commission in 1587 that ushered in the period of storm and stress. This rising ecclesiastic, who had worked his way up the ladder of preferment by sheer ability, had latterly acquired a unique knowledge of the ramifications of puritan activity, parti-cularly of the 'classis' movement, and in February 1589 he delivered a bitter philippic against presbyterianism at Paul's Cross. The sermon raised a storm of protest from Scotland, where the presbyterian type of church government now reigned supreme, and the 'kirk' dominated the state; but Bancroft continued his exposure of the political dangers of the movement, unperturbed by the criticism levelled at him by the irate Scot-tish calvinists. Not only was the sermon itself published, but two other publications entitled *A Survey of the Holy Discipline* and *Dangerous Positions* followed in quick succession, showing that Bancroft could quote chapter and verse for the accusations he made. Soon after the February diatribe the presbyterian leaders, including Cartwright himself, who had been lured back to England in the hope of better times, were summoned before the commission, and on their refusal to take the oath were sent to prison. They remained there until 1592, when the star chamber, as a consequence of attorney-general Sir John Popham's advice, decided to take no further action in the matter, because no overt breach of the law could be proved against them.

Meanwhile the full rigour of the law was being felt by the extremists. John Udall, the martinist (or supposed martinist), was imprisoned under sentence of death in 1590, and died in prison before the sentence could be executed. Three puritan fanatics—Coppinger, Arthington, and Hacket—were accused of plotting to take the queen's life, and Hacket was executed in 1592. In the following year Barrow and Greenwood, the separatists, were put to death for seditious words: John Penry suffered the same penalty for his share in the marprelate movement; and the drastic conventicle act (35 Eliz., cap. 1) was passed punishing, in the last resort with exile or a felon's death, all who obstinately refused to attend the anglican church or attended separate conventicles.

By these means the back of the puritan movement was effectually broken. The queen, in the meantime, kept a rigorous control over any attempt of parliament to come to the rescue of the suppressed party. When Anthony Cope introduced a bill, in 1587, for the repeal of all existing ecclesiastical laws, and the introduction of a new form of common prayer, both the bill and the book were impounded, and Cope was sent to the Tower for disobeying the royal injunction concerning the submission of all such projects to the bishops. A similar sharp punishment was meted out to Attorney Morrice of the court of wards, who took up the cudgels on behalf of the suffering puritans in 1593, and introduced two bills for the abolition of oaths and subscriptions, and against unlawful imprisonments and restraints of liberty. He was deprived of office, suspended from practising as a lawyer, and imprisoned for several years in Tutbury castle.

Note. Elizabethan England was comparatively free from the more fanatical protestant sects, whose politico-religious doctrines, subversive of both church and state, had caused widespread disturbances and bloodshed in many parts of Germany, Switzerland, and the Netherlands during Henry VIII's reign. The anabaptists, in particular, were a relatively small body, mostly Dutch, who had mingled with the stream of refugees fleeing from Spanish tyranny.

In all probability they would have lived peaceably in their English settlements; but the queen, like her father, set her face flintlike against their dangerous creed. In 1561, says Camden, a royal proclamation ordered 'anabaptists and such like heretics' who have come to England under colour of avoiding persecution, to leave the realm within twenty days under penalty of imprisonment and confiscation of goods. In 1568, owing to a fresh wave of immigration, Parker was commanded to undertake, through the bishop of London, an inquiry into the character of the immigrants, and to compile a register. In the same year a second proclamation refers to secret conventicles by which many English people have been corrupted, and orders all who hold anabaptist principles, and refuse to recant, to leave the country within twenty days under penalty of death. But apparently, in spite of all measures and precautions, the danger from anabaptism continued. In 1575, at Easter, a congregation of the much-feared sect was discovered in Aldgate. The culprits, who were Dutch, were tried before the bishop of London at St. Paul's: five recanted their heresies, fifteen were shipped abroad, and five were condemned to death. Eventually only two of the five were burnt at the stake in Smithfield 'in great horror with roaring and crying'. Extermination of the sect, however, was impossible. In 1589 there were said to be several anabaptist conventicles in London and elsewhere; and in 1591 occurred the Hacket affair mentioned in the text. 'Men talk of it,' wrote a contemporary, 'resemble it to the matter of John of Leyden, who took upon himself the kingdom of the anabaptists, and think this fool [i.e. Hacket] plotted some such kingdom as these prophets might have assembled.' Evidently John of Leyden and the Münster atrocities (1534–5) were still a vivid memory even in Elizabethan England.

VI

THE CONSTITUTION

ELIZABETHAN England produced no book on political theory fit to be compared with the writings of Bodin, Hotman, Duplessis-Mornay, or the other great thinkers of contemporary Europe. Presumably Englishmen were, on the whole, too well satisfied with the form of government under which they lived to probe deeply into first principles, or to question the grounds of their political obligation. If they wrote about the state, they did so in a perfunctory manner, explanatory or laudatory rather than critical or controversial.[1] To the political theorist, therefore, the reign is to a large extent devoid of interest. When we turn from theory to practice, however, a somewhat different picture presents itself. We become aware that the apparent complacency is shot through with a good deal of discontent, and that the Tudor ideal of government, now at its zenith, was already entering upon a period of disintegration. True, the question of defining where the sovereign power lay in the constitution had not yet arisen, for the great corporation and body politic of the kingdom, consisting of king and parliament, was still one and undivided. But to the observant eye the germs of the later conflict, that made definition a necessity, were not only latent but active in Elizabeth's reign. For this reason the constitutional historian will find the period a valuable introductory study to the seventeenth-century struggle: it supplies, in fact, an indispensable connecting link between two epochs.

It has been pointed out[2] that if a strong and vigilant executive was justified in the earlier part of the sixteenth century as a preservative of law and order and national security, it was more than ever justified during the latter part, when the renewal of the breach with Rome, the uncertainty of the succession, the Stuart menace, and the constant threat of invasion combined to create a state of emergency without parallel in English history. Every one realized that if the queen were to die before Mary Stuart, the country would be plunged into the vortex of civil war and all the other evils of a disputed succes-

[1] e.g. Sir Thomas Smith's *De Republica Anglorum*: Hooker's work will be referred to later in another connexion.

[2] Prothero, *Statutes and Documents*, Preface, p. xix.

sion, coupled, in all probability, with foreign intervention. Consciousness of this fact provided Elizabeth with the lever by which she moved her world, and at the same time it enabled her to maintain in substantial working order the mechanism of government she inherited from her predecessors. On the other hand, when the sense of omnipresent danger grew less or faded from memory, after the execution of the Scottish queen and the defeat of the Spanish armada, it is noticeable that a spirit inimical to the existing system of government has spread itself through the community. Criticism of the executive became more and more pronounced in parliament: privy councillors were treated with less respect than formerly: even the popularity of the queen, great as it was, seemed to wane; and a *fin de siècle* feeling, compounded of irritation and lassitude, hovered like a cloud over the political landscape. The glory of the reign was passing away, and a prosaic age, unattended by dangers from without and within, was soon to pick to pieces the governmental machinery that had made that glory possible.

All this is true, but not the whole truth. As we shall see, the effort of parliament to enlarge the scope of its rights and privileges at the expense of the prerogative, to eat its way into the Tudor edifice, and to grope forward to the ideal of popular sovereignty of a much later day began quite early in the reign, and continued unabated for more than thirty years before the queen's death. The transition, therefore, was not sudden, nor was it, strictly speaking, a consequence of the removal of public apprehension: on the contrary, the friction between Crown and parliament was itself generated by the emergency in which the country stood. And this friction acted like a forcing-house for the development of parliamentary technique, as well as of a corporate feeling and continuity of tradition that must have proved useful once the real contest began under the Stuarts.

The central organ of the Elizabethan administration was the privy council, a body consisting of seventeen or twenty carefully chosen, experienced men, mostly the great ministers of state and officials of the royal household, through whose hands passed all the routine business of the state. Unlike the modern cabinet, which is responsible to parliament, the council of the sixteenth century was simply and solely the executive instrument of the sovereign, by whom it was appointed and from whom it derived its authority. It had no permanent abode, but met where the queen happened to be in residence at the moment. Owing to its size it worked by committees, and only occasionally, when matters of exceptional interest or importance were remitted to it by the queen, did it meet as a body. As a rule its meetings, which were held in secret, were attended by eight or ten members, and not infrequently business was transacted by only half that number.[1] The queen herself never attended, being content to let her wishes be known through one or the other of those present or through the secretary, who was in constant communication with her. Normally she accepted the council's decisions for better or worse, but she did not regard herself as bound to follow its advice, nor did she scruple, when questions of policy were in debate, to exercise an over-riding authority if she thought her own or the national interests demanded it.

There is no official evidence to show that the resolutions arrived at round the council table were other than unanimous; but from unofficial sources it may be gathered that when great affairs of state were under consideration differences of opinion bred factions,[2] and factions might easily become feuds.[3] Moreover, there was always a latent rivalry among the members for a leading place in the queen's favour. It is noticeable also that councillors were free to communicate individually and privately, as well as publicly, with any one they pleased, for the purpose of acquiring information, providing always they observed their oath of fealty to the Crown. Ambassadors and dignitaries of all sorts consulted them as to the meaning or implication of their instructions, or supplied them with information which they dared not incorporate in their official reports to the queen and the council as a whole. Knowledge as well as wisdom was a necessary equipment of any one who aimed at taking a leading part in debate; and all the great

[1] Mendoza reported to Philip II, in 1578, that although there were seventeen councillors, the bulk of the business was transacted by Leicester, Walsingham, and Burghley; and towards the close of the reign Robert Cecil, Nottingham, and Buckhurst took their places.

[2] e.g. the marked division between Burghley supported by Sussex, Hunsdon, and Lincoln, on the one hand, and Walsingham and Leicester supported by Warwick, Bedford, and Knollys, during the years 1578–85.

[3] e.g. the Essex-Cecil feud in the nineties.

figures of the Elizabethan council—Burghley, Walsingham, Leicester, Essex, Robert Cecil—spared themselves no effort to
acquire, by an elaborate system of espionage, every scrap of
information likely to be of use to them in the discharge of their
duties.[1] Elizabeth had a curious way of showing her approval or
disapproval of those who supported or opposed her own will;
but as a rule she appreciated and respected the man whose
opinions were based upon a rational analysis of facts, or who
could quote chapter and verse for the advice he tendered. The
system was admittedly amateurish and unscientific. The field
of national interests was too vast to be spanned effectively by
any one person however gifted, the accumulation of state papers,
official and unofficial, was wasteful and often futile, but a
more vigilant, hard-working, and loyal body of men than the
privy council in Elizabeth's reign it would be hard to find.

The business that devolved upon the council was widely
varied, enormous in bulk, and extremely difficult to classify.
Besides the highest matters of state, comparatively trivial concerns affecting the private life of individuals were brought
under consideration. Decisions about the army and navy,
seminary priests and jesuits, recusants, foreign affairs, commerce, industry, and piracy were interspersed with pronouncements on civil suits, family disputes, and petitions for relief
from oppressed persons who could not obtain justice in the
ordinary law courts. There was, in fact, no limit to the com-
petence of the council: it could handle any question involving
the security of the state or the welfare of the subject; and it
could issue proclamations with the validity of law. In Hallam's
opinion,[2] England was managed 'as if it had been the household
and estate of a nobleman under a strict and prying steward'.

As the supreme centre of authority under the Crown, the
council stood in the closest relation with the judicial system,
the church, and parliament, and exercised a directing, controlling, and supervisory authority over all officials—bishops,
judges of assize, lords lieutenant, sheriffs, justices of the peace.
A constant stream of letters, reports, and petitions poured into

---

[1] Walsingham, for example, received intelligence not only from every quarter
of England, but also from thirteen towns in France, seven in the Low Countries,
five in Italy, five in Spain, nine in Germany, three in the United Provinces, and
three in Turkey. (F. M. G. Evans, *The Principal Secretary of State* (1923), p. 286.)
[2] *The Constitutional History of England from the accession of Henry VII to the death
of George II* (1827), i. 265.

it from all parts of the country, to be answered by an equally voluminous outgoing stream of writs, instructions, and summonses. To some extent its labours were lightened by the concurrent and subordinate jurisdictions wielded by the council of the north, which was responsible for the administration of the counties lying beyond Trent, and by the similarly constituted council of Wales and the marches. Purely judicial cases, again, were as a rule relegated to the court of star chamber. Sometimes matters capable of being handled by the ordinary law courts were ordered to be dealt with there; and there was a practice of devolving certain defined pieces of administrative or judicial work on commissioners, not privy councillors, who were specially appointed for the task. But the council itself remained the final court of appeal and the co-ordinating centre of the political universe.[1]

Particularly important, perhaps, were its duties and responsibilities as guardian of the state against all manner of disturbances. In order to strengthen its hands for the investigation and castigation of offences under this head, especially with regard to suspected treason, it possessed the unique privilege of employing torture and forced confessions. In the troubled times of the sixteenth century, it was useful to have some body of men empowered to 'bolt out' the truth from suspects by extraordinary means; and it was right that this power, which might give rise to abuses, should be vested in the highest and most respected court in the land, where it would be used with moderation and discretion. It will be apparent, however, that men entrusted with such far-reaching powers and a jurisdiction extending over all officials, local as well as central, might easily develop a conception of the state very different from that entertained by parliament and the common law courts, and by treating them all as its agents work for the transformation of England into a centralized bureaucratic state. This did not happen, but circumstances were undoubtedly tending in that direction until the end of the Tudor régime.

The judicial annex or extension of the council was the court of star chamber. In the reign of Elizabeth this court, albeit it was officially styled 'the lords of the council sitting in the

---

[1] For instances of its supervision and control of local affairs see Sir W. H. Holdsworth, *History of English Law*, iv. 78–80: of ecclesiastical affairs, ibid. 81, 82; of the judicial system, ibid. 84, 85.

star chamber', had a separate organic existence, kept its own records, and followed its own practices and precedents. All privy councillors attended its meetings, together with some of the judges who were not privy councillors, and such peers and bishops as were summoned by the Crown in the capacity of assessors. In contemporary phrase, the star chamber was 'the curious eye of the state and the king's council prying into the inconveniences and mischiefs which abound in the commonwealth'. As such, the range of its activities was practically unlimited; but for convenience' sake the cases with which it dealt may be regarded as falling into two main categories: first, those which involved a breach of public order, or might be construed as provocative of such a breach; secondly, those which arose out of a direct violation of royal ordinances, proclamations, grants, or commands. In the first category might be grouped riots, conspiracies, challenges to duels, forgeries, libels and defamation of character, disturbances arising out of the game laws, hunting, or inclosures, disputes as to the ownership of land, or between lord and tenant, assaults, and disorderly behaviour of all kinds. When justice Shallow saw Falstaff ill-treating his men, killing his deer, and kissing his keeper's daughter, he exclaimed: 'I will make a star chamber matter of it; if he were twenty Sir John Falstaffs he shall not abuse Robert Shallow, Esquire. . . . The council shall hear of it; it is a riot.'[1] It was a 'star chamber matter', also, to write ballads about one's neighbour, to write abusive letters to justices, or to speak scandal of men in high position. It was an outspoken and scurrilous age, when the 'filthy ballad' set to some trite tune became a regular means, even in polite society, of humiliating a rival; and Sir John Harington was in good company when he noted in his diary: 'I will write a damnable story and put it in goodly verse about Lord A.'[2] Next to rioting and forgery, the 'infamous libel' was more often the cause of action by the star chamber than any other offence.

But the peculiarity of the star chamber was not so much the rich variety of cases it dealt with as the informality of its procedure and the swift and summary punishments it inflicted. As opposed to the courts of common law, where accused persons were tried by a jury, the star chamber was a court

[1] Quoted by E. P. Cheyney, *A History of England from the Defeat of the Armada*, &c., p. 90.   [2] Ibid., p. 93.

composed entirely of judges, who proceeded on written deposi-
tions and predigested evidence—sometimes even on verbal confessions (*ore tenus*)—and only summoned the accused before its
bar in the final stage of the trial. The method was similar in
many respects to that of a modern court martial. The penalties
inflicted were not according to statute, but at the discretion
of the court, and adapted to the nature of the offence, being
often of an ingenious and bizarre nature. Capital punishment,
it is true, was beyond its province; but heavy fines, imprisonments, corporal punishment, whipping, branding, the pillory,
cutting of ears, slitting of noses, and other forms of public
humiliation were freely used *in terrorem populi*. On the other
hand, it is important to note that the proceedings in the court
of star chamber were not conducted in secrecy, but in full view
of the public, who were admitted freely, and 'flock thither in
great abundance when causes of weight are there heard and
determined'. Not only so: all classes might apply to it for
redress of wrongs, and its diligence in the discharge of business,
together with the rigorous justice it dispensed, made it the
most popular court in the realm. Lambarde, speaking of it in
1591, alludes to it as 'this most noble and praiseworthy Court,
the beams of whose bright justice, equal in beauty with Hes-
perus and Lucifer, ... do blaze and spread themselves as far as
the realm is long or wide'. Coke was no less complimentary
when he wrote of it as 'the most honourable court (our parliament excepted) that is in the christian world'.

Of the local administration only a brief description can be
given. Here again, however, the queen made no innovation
in the system she inherited from her predecessors, although two
of the offices, the justiceship of the peace and the lord lieutenancy of the shire, took on a greater significance owing to the
expanding activities of the central government and the political
dangers of the time. As a rule all the chief officers of the local
administration—justices of the peace, sheriffs, lords lieutenant,
vice-admirals, coroners, and high constables—were drawn from
the ranks of the rural gentry, who performed the duties assigned
to them, not as a professional class trained for the purpose, but
as amateurs. To each was entrusted a definite function or
series of functions in the preservation of order, the execution
of justice, and the military security of the community; but the
diffusion of authority and the exact apportionment of respon-

sibility in particular cases is not always easy to determine. Frequently offices seem to overlap or run into one another, and there was a tendency on the part of the central government to modify the system or interfere with its working without paying too much deference to formality. Efficiency rather than formality was the guiding principle.

It must have been difficult, for example, for an Elizabethan justice of the peace to discover just where he stood with regard to his various responsibilities at any particular time: he was the beast of burden on whose broad shoulders the government continually devolved new tasks.[1] Besides the 'commission of the peace', which implied authority to hear and determine all manner of criminal charges, except those which were indictable, that is, subject to trial by jury in the ordinary courts of law, he was the principal officer through whom council ordinances and parliamentary statutes were brought into touch with local life. The quarter sessions, or general assembly of the justices, held in each shire in October, January, spring, and midsummer, display a remarkable variety of business, covering criminal offences such as murder, assault, burglary, vagabondage, witchcraft, breaches of industrial laws and game laws, and recusancy; and the punishments inflicted ranged from fine and imprisonment to flogging, branding, and hanging. But the justices were also responsible for the repair of roads, licensing of taverns, binding of apprentices, relief of the honest poor in times of inordinate distress owing to famine, fire, or pestilence, fixing of the wage level for the district, regulation of the price and the export of grain, and the making of provision for redeeming English sailors from the Barbary pirates. They did not deal with civil litigation, but in certain circumstances they co-operated with the bishops in the suppression of irregularities in the church, and gradually the responsibility for the entire administration of the poor law was deputed to them. In short, there was no more useful officer in the local administration than the justice of the peace. He was its pivotal point and the guarantee of its efficiency, and his integrity was carefully watched by the council.

The office of the sheriff, on the other hand, was becoming more honorific, and was certainly less arduous than the justice-

---

[1] W. Lambarde in *Eirenarcha* (1581) takes over 600 pages to describe the powers and responsibilities of the J.P.

ship. As the immediate representative of the Crown in the shire, the sheriff was charged with the provision of hospitality in the event of a royal visit, or the sojourn of distinguished strangers from abroad, and the periodical visit from the justices of assize. He had also important financial responsibilities in connexion with the collection of debts due to the Crown and of fines imposed by the law-courts, either locally or at Westminster, together with the enforcement of the queen's right to purveyance. He had nothing to do, however, with the collection of taxes levied by the authority of parliament; but in the county court, over which he presided, he supervised what was by far the most important function of this decaying assembly, the election of members of parliament. In the same place he proclaimed the laws, published outlawries, decided civil suits involving less than 40s., and awarded damages for breach of contract and non-payment of wages. After the close of each parliamentary session writs were sent to him authorizing the payment of wages and travelling expenses to members of parliament, and these were assessed upon the hundreds and townships at the next meeting of the county court. The sheriff also attended the quarter sessions of the justices of the peace and the assize courts for the impanelling of juries, and the due presentment of prisoners from jail, whose safe keeping was part of his duty. These courts could not be properly constituted without his presence, nor could the sentences imposed be discharged without his co-operation; and all writs and summonses required his sanction. Certain other powers were also exercised by the sheriff, either on his own initiative or in conjunction with other Crown officials. Such were the suppression of riots and unlawful assemblies and the supervision of the county musters, but in these matters his responsibility was overshadowed by that of the lord lieutenant.

The lord lieutenant[1] was, like the justice of the peace, a typically Elizabethan functionary. The constant fear of internal rebellion and invasion from abroad combined to elevate the lord lieutenancy to a position of great importance in the local administration. On this office fell the supreme responsibility for holding the musters in the shire, selecting officers and men for the trained bands or local militia of the time, purchas-

[1] Appointments to this office were temporary until 1585; after that they became permanent.

ing and listing armour and munitions of war, keeping up the beacons, and generally speaking making all local arrangements for the defence of the country. In case of invasion or rebellion it was his duty to lead the county levies; and if men were wanted for foreign service he supplied them, provided for their equipment, and saw to their transport.

In the maritime counties there was another typical functionary, the vice-admiral, whose jurisdiction covered all matters arising out of the naval and commercial exigencies of the age. He saw to the suppression of piracy, the impressment of seamen for the navy, the enforcement of embargoes, the registration of captures at sea, and the salvage of wrecks, together with the settlement of disputes connected with these matters. He was directly responsible to the lord admiral in London.

When the wide area covered by prerogative government is considered, it will be apparent that the Elizabethan parliament enjoyed a very restricted sphere of activity, and a very imperfect influence over the executive.[1] The queen in council could issue proclamations modifying the law without its sanction, except for the infliction of new penalties, and in finance was dependent upon it only for the 'extraordinary' revenue, the 'ordinary' or main part of the revenue being entirely at her own disposal and derived from sources beyond parliamentary control. In spite, therefore, of the showy title accorded to the national assembly by Sir Thomas Smith, it remained constitutionally subordinate to the council, and was frowned upon by the queen if it ventured to assert an independent judgement in regard to policy. It is a striking testimony to the relatively humble position it occupied that in a reign of forty-four and a half years the total time spent in parliamentary discussions was only some thirty-five months. Sessions were short, and the intervals between them sometimes long and always irregular—circumstances that made it difficult for parties to develop or for leaders to arise, who might focus and direct public opinion or become the centre of opposition to the executive. The only member of parliament who showed the makings of a great parliament man was Peter Wentworth,[2] a member for the Cornish borough of Tregony;

[1] It had not even obtained full control over its own elections, controversies in regard to which were in many cases dealt with by the council.
[2] For Wentworth's parliamentary career see Professor Neale in *Engl. Hist. Rev.* xxxix. 36–54, 175–205.

but his courageous independence of mind and fervid political genius were largely wasted on the sixteenth century. The conservative force of tradition, operative even in parliament itself, was too strong for men of his stamp to take command. Moreover, there was always a number of privy councillors with seats in the house who assumed an informal leadership, and by co-ordinating their actions made it certain that any organization of opinion that existed would be in favour of, rather than against, the royal policy. These councillor members were always men of experience, who did excellent work in committees and by acting as a *liaison* between parliament and the executive; but their weight was invariably thrown on the conservative side when disputes arose concerning the rights and privileges of the house.

The same criticism applies, with reservation, to the Speaker. Nominally he was the freely elected spokesman or prolocutor of the commons; but in actual fact his nomination usually came from one or other of the privy councillors present, and it would be no exaggeration to say that he was as much a Crown official as the 'mouthpiece' of the popular assembly. His conflicting duties in this dual capacity sometimes led to trouble; for, as chairman of the house, he had to allow a certain liberty of discussion, while, at the same time, as an official responsible to the Crown he had to exercise care that members did not trespass upon 'points of sovereignty'. True, he received no special instructions from the queen about the discharge of his duties; but in the speech delivered by the lord keeper or the lord chancellor at the opening of parliament, the purposes for which it was summoned were more or less clearly indicated, and it was the Speaker's task to see that these terms of reference were adhered to. If he permitted an undue liberty of criticism, he might be summoned by the queen and brought to book for his laxity; and generally speaking he was subject to direction as to the receiving and reading of bills. 'Messages from Caesar', or, as Peter Wentworth described them in his speech of 1575, 'rumours and messages', were a common occurrence in Elizabethan parliamentary history.

Nor was this all. It may be assumed that although there was no discrimination against private members' bills, the initiation of all the important legislation lay not with parliament but with the council, where it was carefully prepared beforehand, and the councillor members were the men who piloted it through the house. Sometimes, too, bills were brought from the upper house to the commons 'specially recommended from the queen's majesty'; and if bills were introduced which appeared to touch the royal prerogative, there were always members who 'moved that petition be made by this house to her majesty for her licence and privity to proceed'. Finally, it is to be noted that the summoning and dissolution of parliament, and the validity of the measures it passed, were all dependent upon the will of the sovereign alone. 'It is in me and my power', said Elizabeth to the Speaker, in 1593, 'to call parliaments: it is in my power to end and determine the same: it is in my power to assent or dissent to any thing done in parliaments.' That she made full use of this power may be gathered from the fact that she summoned her parliaments only when it suited her convenience, and dismissed or prorogued them when she thought they had sat long enough. Her hatred of long parliaments and lengthy debates was notorious. Again and again the lord keeper was instructed in his speech on the opening day of the session to emphasize the importance of making existing laws more effective rather than of multiplying new laws. To curtail discussion whenever and wherever possible, to expedite the passing of subsidy bills and other necessary measures, and to send the members back to their constituencies with the minimum of delay—this seems to have been Elizabeth's settled policy in the conduct and control of parliament.

The question therefore arises: how much liberty was allowed in the discussion of public affairs?[1] In the first place, we must note that it was customary for the speaker of the house of commons to petition the Crown at the commencement of the session for recognition of certain privileges: (1) freedom of access for the house to the royal presence on all matters of urgency and importance; (2) favourable construction of the Speaker's words when reporting resolutions of the house; (3) freedom from arrest of members and their servants during the continuance of parliament; and (4) freedom of speech 'in whatsoever was treated, propounded, and debated in the house'. All four liberties were established in principle before Elizabeth's

[1] For this subject see Neale's excellent chapter in *Tudor Studies*, ed. R. W. Seton-Watson (1924), together with other studies by the same author referred to in the Bibliography below.

accession, and the queen found no difficulty in giving her
assent to them in 1559 'as largely, as amply, and as liberally
as ever they were granted by her noble progenitors'. This did
not mean, of course, that they were recognized in an absolute
sense, or without qualification as to the use that might be made
of them: on the contrary, it rested with the Crown to determine
their due limits, and to apply restrictions if liberty threatened
to degenerate into licence. Thus freedom of speech was granted
to the first parliament of the reign with the condition that
members were 'neither unmindful nor uncareful of their duties,
reverence, and obedience to their sovereign'. But it was found
necessary later to define the condition more narrowly and
specifically, for what constituted 'duty', 'reverence', and 'obedience' was apt to be very differently interpreted by Crown and
parliament. Consequently, in 1593, after thirty-eight years of
more or less continuous friction, the lord keeper, in his reply
to the Speaker's petition, summed up the attitude of the Crown
as follows:

'For even as there can be no good consultation where all freedom
of advice is barred, so there will be no good conclusion where every
man may speak what he listeth, without fit observation of persons,
matter, times, places, and other needful circumstances. . . . For
liberty of speech her majesty commandeth me to tell you that to say
yea and no to bills, God forbid that any man should be restrained
or afraid to answer according to his best liking, with some short
declaration of his reason therein, and therein to have a free voice,
which is the very true liberty of this house; not as some suppose to
speak there of all causes as he listeth and to frame a form of religion
or a state of government as to their idle brains shall seem meetest.
She saith that no king fit for his state will suffer such absurdities. . . .'

From the queen's standpoint the question was not so much
one of free speech as of the topics to which this freedom might
be applied. Certain matters she regarded as definitely outside
the scope of popular debate, and fit to be handled by herself alone, or by herself in conjunction with the council, viz.
the succession, religion, foreign policy, and trade. Since parliament was not prepared to accept all of these extensive
reservations, which covered every important national interest,
a struggle ensued; and it was only by adroit tactics, or by coercive measures, that the queen was able to obtain the ultimate
victory.

The first clash occurred over the succession, in the autumn of 1566. Already petitions had been addressed to the queen either to marry (1559) or to marry and declare a successor (1562-3); but in each case the supplication was met by an evasive answer or by the non-committal statement that the matter was weighty and required further consideration. As time passed, however, and nothing was done, the anxiety of the commons increased; and in October 1566, when the second parliament met for its second session, Mr. Molineux moved that their former suit to the queen be revived. Sir Ralph Sadler and others of the council who were present tried to 'stay' the house; but it was resolved to go on with the petition, in conjunction with the lords. Whereupon Cecil and the vice-chamberlain intimated a message from the queen that 'by the words of a prince, she by God's grace would marry, and would have it believed', but that 'touching the limitation of the succession, the perils be so great to her person . . . that the time will not yet suffer to treat of it'. Despite this assurance and warning, the house went on with the matter in hand. The queen then tried to forestall it by summoning a deputation of both lords and commons before her. She would not, she said, be moved by violence to do anything, for it was 'monstrous that the feet should direct the head', but she would limit the succession when she conveniently could without peril. The report of this speech was communicated to the house by Cecil on the following day (6 Nov.); but the ferment continued. On the 8th Mr. Lambert moved that a further attempt be made. Next day the vice-chamberlain, Francis Knollys, 'declared the queen's majesty's express command to the house that they should proceed no further in their suit, but satisfy themselves with her majesty's promise to marry'. In the debate that took place on receipt of this inhibition, Paul Wentworth desired to know whether the action of the Crown was not against the liberties and privileges of the house. Tempers were obviously rising. Two days later the Speaker was summoned to court, and returned with the news 'that he had received a special command from her highness . . . that there should be no further talk of the matter . . . and that if any person thought himself not satisfied let him come before the privy council and there show them'. Even this peremptory veto did not stop the discussion; but later in the month the queen, confronted with the

possibility of having to dissolve parliament and lose her supplies, voluntarily 'revoked her two former commandments'—
a concession that was received with great joy by the commons.
But when the jubilant house then tried to turn its 'victory' to
account by incorporating the queen's promise of 5 November
in the preamble to the subsidy bill—a significant move—Elizabeth became very angry: 'as if', she scribbled on the draft, 'my
private answers to the realm should serve as prologue to a
subsidies book'.[1] At the dissolution, on 2 January 1567, the
lord keeper seized the opportunity to rebuke the house through
the Speaker for 'bringing her majesty's prerogative in question';
and the queen, who was present as usual in her place, delivered
them a short but caustic lecture on the same text.

'I have in this assembly', she said, 'found so much dissimulation,
where I always professed plainness, that I marvel thereat, yea two
faces under one hood, and the body rotten, being covered with two
vizors, succession and liberty, which they determined must be either
presently granted, denied or deferred. . . . But do you think that
either I am unmindful of your surety by succession, wherein is all
my care, considering I know myself to be mortal? No, I warrant
you. Or that I went about to break your liberties? No, it was never
in my meaning, but to stay you before you fell into the ditch. For
all things have their time . . . yet beware however you prove your
prince's patience, as you have now done mine.'

At the opening of the third parliament in April 1571, when
Christopher Wraye, the Speaker, made his petitions as usual, the
lord keeper reminded him 'that her majesty having experience
of late of some disorder and certain offences, which though they
were not punished, yet were they offences still', and 'that they
should do well to meddle with no matters of state but such as
should be propounded unto them, and to occupy themselves in
other matters concerning the commonwealth'.

In spite of this warning off 'state affairs', the parliamentary
session had scarcely begun when Mr. Strickland, 'a grave and
ancient man of great zeal', broached the question of reforming
the prayer book so as to 'have all things brought to the purity
of the primitive church and the institution of Christ'. A few
days later he introduced a bill 'for reformation of the book of
common prayer'; and the Speaker, having no authority to
refuse it, allowed it to be presented and read for the first time.

[1] See Neale in *Eng. Hist. Rev.* xxxvi. 497–520.

Mr. Treasurer and Mr. Comptroller pleaded that it was a matter for the queen alone as head of the church, and after 'divers long arguments' the house agreed to petition her majesty for licence to proceed.

Strickland's bill, it should be noted, was only one of a series introduced into the commons at this time for the regulation of ecclesiastical affairs, all of which were subsequently 'dashed' by the queen, on the ground that 'she could not suffer these things to be ordered by parliament'. But as the unfortunate Strickland had taken the lead and was evidently the prime mover, Elizabeth determined to make an example of him. During the Easter recess he was summoned before the council, reprimanded, and commanded to refrain in the meantime from attendance at the house. The consequence was that soon after business was resumed a debate ensued as to the legitimacy of his detention. Mr. Carleton represented it as a breach of parliamentary liberty, and urged that the offending member be sent for to the bar of the house, 'there to be heard, and there to answer'. Mr. Treasurer, on the other hand, advised caution and showed that Strickland 'was in no sort stayed for any word or speech by him in that place offered, but for the exhibiting of a bill into the house against the prerogative of the queen; which was not to be tolerated'. Whereupon Mr. Yelverton, taking his cue from the treasurer's remarks, boldly attacked the prerogative. Characterizing Strickland's detention as a 'perilous precedent', he proceeded to vindicate the authority of parliament in these words: 'It was fit', he said, 'for princes to have their prerogatives; but yet the same to be straitened within reasonable limits.' The prince, he showed, could not of herself make laws, neither might she by the same reason break laws.

The councillor members were obviously uneasy at the turn the discussion was taking, and were observed to 'whisper together': so the Speaker, realizing that matters had gone far enough, moved 'that the house should make stay of any further consultation thereupon'. On the next day, to the joy of the commons, Strickland reappeared in his place, and the committee on religious affairs resumed its discussions. For the remainder of the session there was no further trouble; but when parliament was dissolved on 29 May the lord keeper took occasion, in his reply to the Speaker's speech, to animadvert upon the 'audacious, arrogant, and presumptuous folly' of the house 'thus by superfluous speech spending much time in meddling with matters neither pertaining to them nor within the capacity of their understanding'. A year later, during the first session of the fourth parliament, the commons were informed through the Speaker that 'her highness' pleasure is that from henceforth no bills concerning religion shall be preferred or received into this house, unless the same should be first considered and liked of the clergy'. Coupled with this inhibitory message came another commanding the delivery to the queen of two bills concerning rites and ceremonies, which were then pending in parliament. The house had perforce to surrender them, and they were duly impounded.

Although it must now have been evident that the queen was inflexible, and the vast majority of members disinclined to continue the struggle, there was one man, Peter Wentworth, who determined to protest vigorously against the treatment meted out to parliament. Wentworth was one of the younger members, and perhaps on that account less cautious than his fellows; but his sincerity, eloquence, and courage were of a high order, and his grasp of principle still more remarkable. His speech in the house on 8 February 1575 was as much an invective against the subservience of parliament as it was a calculated indictment of the royal policy.

'There is nothing so necessary', he said, 'for the preservation of the prince and state as free speech, and without it is a scorn and mockery to call it a parliament house, for in truth it is none, but a very school of flattery and dissimulation. . . . Amongst other, two things do great hurt in this place, of the which I do mean to speak: the one is a rumour which runneth about the house, and this it is, take heed what you do, the queen's majesty liketh not such a matter, whoever preferreth it, she will be offended with him. . . . The other: sometimes a message is brought into the house either of commanding or inhibiting, very injurious to the freedom of speech and consultation. I would to God, Mr. Speaker, that these two were buried in hell, I mean rumours and messages. . . . So that to avoid everlasting death and condemnation with the high and mighty God, we ought to proceed in every cause according to the matter, and not according to the prince's mind. . . . Free speech and conscience in this place are granted by a special law, as that without the which the prince and state cannot be preserved or maintained. . . . It is a dangerous thing in a prince to oppose or bend herself against her nobility and people.

... I beseech the same God to endue her majesty with His wisdom, whereby she may discern faithful advice from traitorous sugared speeches, that will may not stand for a reason; and then her majesty will stand when her enemies are fallen, for no estate can stand where the prince will not be governed by advice. . . . For we are incorporated into this place to serve God and all England, and not to be time-servers, as humour-feeders, cancers that would pierce the bone, or as flatterers that would beguile all the world, and so worthy to be condemned both of God and man. . . .'

The commons were so surprised at the vehemence of the speech that 'out of a reverent regard of her majesty's honour' they stopped him before he had fully completed his remarks, and presently committed him to the serjeant's ward, to be examined the same afternoon by a committee of the house. The upshot was that Wentworth was sent to the Tower by a resolution of the house of commons itself and remained there a prisoner until 12 March, when he was released by 'special favour' of the queen.

When the fourth parliament met for its third session in January 1580 puritan feeling ran high in the commons; and on the initiative of Paul Wentworth, brother of the redoubtable Peter, it was resolved to hold a public fast and a daily preaching, so that God 'might the better bless them in all their consultations and actions'. It was agreed that the preachers to conduct the first service should be selected by such of the privy council as were members of the house 'to the intent that they may be discreet persons and keep convenient proportion of time, without intermeddling with matter of innovation or unquietness'. But the queen was annoyed at the innovation being made without her 'privity and pleasure first known', and a message was sent through Mr. Vice-chamberlain to that effect. The reprimand, however, was tempered by the remark that the 'error' was not 'wilful and malicious', but simply 'rash, ill-advised, and inconsiderate'. The house therefore agreed, on the suggestion of the vice-chamberlain, to make 'humble submission unto her majesty, acknowledging the said offence and contempt'. On the following day answer was brought from the queen 'of her most gracious acceptation of the submission', accompanied by the comment that the cause of the royal displeasure was 'not for that they desired fasting and prayer, but for the manner in presuming to indict a public fast without order and without her privity, which was to intrude upon her authority ecclesiastical'.

Meanwhile the yeast of puritanism continued to ferment beneath the surface, and in the succeeding parliament, on 27 February 1587, Anthony Cope introduced into the house a 'bill' petitioning for the repeal of all existing laws touching ecclesiastical government, and for the authorization of a book containing a new form of public prayer and administration of the sacraments. The Speaker, mindful of the queen's previous command not to meddle with such matters, asked to be spared the reading of the book; but the house insisted, and but for the lack of time the Speaker would have been overborne. When the queen heard of the matter, she sent for the Speaker and ordered him to deliver up the bill and the book. On the following day (1 March) Peter Wentworth presented a number of articles concerning the liberties of the house, which he desired should be read and put to the question by the Speaker. He was asked to wait until the queen had signified her pleasure with regard to Cope's bill and book; but Wentworth refused, and the Speaker retorted that he must first peruse the articles before he took any further step. The articles were as follows: 'Whether this council be not a fit place for any member of the same here assembled freely and without controlment of any person or danger of laws, by bill or speech to utter any of the griefs of this commonwealth whatsoever touching the service of God, the safety of the prince, and this noble realm? Whether that great honour may be done unto God, and benefit and service unto the prince and state without free speech in this council, which may be done with it? Whether there be any council which can make, add to or diminish from the laws of the realm, but only this council of parliament? Whether it be not against the orders of this council to make any secret or matter of weight, which is here in hand, known to the prince or any other, concerning the high service of God, prince, or state, without the consent of the house? Whether the Speaker or any other may interrupt any member of this council in his speech used in this house, tending to any of the aforesaid high services? Whether the Speaker may rise when he will, any matter being propounded, without consent of the house or not? Whether the Speaker may over-rule the house in any matter or cause there in question; or whether he is to be ruled or over-ruled in any matter or not? Whether the prince and state can continue, stand, and be maintained without this council of parliament, not altering the government of the state?' The Speaker 'pocketed up' the questions and showed them to Sir Thomas Heneage, one of the privy council, 'who so handled the matter that Mr. Wentworth went to the Tower, and the question not at all moved'. Cope and others who had supported him were likewise committed to the Tower, by order of the lord chancellor, on the following day. No explanation was officially given for these arrests, nor did the commons take any steps, either by petitioning or otherwise, to secure the liberation of the prisoners. The conjectural statement made by the vice-chamberlain 'that they might perhaps be committed for somewhat that concerned not the business or privilege of the house', which was accepted, seems to indicate that they were punished not for anything they said or did in parliament but for their actions outside.

But Wentworth was nothing if not pertinacious. For some time he had been brooding over the still unsettled succession question, which most men were now inclined to leave alone, and in August 1591 was again sent to prison for writing a pamphlet entitled *A Pithy Exhortation to Her Majesty for Establishing the Succession*. On being liberated he proceeded to work up a party in support of his plan; and when the eighth parliament of the reign met, in February 1593, he and Sir Henry Bromley presented a petition to the lord keeper, Puckering, in which they urged the lords of the upper house to join with the commons in a supplication to the queen for 'entailing' the succession, 'whereof a bill was ready drawn up by them'. For this offence against the 'former strait commandment' relating to matters of state they were 'presently sent for' by Sir Thomas Heneage, and ordered not to attend parliament or leave their lodgings. But the queen was so offended at their conduct that she gave instructions for their imprisonment, and Wentworth and Bromley along with others consentient to their proposal were severally sent to the Tower and the Fleet Prison. When the question was raised in the commons, and Mr. Wroth moved that a petition be made to her majesty for their release, on the ground that since they were absent from the debates on the subsidy their constituents might complain of the heavy taxation about to be imposed on them, the councillor members in the house replied that 'her majesty had committed them for

causes best known to herself, and for us to press her majesty
with this suit, we should but hinder them whose good we
seek'.

Meanwhile, two days after the imprisonment of Wentworth
and Bromley, Mr. Attorney Morrice of the court of wards
raised the question of tyranny in the ecclesiastical courts,
pointing out that the 'hard courses' adopted by the bishops,
ordinaries, and other ecclesiastical judges towards 'the godly
ministers and preachers of this realm' were 'contrary to the
honour of God, the regality of her majesty, the laws of the
realm, and the liberty of the subject'. He then presented two
bills to be read: the first concerning unlawful 'inquisitions,
subscriptions, and offering of oaths'; the second concerning the
imprisonments inflicted for refusal to take oaths. Various members spoke in favour of the mover; but Sir Robert Cecil reminded the house that the queen had forbidden its interference
in such matters, and advised that as the bill 'seemed to contain
matters needful', it should first be 'commended to her majesty',
with a view to obtaining her approval. This, however, did not
meet with any support; but on the Speaker's protest that the
bill was weighty and complicated, and that he could not 'open'
it as he should until he understood it, it was agreed to let him
have it for his private perusal. The same afternoon he was
sent for to the court and questioned as to the content of the
bill. He returned to the commons with the information that
the queen was highly offended that, in spite of her express
prohibition, the house had intermeddled with ecclesiastical
affairs. 'Her majesty's present charge', he said, 'and express
commandment is, that no bill touching the said matters of
state or reformation in causes ecclesiastical be exhibited. And
upon my allegiance I am commanded, if any such bill be
exhibited, not to read it.'

This really terminated the efforts of the house of commons
to legislate in religious matters; for although the queen removed
her prohibition in 1597, and a bill against pluralities was read
twice and 'committed' in 1601, there is no record that it
reached the statute-book. As for the succession question, it was
discreetly dropped, as being a subject in which the house had
no further interest.

So far no reference has been made to the financial aspect of
the relations between Crown and parliament, for, strictly speak-

ing, finance did not enter as a factor into the constitutional struggle of the reign. The permanent hereditary revenue was derived from the income from the Crown lands, feudal incidents, the customs, ecclesiastical first-fruits, the proceeds of justice, recusancy fines, and the right to purveyance and preemption. It was sufficiently large to make the queen in ordinary years more or less independent of parliament, and able to 'live of her own'. Indeed the revenue granted by parliament in the shape of subsidies, tenths, and fifteenths was intended to meet only exceptional expenditure.

Until 1589 it was customary for the subsidy bill to consist of one subsidy and two tenths and fifteenths—roughly £140,000; and although the yield of these taxes had fallen greatly since Henry VIII's time, this was ample enough—thanks to the queen's frugality—not only for the needs of the government, but for the remission of the third instalment of the subsidy, in 1566. But a great change came over the situation after 1588, owing to the intensification of the war with Spain, the commitments in France and the Netherlands, and the growing expenditure in Ireland; and for the first time we hear of parliament being summoned primarily for the provision of 'treasure'. In 1589 the committee for the subsidy in the house of commons, stimulated by the arguments of Sir Walter Mildmay, chancellor of the exchequer, recommended two subsidies and four tenths and fifteenths. This was agreed upon by the house, subject to the proviso that the double subsidy would not become 'the occasion of a precedent to posterity'. But when the next parliament met, in 1593, it was confronted with a fresh request for liberality. The foreign peril and the military and naval needs of the country were strongly emphasized by the lord keeper in his opening speech, and by Sir Edward Stafford, Sir John Fortescue, and others in the commons. It was a question, said Fortescue, 'non utrum imperare sed utrum vivere'. In spite, therefore, of its previous decision not to allow a precedent to be established, the house of commons agreed to repeat the double subsidy and four tenths and fifteenths of 1589, and gave instructions for the bill to be drawn up accordingly.

At this point the lords of the upper house requested a conference with the commons about the subsidy, and the lord treasurer pointed out that whereas the last parliamentary grant brought in only £280,000, the queen had spent over £1,000,000 since the grant had been made. He also stated that the lords could not assent to less than three subsidies payable in the next three years. This at once gave rise to a somewhat acrimonious discussion in the commons on the question of privilege, it being asserted that the amount of the subsidy was a matter for the lower house alone to determine. Eventually, however, it was decided to revise the grant in the light of the treasurer's statement, and a lengthy debate ensued as to the capacity of the country to pay a triple subsidy. Sir Francis Bacon avowed himself in favour of three subsidies, but would have them spread over six years, because 'gentlemen must sell their plate, and farmers their brass pots ere this will be paid'. Mr. Heyle, on the other hand, affirmed 'that of his own knowledge from the Mount to London the country was richer many thousand pounds than heretofore'. Sir Walter Raleigh was of the opinion that 'for the difficulty in getting this subsidy, I think it seems more difficult by speaking than it would be in gathering'. 'Let us let the people blood', remarked Mr. George Moore, using a surgical metaphor, 'and so prevent the danger.' Once again the bogy of undesirable precedents came up, to be quashed by Sir Robert Cecil, her majesty's principal secretary, with the remark that 'precedents have never been perpetual, but begun and ended with the causes. . . . We have no reason to give prejudice to the best queen or king that ever came, for fear of a worse king than ever was.' After 'many long and grave speeches' the house agreed to three subsidies and six tenths and fifteenths. The bill passed the house on 22 March 'very difficultly . . . by reason of the greatness thereof', but the Speaker 'did overreach the house in the subtle putting of the question'. From start to finish the discussions had lasted twenty-four days.

Despite the increased revenue the financial situation had in no way improved when, in October 1597, the queen summoned her ninth parliament. The lord keeper laid great stress on the fact that Crown lands had been sold to maintain the necessary forces by sea and land, 'whereby with such small helps as from her subjects have been yielded, she hath defended and kept safe her dominions'.[1] So serious was the position of the country, he said, that 'he that would seek to lay up treasure and enrich

---

[1] Loans and benevolences were levied on the wealthy by privy seal, and ship money was imposed on inland towns, while the customs were augmented; but still there was a deficit. For ship money see *A.P.C.* xvi. 304, 353–4.

himself, should be like to him that would busy himself to beau-
tify his house when the city where he dwelleth were on fire or
to deck up his cabin when the ship wherein he saileth were
ready to drown'. It was impossible for the commons to do less
than repeat the three subsidies of the previous parliament and
they did so apparently without any serious criticism. But their
troubles were not yet at an end.

The last parliament of the reign, which sat from October to
December 1601, was in many respects the most interesting, as
it was certainly the most turbulent, of all. The purpose for
which it was called was to provide ways and means of financing
the costly war in Ireland; but trouble was in the air almost
from the very start of the session. When the commons, as was
their wont, repaired to the upper house to hear the opening
speech of the lord keeper, they found the door of the chamber
accidentally barred against them; and a day or two later (30
Oct.), when the queen left the parliament house, after the
Speaker's petitions had been granted, there were few who
greeted her with the customary 'God save your majesty'.[1] It
was a bad beginning, but worse followed, as soon as the commons settled down to business. Acrimony, noise, and confusion
characterized the debates. Speakers were 'cried or coughed
down', or otherwise interrupted in their remarks: voting was
interfered with, members being, on one occasion, pulled back
into the house or pulled out into the lobby against their will;
and even the precincts of the house were disturbed by brawling
and disorder. The two principal matters in dispute were the
subsidy bill and monopolies. In regard to the former it was
pointed out by Cecil, who repeated the substance of the lord
keeper's speech in the house of commons on the following day,
that the rebellion in Ireland and the war with Spain had
created a deficit of £140,000; while the government would
require £300,000 for fresh outlay before Easter. In order to
meet this, it would be necessary for the house to offer no less
than four subsidies, or twice the sum granted under protest
in 1589. The commons were greatly perturbed, and in the

---

[1] 'And the throng being great, and little room to pass, she moved her hand to
have more room; whereupon one of the gentlemen ushers said openly, "Back
masters, make room." And one answered stoutly behind, "If you will hang us,
we can make no more room": which the queen seemed not to hear, though she
heaved up her head and looked that way towards him that spake.' (H. Townshend,
Historical Collections, &c., pp. 178, 179.)

debates that followed many raised their voices on behalf of the
poor, who, it was said, were compelled to sell their 'pots and
pans' to meet the already heavy taxation; and a desire was
expressed that the incidence of the taxes should be made more
equitable by higher assessment of the wealthy. Some thought
that the 'three pound men' should be spared; others that the
'four pound men' should pay double, with a correspondingly
increased charge on the 'rest upwards'. Cecil reminded the
house that 'when Hannibal resolved to sack Rome, he dwelt
in the cities adjoining, and never feared or doubted of his
enterprise, till word was brought him that the maidens, ladies,
and women of Rome sold their ear-rings, jewels, and all their
necessaries to maintain the war against him'. Serjeant Heyle
averred that the queen 'hath as much right to all our lands
and goods as to any revenue of her crown'—a remark that
caused the house to 'hem, laugh, and talk'. Eventually, however, after a protracted debate, it was agreed to grant the four
subsidies and the eight tenths and fifteenths, and a committee
was appointed to draw up the necessary bill.

The climax of the session was reached when the question of
monopolies was brought before the house by Mr. Lawrence
Hide, for this was a subject that touched every one closely, and
on which every one believed he could speak with authority.
A monopoly, said Mr. Spicer of Warwick, is 'a restraint of any
thing public in a city or commonwealth to a private use, and
the user called a monopolitan'. The practice of granting monopolies by royal patent for the manufacture, distribution, or
sale of certain articles of commerce had grown up since the
sixteenth year of the reign, with disastrous effect on the cost of
living and the general welfare of the community. 'I cannot
utter with my tongue', said Mr. Francis Moore, 'or conceive
with my heart the great grievances that the town and country
which I serve suffereth by some of these monopolies; it bringeth
the general profit into a private hand, and the end of all is
beggary and bondage of the subject.' When Sir Robert Wroth
read out the list of fresh patents issued since the last parliament
—for currants, iron, powder, cards, ox shin-bones, train oil,
oil of blubber, transportation of leather, cloth, ashes, aniseeds,
vinegar, sea-coals, salt-petre, lead, calamine stones, pilchards,
&c., Mr. Hakewell of Lincoln's Inn stood up and asked: 'Is
bread not there?' 'Bread?' quoth one. 'Bread?' quoth another.

'This voice seems strange,' quoth a third. 'No,' quoth Mr. Hakewell, 'if order be not taken for these, bread will be there before the next parliament.'

There was no doubt about the evils of the system: they were admitted by every speaker; but the question was how to proceed towards their amendment. As Francis Bacon pointed out, 'the queen, as she is our sovereign, hath both an enlarging and restraining power. For by her prerogative she may first set at liberty things restrained by statute law or otherwise; and secondly, by her prerogative she may restrain things that be at liberty.' Wherefore, he concluded, 'I say and I say again, that we ought not to deal, to judge or meddle with her majesty's prerogative.' But the house found it difficult to reach unanimity as to procedure. Some demanded a bill: others insisted that all patents should be brought before the house and cancelled: others, again, were for petitioning the queen. The confusion was so great that Cecil exclaimed: 'This is more fit for a grammar school than a court of parliament. I have been a counsellor of state this twelve years, yet did I never know it subject to construction of levity and disorder.' Outside the house, in the streets, there were ominous mutterings and cries of 'God prosper those that further the overthrow of these monopolies, God send the prerogative touch not our liberty'. 'Let me give you this note,' said Cecil, 'that the time was never more apt to disorder and make ill interpretation of good meaning: I think those persons would be glad that all sovereignty were converted into popularity.'

Fortunately at this juncture the queen, who had been informed of the debate by her council, and had received many petitions 'both going to chapel and also to walk abroad', intervened with a drastic solution of the problem that took the house by storm, and won for her one of the greatest triumphs of her reign. Summoning the Speaker, she informed him that 'she herself would take present order of reformation'; that 'some [of the monopolies] would presently be repealed, some suspended, and none put in execution, but such as should first have a trial according to the law for the good of the people'. When the news was communicated to the commons they agreed to send a deputation, headed by the Speaker, to express their gratitude to her majesty; and on 30 November, in the council chamber at Whitehall, was transacted one of the most memorable and glamorous scenes in Tudor history. After receiving the thanks of the Speaker, which were couched in a language that must have sounded sweet to her ears, the aged queen replied with a speech full of gracious condescension, fervour, and evident sincerity:

'Though God hath raised me high,' she said, 'yet this I count the glory of my crown, that I have reigned with your loves. This makes me that I do not so much rejoice that God has made me to be a queen, as to be a queen over so thankful a people. . . . Neither do I desire to live longer days, than that I may see your prosperity, and that is my only desire. . . . Of myself I must say this, I never was any greedy, scraping grasper, nor a strait fast-holding prince, nor yet a waster; my heart was never set on worldly goods, but only for my subjects' good. . . . Yea mine own properties I count yours to be expended for your good.'

Hereupon she bade the kneeling commons to stand up, and thanked them for preventing her from falling 'into the lap of an error, only for lack of true information', and for their zeal for the common weal.

'Since I was queen', she said, 'yet did I never put my pen to any grant, but that upon pretext and semblance made unto me that it was both good and beneficial to the subjects in general, though a private profit to some of my ancient servants who deserved well.'

But if her 'kingly bounty' had been 'abused', and her grants 'turned to the hurt of my people', she hoped that 'God will not lay their culps and offences to my charge'. 'To be a king and wear a crown', she assured them, 'is more glorious to them that see it, than it is a pleasure to them that bear it.' For herself, she continued,

'I was never so much enticed with the glorious name of a king, or royal authority of a queen, as delighted that God hath made me His instrument to maintain His truth and glory, and to defend this kingdom from peril, dishonour, tyranny, and oppression. . . . I speak it to give God the praise as a testimony before you, and not to attribute anything to myself; for I, O Lord, what am I, whom practices and perils past should not fear! O what can I do that I should speak for any glory! God forbid.'

On this note of humility Elizabeth ended her speech; and having bade them all kiss her hand, she dismissed the deputation. It was the last occasion on which she publicly addressed her subjects, and the most notable.

VII

THE EXPANSION OF ENGLAND, AND THE ECONOMIC AND SOCIAL REVOLUTION

T HE Englishman of the Tudor period was not by nature or tradition an explorer or a conquistador. The cult of the map and the flag was unknown to him: he had no desire to search out the distant places of the earth, or to found a new England beyond the seas. After the nine days' wonder of the Cabot voyages had subsided, leaving nothing behind it but the memory of a new fishing-station for cod off the Canadian coast, sixty years elapsed during which England made no contribution whatever to the extension of geographical knowledge, or to the opening up of new sea routes. Neither the fabled wealth of Cathay, nor the epoch-making achievements of the Spaniard in central and south America, nor the study of the new geography, which created a ferment in all European countries, from Cracow to Seville, could tempt the sea-girt Englishman out of his isolation. To all intents and purposes he remained a passive spectator of the world's far-flung maritime activity. Not even a book appeared, with an English superscription, to commemorate the greatness of human endeavour in the most momentous half-century of the world's history.[1] It was only when the magical movement of trade began, after 1550, when the expansion of industry at home made distant markets a necessity, and when, in Elizabeth's reign, spoliation of the Spaniard became a patriotic duty, that Englishmen took the lead in maritime enterprise, and entered the field as competitors for geographical honours. In other words, it was trading and buccaneering, rather than the lure of the unknown, or the thirst for knowledge, or the vision of empire, that gave birth to the greatest period of English exploit on the sea.

In 1558 the population of England and Wales was probably not more than from two and a half to three millions, with an average density approximating to that of European Russia in the nineteenth century; and the wealth of the richest London merchants was estimated by the Venetian ambassador, in 1557,

---

[1] The first original English work on cosmography was W. Cunningham's *Cosmographical Glasse* (1559): 'I am the first', wrote the author, 'that ever in our tongue have written of this argument.'

as between fifty and sixty thousand pounds—a figure that compares very unfavourably with the three and a half million ducats of capital (£875,000) possessed by a first-class continental business firm like the Fuggers of Augsburg. England, in fact, was a relatively poor country, and the standard of living was low. In the production as well as in the consumption of wealth, in industrial technique and scientific skill, she was markedly behind her nearest neighbours on the Continent. Her extractive industries, with the exception of lead and tin mining, which were of ancient standing, were so backward that it was necessary to import Germans to run the copper mines at Keswick; and coal was only beginning to come into its own as a fuel, either for the forge or the domestic hearth.[1] Nor was agriculture in a much better state, for only one-quarter of the soil was under grain crops. One economic asset, however, she did possess, and it was of priceless value, viz. the cloth and woollen industry. Here, indeed, she enjoyed a pronounced advantage over all rivals, for English sheep and wool were

[1] According to a reliable estimate, the total annual production of coal in the British isles, about the beginning of Elizabeth's reign, was slightly over 200,000 tons, or, if we exclude Scotland and Ireland, 160,000 tons. The chief coalfields were situated in Northumberland and Durham, the Midlands (Yorkshire, Lancashire, Cheshire, Derbyshire, Shropshire, Staffordshire, Nottinghamshire, Warwickshire, Leicestershire, and Worcestershire), South Wales, Cumberland, Kingswood Chase (near Bristol), Forest of Dean, and Devonshire. Only some 12,000 tons of the whole output were exported, the rest being consumed mainly in the coal-producing districts and in London and the south-east generally. From this it may be concluded that (1) although the coal industry was not a negligible quantity in the national economy, the consumption of mineral fuel per head of the population was relatively small, probably about one hundredweight; (2) England had not yet begun to use her coal to any appreciable extent as a means of purchasing food and raw materials from other countries. It cannot be too strongly emphasized, therefore, that Elizabethan England was still in the timber age. Every industry of importance, excepting that of the smith and the lime-burner, in which the use of coal was well established, drew on the timber supplies of the country to a degree almost inconceivable to-day: so much so, indeed, that, as the reign advanced, the conservation of woods and forests became a serious problem for the government, and the cost of firewood almost trebled. It was this timber shortage—one might call it a crisis— that really gave coal its opportunity: this, and the growing demand of an expanding population for manufactured goods of all kinds. The noxious fumes which rendered coal unpopular at first, both for domestic and industrial purposes, gradually assumed less significance as the potentiality of the new fuel was discovered, and a better technique was invented for handling it in furnaces. Finally, it would be no exaggeration to say that the substitution of coal for wood (or wood charcoal) in industry, which began in Elizabeth's reign, made possible its great industrial development; for the rapid diminution of timber-supplies and the consequent enhancement of prices for wood fuel might easily have nipped such a movement in the bud.

admitted to be the best in the world; and although 42 per cent. of the export trade in woollen products was still in the hands of the foreigner, and the finer branches of the manufacture were practised only in Flanders, the paramount importance of the industry in the national economy is clearly reflected by the customs statistics. During the year 1564–5 the total value of all goods exported from England amounted to the round sum of £1,100,000, of which no less than 81·60 per cent. accrued from cloth and woollens: the remaining 18·40 per cent. being divided between raw wool, woolfells, lead, tin, corn, beer, coal, and fish in that order of value. It is impossible, therefore, to be mistaken as to the basic source of the national wealth. Not only so: the cloth trade was so far ahead of the others, and so powerfully organized under government favour, as to constitute a 'phalanx of irresistible momentum, to which other nations had nothing similar to oppose'.

When Elizabeth's reign began, the export trade in cloth, so far as it was conducted by English merchants, had a very restricted market, being confined chiefly to Antwerp, where the Merchant Adventurers had their 'staple'. The Muscovy Company, founded in 1555, had just begun to explore the possibilities of the Russian market. And there was a group of London and west country merchants who plied a lucrative if somewhat precarious traffic with Barbary and the Guinea coast. But with the exception of Africa England had no markets of value outside Europe. The habitable part of the north American continent from Newfoundland to Florida was unknown to English navigators, and remained so for another generation: central and south America lay within the monopolist area of the Spaniards; and the Near and Far East were preserves of the Venetians and Portuguese respectively. Nevertheless a beginning had been made by the Muscovy Company to penetrate beyond Russia into Asia by sea and by land. In 1556 Stephen Burrough was sent in the *Searchthrift* to force a passage by the North Cape to Cathay, following the line taken by the ill-fated Willoughby and Chancellor expedition three years before. He succeeded in passing the Kanin peninsula and the island of Kolgiev, and reached the bay of Petchora, but was eventually compelled to turn back by drift ice, fogs, and contrary winds at the straits of Waigatz. In the next year Anthony Jenkinson, another agent of the company, set out on his first journey overland into Asia via the great rivers of Russia, Astrakhan, the Caspian sea, the Transcaspian desert, and the Oxus, reaching Bokhara about Christmas. Meanwhile Richard Eden had published his *Decades of the New World* (1555), the first compendium in English of information concerning the Spanish exploits in America; and a few years later Burrough placed his countrymen in touch with the latest developments in navigation by inducing the Muscovy Company to publish a translation of Martin Cortes's *Art of Navigation*, a notable product of the Spanish school of seamanship at Seville.

Although it was now clear that Englishmen were now acquiring knowledge and experience in exploration, so far no positive gains had resulted from the efforts of the Muscovy Company, beyond the opening up of the Russian market. The barren tundras of Siberia, and the semi-cannibal tribes of Samoyeds who inhabited them, offered little prospect of trade in this direction; and Jenkinson at Bokhara found that the main import and export traffic of Asia was handled by Venetians, who traded with Persia through the Turkish empire, and by Portuguese from their stations at Ormuz and Malacca. The company therefore suspended its interest for the time being in the north-east passage, but decided to make a bid for the Persian trade. In 1561 Jenkinson was dispatched on his second journey, with instructions to negotiate a commercial agreement with the Shah. Following the same route as on the previous occasion as far as Astrakhan, he crossed the Caspian in a more southerly direction to Derbend, and entered Old Persia, where he met the Shah at Kasbin. In spite of opposition from the Turk, who sponsored the Venetians, and was unwilling that the fertilizing stream of traffic should be diverted from the Turkish empire to the northerly route through Russia to England, an understanding was reached; and during the next twenty years no fewer than six other commercial missions were sent from England to Persia. Thus a considerable quantity of the finer English cloths, dyed with bright colours to suit the oriental taste, found its way to the markets of Tabriz, Shamakha, and Kasbin. This route to Persia via Russia and Astrakhan continued to be frequented until 1581, when it was superseded by the Mediterranean route used by the Levant Company.

In the meantime the Muscovy Company strengthened its position in Russia. Under the protection of the tsar, Ivan the

Terrible, exclusive rights were obtained to trade in the northern
regions of the empire, and factories were established linking up
Rose Island in the White Sea (the company's head-quarters)
with Cholmogory, Moscow, Jaroslav, Vologda, and even far
distant Kazan and Astrakhan. It was the hope of the promoters to acquire a monopoly of Russian exports and imports,
thereby enhancing profits; but this was prevented, partly by
the presence of numerous English 'interlopers', partly by Russian resentment at the high prices charged by the company,
and partly by the competition of the Dutch. The loss of ships
and cargoes on the hazardous sea passage, the dangers to which
merchants were exposed in a semi-barbarous land, the plunder
of caravans either by the wild tribes in Russia itself or on the
farther side of the Caspian, and the dishonest practices of
the company's own factors also added to the difficulties of the
enterprise. Still the profits were satisfactory for a number of
years, and the exports from Russia—train oil, tallow, hides,
tar, timber for masts, hemp, cordage, and wax—were of the
greatest value to England as raw materials and supplies for the
navy and ship-building.

One notable advantage the Muscovy Company enjoyed was
its comparative immunity from the political disturbances of
western Europe, and the evil consequences of Anglo-Spanish
rivalry, which played havoc with the Netherlands market, and
compelled the Merchant Adventurers to remove their staple
from Antwerp to the German cities of Hamburg, Emden,
Stade, and latterly to Middleburg in Zeeland. Nor was it
troubled by the severe competition of the Hanse merchants,
who fought a long and desperate struggle with the Adventurers
for the economic mastery of the German market, and the maintenance of their privileged position in the English carrying
trade—a struggle that ended with the defeat of the Hanse and
the evacuation of the Steelyard in London, but told heavily on
the finances of the English company.

While the Adventurers were endeavouring to establish themselves on the Rhine and the Elbe, a new company, the Eastland
Company, was formed in 1579, for the purpose of still further
weakening the Hanse by capturing the trade of Scandinavia,
Pomerania, and Poland. Its depot was at Elbing in Poland,
and later at Danzig. Like the Muscovy Company, it dealt in
materials for shipping, which were exchanged for English cloth.

The lure of Cathay, however, was irresistible. During the seventies a project began to take shape for the opening up of a passage by the north-west instead of the north-east. For some years the fertile and adventurous brain of Humphrey Gilbert had been busy with the technical aspect of the problem, working over geographical treatises, ancient as well as modern, and old chronicles, and collating every scrap of information available on ocean currents and prevailing winds. The outcome was the celebrated *Discourse to Prove a North-west Passage*. The thesis of this remarkable book, 'that there lieth a great sea between it [i.e. America], Cathaia, and Greenland, by the which any man of our country that will give the attempt, may with small danger pass to Cathaia, the Moluccae, India, and all the other places in the east, in much shorter time than the Spaniard or Portugal doth', was doubtless something of an academic *tour de force*; but it was stated with such fervour and plausibility that it became the signal for a fresh onslaught on the Arctic. Other men had also been thinking on similar lines, notably Martin Frobisher and Michael Lock—the former a practised navigator and the latter a London merchant and member of the Muscovy Company who dabbled in the new geography. Sanction was obtained from the queen to make the venture, and the Muscovy Company waived its monopoly of the 'search for new trades to the northward, north-eastward, and north-westward'. Thus, under the patronage of the earl of Warwick, and with the financial backing of the queen, Burghley, Walsingham, Leicester, Sir Thomas Gresham, and others, Frobisher set out on three voyages to the unknown lands north of Labrador (1576–8). By some mischance he missed the entrance to Hudson strait, which would have taken him a long way on his journey; but he succeeded in destroying the legend of a *mare glaciale* or frozen sea, and planted the flag on the 'Countess of Warwick Island', in the strait that subsequently became known as Frobisher Strait. Unfortunately the report got abroad that some black stones, which one of the crew brought home with him from the first voyage, contained gold: with the result that the geographical interest of the enterprise was quickly subordinated to mining operations on the Canadian coast. In the end the alleged ore turned out, on being assayed, to be worthless, and was thrown out, says Camden, to mend a road at Dartford in Kent. On the other hand, although Frobisher's

'Cathay Company' failed of its object, a milestone had been set up in Arctic exploration, and England's faith in the existence of a north-west passage was not shaken. The anxiety of the government to keep the achievement a secret from the Spaniard shows the importance attached to it.

Two years later the Muscovy Company revived its interest in the north-east passage; and an expedition was organized (1580) for another attempt to break through the straits of Wai-gatz. Of all the voyages of the period none was more elabo-rately prepared or more carefully thought out. The knowledge gained on previous expeditions was utilized, and the leaders, Arthur Pet and Charles Jackman, bore with them a mass of instructions and advice, in which several of the best-known geographers of the time, Mercator, Dr. Dee, Stephen Burrough, and Richard Hakluyt, had a hand. They were to proceed to the straits of Waigatz, and push on through the Kara sea to the 'river of Ob' (Obi); thence 'in God's name proceed eastwards with the land on the starboard side until Cathay is reached'. It was believed that if once the northern trend of the Siberian coast could be circumnavigated, the land would fall away rapidly to the southward, and Cathay would soon be in their grasp. But alas for the hopes of the promoters! After a weary battle for two months with the ice, fog, rain, and contrary winds that had barred Burrough's progress in 1556, Pet and Jackman found it impossible to penetrate into the Kara sea, and gave up the project in despair. It was the last Elizabethan voyage to the north-east. The task of discovering the passage to Cathay by the North Cape had to be left to the better tech-nique and more scientific equipment of a later age.

So far the only contact which England had established with Asiatic markets was the Muscovy Company's trade with Persia. The older Levant trade, which had flourished in the fifteenth century and given England a share in Turkish markets, and the means of access to the spices, drugs, and gems that came from farther east, had practically died out. In 1578, however, two London merchants, Sir Edward Osborne and Richard Staper, 'seriously considering what benefit might grow to the common-wealth by renewing the aforesaid discontinued trade', sent out an envoy, William Harborne, to the court of Murad III; and in June 1580 this able negotiator secured a *firman* from the Sultan, granting full rights and privileges to English merchants

trading in all parts of the Turkish empire. Immediately afterwards the Levant Company was formed and received a charter
from the queen, who hoped to turn the commercial connexion
with the Porte to political advantage in the struggle with Spain
(1581). Depots were founded at Aleppo, Damascus, Tunis,
Alexandria, Tripoli, and other Turkish ports, and an ambassador took up residence at Constantinople, at the company's
expense. Prosperity came quickly to the Levant merchants,
and also to the Venetian Company, which was formed a few
years later for trade with Venice and her possessions in the
eastern Mediterranean; for the sweet wines of Candia, olive
oil, and dried currants were in great demand in England.
Furthermore, the friendship of Turkey opened up an entrancing
vista of trade with the countries lying beyond, as far as the
Persian gulf and perhaps even India.

Thus, in 1583, the same two merchants who had floated the
Levant Company planned the crowning adventure of the reign
into the east by land. In that year a small party of Englishmen
led by John Newbery and Ralph Fitch made its way to Jaffa
by sea, crossed the Lebanon mountains to Aleppo, and joined
a caravan for the Euphrates. Sailing down this river to Basra
and the Persian gulf, they took ship to Ormuz; and from
Ormuz crossed the Indian ocean in a Portuguese vessel to Goa,
where the authorities threw them into prison as spies. But they
escaped through the good services of an English jesuit, Father
Stevens, who had settled in Goa in 1579, and picking their way
eastward to Golconda, turned north across the great river and
mountain system of India, and eventually arrived at Agra, the
residence of the Moghul, to whom Newbery presented a letter
of credence from Queen Elizabeth. Thereafter the little party
broke up. Newbery set out on a perilous return journey to
Europe through the Punjab, promising to come back with a
ship and pick up Fitch in Bengal in two years. Another member
of the expedition took service with the Moghul. But Fitch,
whose appetite for adventure was only whetted, pursued his
way down the Jumna and the Ganges to Bengal, from Bengal
to Pegu (Burma), and from Pegu to Malacca. Returning to
Bengal in time to keep his appointment with Newbery, he
found no trace of him—he had perished on the homeward
journey: whereupon he took ship for Goa; and at long last, after
nine years of wandering in the east, arrived back in England by

approximately the route he had used on the outward expedition.
From many points of view he had accomplished the most remarkable journey of the period; and, above all, he had opened communications with India, Burma, and the Malay peninsula.

While these events were taking place in India, John Davis
was making the last and greatest attempt of the reign to prove
the existence of the north-west passage. Supported by Walsingham, Adrian Gilbert,[1] and William Sanderson, a member
of the merchant community in London, this prince of Elizabethan navigators made three voyages into the waters between
Greenland and the Canadian coast. In 1585 he reached latitude 66° 40', to the north of Frobisher Strait, and discovered
Northumberland Inlet, which he took to be the passage he was
in search of. Further investigation, on the second voyage,
however, revealed the fact that Northumberland Inlet was
merely an arm of the sea. On the third voyage (1587) Davis
held on to latitude 73° on the Greenland side, the 'farthest
north' of the period, which he named 'Sanderson His Hope'.
To the north of this point the sea was invitingly open; but when
he turned westwards to look for the passage, he ran into pack
ice, and was compelled to return southwards by the Canadian
coast to Labrador. Curiously enough he too, like Frobisher,
passed Hudson Strait without noticing it. But he was greatly
impressed by the 'great sea, free, large, very salt, blue, and of
unsearchable depth', which lay northwards from Sanderson's
Hope. 'I have been in 73°,' he wrote, 'finding the sea all open,
and forty leagues between land and land': wherefore, he concluded, 'the passage is most probable, the execution easy'.
Subsequent events hardly justified this optimistic forecast; but
Davis had rendered a great service to geography. Besides discovering the strait that now bears his name, he had scattered
a number of new names over the coasts of Greenland and
North America, such as Mount Raleigh, Dier's Cape, Cape
Walsingham, Lumley's Inlet, Warwick's Foreland, Cape Child-
ley, and Darcy's Island. Although he remained confident of
ultimate success, and would doubtless have pursued it, if chance
had offered, the numerous distractions of the later years of the
reign, the war with Spain, and the diversion of the nation's
energies into other channels made any resumption of the quest
for Cathay impossible.

[1] A younger brother of Sir Humphrey.

Although the search for a passage to the Pacific was the lodestar of Elizabethan navigators, it must have been borne in on the mind of the merchant community, after 1580, that the exploration of the Siberian, Greenland, and American coasts, and the increase of geographical knowledge resulting therefrom, was but a poor and unproductive outcome of the effort involved. The need for markets beyond the seas, free from interference by the foreigner, was now more urgent than ever; for English trade, though still prosperous in Europe, was being subjected to increasing restrictions, dangers, and interruptions. Baltic shipping was at the mercy of Danish tolls: the Muscovy Company had likewise to pay an annual tribute to Denmark for the right of passage by the North Cape: the Merchant Adventurers and the Eastland Company were fighting the Hanse for the control of markets in Germany, Norway, Sweden, and Poland; and English ships bound for Mediterranean ports had to run the gauntlet of Barbary and Spanish privateers, which haunted the straits of Messina and Gibraltar. Indeed the Mediterranean became so unquiet a region that merchantmen had to go in convoys heavily armed. Moreover, apart from troubles of this nature, England had a surplus population, always increasing, which her industries could not absorb, and a considerable recusant population whose existence taxed her jails and provided the government with a continual source of anxiety. The question therefore arose whether these various problems and difficulties could not be met by 'planting' Englishmen on some suitable part of the American seaboard, where they could develop and exploit the resources of a great continent, beyond the Spanish sphere of influence. Already Gilbert had hinted at some such plan in his pamphlet of 1567; and Richard Hakluyt had suggested to some of Frobisher's men that they should endeavour to find out whether it was possible to set up a 'stapling place' on the other side of the Atlantic—a sort of American Calais—with an English population to maintain it. Gradually the idea was born of transferring the nation's principal economic activities to America, and a rudimentary 'colonial' policy took shape in the minds of Captain Christopher Carlile and Sir George Peckham, and received the able backing of Hakluyt. All three writers set themselves to show that America could supply all the important raw materials which England obtained from European countries, while her vast native population, together with the English to be settled in their midst, would easily absorb all the cloth England could produce. The fact that the passage was quick and easy, and free from molestation by foreign powers, was an additional advantage of the highest importance. Nor was it overlooked that the American aborigines would profit by the introduction of the christian religion and contact with a more advanced civilization. Even the problem of the sea route to Cathay might be solved if England secured a base of operations on the other side of the Atlantic.

In pursuance of this new policy, Gilbert procured his patent from the queen to 'plant' America, in 1578, and five years later organized his expedition to Newfoundland. If he had been successful, Newfoundland would have become the first English colony in the new world; for the patent gave ample rights to its promoter not only to settle a community but also to provide for its government and administration. Fate, however, decreed otherwise. Gilbert's scheme was wrecked by his own impatient, unstable, and perverse nature. After a hurried survey of the mineral and other wealth of the country, a provisional distribution of the lands round St. John's harbour, and the proclamation of English sovereignty over the island, he determined to push on with a grandiose project for carving out similar settlements farther to the south, and plunged into a hazardous voyage along the American coast, which was just as little known to English navigators at this time as the coast of Siberia or of northern Canada. In the shoals off Sablon island, Cape Breton, he lost his principal ship, the *Delight*, with all his maps and plans and geological specimens, and was compelled to turn homeward with his two remaining ships, the *Squirrel* and the *Golden Hind*. Then his strong sense of duty to his crews, admirable enough in itself but tragic in its consequences, bade him travel in the *Squirrel*—a tiny frigate, heavily gunned and overloaded with deck hamper—in preference to the larger and safer ship, which his pilot advised. North of the Azores the little flotilla was struck by a storm, 'terrible seas breaking short and pyramid wise'; and during the night of 9–10 September the lights of the *Squirrel* disappeared, and all hands were lost, including the 'general'. So ended the first attempt to build up a greater England on American soil.

The death of Gilbert, however, did not affect the continuance of the project of which he was the pioneer: his mantle fell on the shoulders of Walter Raleigh, who sought and received a similar patent from the queen in 1584, and threw himself into a fresh scheme for appropriating and planting the American mainland. A month after the issue of the patent (April 1584) he dispatched Philip Amadas and Arthur Barlow to reconnoitre the coast north of Florida; and when they returned with a glowing report of the natural resources and friendly natives of Roanoak island in Pamlico Sound, he decided to plant there the new colony of Virginia. In the spring of 1585 a body of settlers set sail under Sir Richard Grenville, over a hundred of whom elected to remain in America with Ralph Lane as their governor. Next year a ship was sent out from England with supplies, and Grenville followed with three other ships. But by this time (June 1586) the entire community, threatened with starvation and at cross purposes with the Indians, had decided to repatriate themselves by the kind services of Drake, who paid the settlement a visit on his way home from his campaign in the West Indies. Nothing daunted by this unexpected set-back, Raleigh launched a second attempt in 1587, when another batch of over a hundred men and women were transported to the same locality by John White, under the title of the 'Governor, Assistants, and Company of the City of Raleigh in Virginia'. But once again disaster dogged the path of the luckless pioneers; and when White, who had gone back to England for supplies, returned to America in 1590 no trace of the settlement could be found, nor could any information be procured of its fate. In all probability it had perished at the hands of the Indians.

Doubtless the failure is to be attributed, primarily, to the incompetence and lack of foresight shown by Raleigh himself. To remove a group of men and women from the political and social surroundings of an old civilization, and plant them down in the wilderness without ample preparations being made to maintain them during the hard struggle against natural difficulties, was to court defeat from the very start. 'Planting of countries', wrote Bacon,[1] 'is like planting of woods; for you must make account to lose almost twenty years' profit, and expect recompence in the end.' But the spirit of the age was against this slow and careful nursing. A race of adventurers

[1] Essay *Of Plantations.*

like the Elizabethans, who looked upon the world as an oyster
to be opened by the sword, could not readily adapt itself to the
hard work of pioneering in a virgin country, or realize that
the development of natural resources is a surer way to wealth
than the robbing of treasure-ships and the spoliation of cities.
Spanish loot was a sore distraction to would-be empire-builders.
Moreover, the prevalent idea of the period that plantations
should be used to rid the mother country of its undesirable
population may have contributed its share also, as it did in
the case of the Irish plantations, in rendering the Elizabethan
experiment in America abortive. To quote the wisdom of
Bacon again:

'It is a shameful and unblest thing to take the scum of people and
wicked condemned men to be the people with whom you plant, and
not only so, but it spoileth the plantation; for they will ever live like
rogues, and not fall to work, but be lazy, and do mischief, and spend
victuals, and be quickly weary, and then certify over to their country
to the discredit of the plantation.'[1]

It is easy to be wise after the event: a more generous criticism
would perhaps regard the collapse of the Elizabethan experiment in colonization as one of the more or less inevitable
disasters that accompany the launching of a great idea into the
world of practical affairs.

While the merchants, aided by the geographers and explorers,
busied themselves with the search for distant markets, and men
of vision were dreaming of establishing a new England beyond
the Atlantic, there were adventurers of a different sort who
held that the surest way to strengthen England was to weaken
Spain. The leader of this class was Francis Drake, the stout-
est fighter of the war party in England. Drake's feud with
the Spaniard dates from the disaster at San Juan de Ulloa in the
autumn of 1568; and from that time to his death in 1595 the
fixed and unalterable purpose of his existence was the humiliation of his enemy. 'A single purpose animates all his exploits',
says the late Sir Walter Raleigh,[2] 'and the chart of his movements is like a cord laced and knotted round the throat of the
Spanish monarchy.'

As a seaman Drake had few rivals; but it was in the leadership of men, in the originality of his designs, and in the superb

[1] Ibid.
[2] *English Voyages of the Sixteenth Century* (1906), p. 84.

brilliance with which he executed them, that the genius of the man shone at its brightest. In Europe, where England and Spain were nominally at peace, his talents were lost, at least until nearly thirty years of the reign had passed; but beyond the line, where the feud originally started, and the reign of law was simply the right of the stronger to enforce his will, an ideal theatre presented itself for the display of his singular ability. Doubtless, also, his strategic eye detected in the isthmus of Panama the weakest point in the Spanish empire; for it was through this narrow neck of land that all the commerce between Spain and her south American colonies of necessity passed. The silver from the mines of Peru was shipped at Lima to Panama, carried on mules across the mountains to Nombre de Dios, and there transferred to galleons which brought it to Europe.

Most of the route was perfectly safe from marauders; for European pirates had not yet broken into the Pacific, and once the treasure was afloat on the Atlantic it ran no risk of molestation until it reached the coast of Spain. Even there the danger was small, there being no instance of an attack on the treasure-ships in European waters before 1580. The only danger zone lay in the land passage across the Panama isthmus. It was here that Drake scored his first piratical success against the Spaniards. In 1572 he descended on Nombre de Dios, and after ransacking the town ambushed the treasure convoy some distance inland, escaping with his booty to the coast before the garrison at Panama had time to act. Complete surprise and the help of the Simaroon Indians, who hated the Spaniards, account for his triumph. A few years later (1575), when John Oxenham tried to repeat the achievement, he found that the convoy was now heavily armed, and that the only way to lay hand on the treasure was to intercept it between Panama and Lima. This he accomplished by building a pinnace on the Pacific coast, to the north of Panama, while he hid his own ship in a creek on the Atlantic side. But the adventure ended disastrously, for Oxenham was trapped with his booty on the way back across the isthmus and hanged as a pirate at Lima. Oxenham's failure, however, did not deter Andrew Barker from carrying out an extended raid along the mainland of Central America from Cartagena to Honduras in 1576; but Barker himself was killed in a skirmish with the Spaniards, and the ship on which his booty was loaded sank on the way home. The next step was boldly to transfer piratical activity on a grand scale to the Pacific, where the element of surprise might still prove effective; and here, again, it was the genius of Drake that gave the lead.

The motive, or motives, underlying the famous voyage of circumnavigation (1577–80) are still something of a mystery; but the theory is now gaining ground that the plunder of Spanish settlements and shipping in the Pacific was not its sole, or even principal, object. The occupation of the Californian coast in the queen's name, under the title of New Albion, together with the suggested plan of a great English settlement in North America, stretching from California to Florida,[1] and the prominence which Drake gave to his search for the western or Pacific end of the north-west passage,[2] would seem to show that part of his intention at least was to set up a base for some colonial and commercial project in the Pacific. The subsequent treaty with the Sultan of Ternate, in the Moluccas, also lends corroboration to this view. Yet, when the results of the expedition are summed up, it becomes difficult to believe that Drake seriously regarded himself as an ambassador of commerce or an empire-builder. His search for the place where the passage to Cathay debouched on the Pacific was perfunctory in the extreme, perhaps disingenuous, since he abandoned the quest at latitude 48° on the ground of excessive cold and contrary winds—a strange description, as one writer remarks, of the climate of Vancouver in the month of June! Nor should too much importance be attached to the treaty with the native ruler of Ternate, albeit it gained subsequent notoriety as the title-deed of the East India Company. The truth is, Drake went forth with the queen's authority to avenge the wrongs done to England by the Spaniard, to strike a resounding blow at catholicism, perchance as a counterblast to the reverses sustained by protestantism in Europe, and, lastly, to make good the losses sustained by Hawkins at San Juan a decade before. It would have surprised him had he been designated a pirate, for he had the royal commission to cover his enterprise; but the

[1] See 'New Light on Drake' (*Hakluyt Society*, ser. ii, vol. xxxiv, introduction, p. lv, and accompanying map).

[2] The north-west passage was believed by some geographers to slant south-west from the Atlantic to the Pacific, debouching on the latter ocean between latitudes 40° and 50° north.

fact remains that the substantial result of the voyage, apart from the glory it conferred on its commander, was the colossal treasure the hold of the *Golden Hind* disgorged at Plymouth in the last days of September 1580. It was believed at the time that it amounted to a million and a half sterling, or between a quarter and a half of the whole annual produce of King Philip's American mines.

The principal geographical interest of the voyage lies in the fact that, after passing into the Pacific by the Strait of Magellan, Drake ran into a storm which drove him 'towards the pole antarctic' as far as 57° south, and it was a month before he recovered the mainland of America. Whether he actually discovered Cape Horn is uncertain; but at least he had disposed of the legend that the Strait of Magellan ran between the American continent and a great southern continent stretching unbroken as far as the pole.

Where Drake had sailed others could follow. In 1586-8 Thomas Cavendish carried out a similar voyage of circumnavigation and plunder, arriving back in London in the autumn of the Armada year, his ships gaily bedecked with blue damask sails, and his men each with a chain of gold.

The voyages of Drake and Cavendish may be said to mark the culmination of distant sea-faring in Elizabeth's reign. The men of war had outrun the traders: it remained for the latter to take advantage of the knowledge gained. Raleigh's failure in Virginia had spelt the end of England's interest in America for the time being. It was now apparent that neither Frobisher's Meta Incognita, nor Gilbert's Newfoundland, nor the lands favoured by Raleigh south of the Chesapeake could offer any profitable outlet for trade. There was left India and the Far East, interest in which had been revived by Drake's visit to the Moluccas in 1579, and by the journey of the jesuit father, Thomas Stevens, to Goa in the same year. Accordingly, in 1589, a group of London merchants applied to the privy council for permission to send a trading expedition to the East; and in 1591 James Lancaster took three ships by the Cape route to Sumatra, Malacca, and Ceylon. Although only twenty-five men and officers returned to England in May 1594, the information acquired, coupled with the full report of Dr. Thorne, who resided at Seville, on the advantages of trade with India, led to the projection of a more pretentious expedition, and

eventually to the formation of a company to trade with the
East Indies. Probably the action of the merchants was accelerated by Ralph Fitch's return after his long sojourn in India,
Burma, and Malacca (1594), and the fact that the Dutch were
combining to develop an independent Malaysian trade of their
own. A sum of £30,000 was quickly subscribed, augmented
later to £72,000; and the queen's approval of the voyage was
secured. On 31 December 1599 the charter of incorporation
of the East India Company was granted, being a concession
for fifteen years to George Clifford, earl of Cumberland, and
215 knights, aldermen, and merchants for the discovery of
trade with the East Indies. Lancaster was appointed general
of the fleet, with John Davis of Arctic fame as his pilot-major;
and in April 1600 five ships—the *Red Dragon*, the *Ascension*,
the *Hector*, the *Susan*, and the *Guest*—set sail from Torbay for
Acheen (Sumatra) and Bantam (Java), to launch the Company
on its amazing career as the pioneer of imperial England in
the East.

Economic history cannot as a rule be written in terms of
reigns, for its beginnings and endings do not correspond with
the normal punctuation of the political calendar. There is some
justification, however, for regarding Elizabeth's reign as coincident with an important phase in England's economic development. For one thing, the gradual substitution of free for servile
labour, begun two centuries before, had resulted in the destruction of the economic fabric of the medieval manor. The vast
bulk of the landholding population were now either freeholders,
copyholders, leaseholders, or tenants at will. Villeinage as a
tenure was extinct; and although it still survived as a status in
something like one per cent. of the population, the progress of
manumission would soon account for this last vestige of the old
economy. In this respect at least the transition from medieval
to modern conditions was virtually complete. On the other
hand, the inclosure movement was still running its course. The
conversion of arable land into pasture for the production of
wool and the economizing of wages, the fencing in of the
'commons' by the gentry, either for grazing purposes or to
provide themselves with enlarged parks, warrens, and chases,
and the substitution of 'severalty' (inclosed) for 'champaign'
(open field) farming, together with all the other phenomena

that accompanied and aggravated the change from a feudal economy to an economy based upon individual enterprise—the building up of great estates by the new moneyed interest, rack-renting, eviction, the decay of husbandry, the drift of the rural population to the towns, ruined villages, pauperism, vagabondage—continued to disturb the countryside throughout Elizabeth's reign. Nevertheless there were indications before the reign terminated that at least the acuteness of the crisis was past. A new race of landlords, more enterprising than their predecessors, were finding it more and more profitable to break up fresh ground for tillage, and farming for gain was supplanting the old subsistence farming of the middle ages. So marked was the change that in 1592 Bacon ventured to assert that England was now in a position to feed other nations instead of being fed by them. Even if he interpreted the signs of the times in too favourable a light, as subsequent events showed, his testimony cannot be entirely overlooked. Meanwhile great strides were being made by the application of capital to economic undertakings of all kinds on a scale unprecedented in history. Manufacturers and merchants rivalled graziers in their prosperity. Indeed it would be no exaggeration to say that the grievances of society were now beginning to veer round from the miseries accompanying the dissolution of the old order to the vices incidental to the introduction of the new. In this connexion it is not without significance that the last great parliamentary controversy of the reign was concerned with patents and monopolies. By this time, too, the terrible problem of pauperism, that had perplexed and baffled earlier Tudor statesmen, was solved, or sufficiently solved to warrant the assumption that the worst menace to social order had been successfully disposed of.

We must be careful, however, not to credit the reign of Elizabeth with more than its due. The most striking feature of rural history between 1558 and 1603 was the prevalence of unrest—an unrest provoked, in large measure, by the continued activity of the inclosers for pasture. Complaints on this subject reached the government from many districts, from Lincolnshire, Derbyshire, Norfolk, Hertfordshire, Bedfordshire, Kent, Monmouthshire, Gloucestershire, and Worcestershire. William Harrison, writing about 1577, alludes bitterly to those who 'for the breed and feeding of cattle do not let daily to take in more, not sparing the very commons, whereupon many townships . . . do live', alleging in excuse that 'we have already too great store of people in England'. 'These inclosures', said Philip Stubbs[1] in 1583, 'be the causes why rich men eat up poor men as beasts do eat grass. . . . They take in and inclose commons, moors, heaths, and other common pastures, where-out the poor commonalty were wont to have all their forage and feeding for their cattle, and (which is more) corn for themselves to live upon.' In 1596, after several bad harvests, there were inclosure riots in various districts. In south Oxfordshire, where a rising of the peasantry was only checked by the timely intervention of the council, there was talk of throwing down the inclosures, killing the gentry, and seizing their corn to alleviate the sufferings of a hunger-stricken multitude. Even the north, hitherto exempt from the movement, was beginning to feel its effects. In 1597 the dean of Durham reported that 500 ploughs had decayed in the bishopric within a few years, and of 8,000 acres lately in tillage, not 8 score were now tilled. Needless to say, contemporary literature continued its wail against the iniquities of the incloser: he was a 'cormorant', a 'greedy gull', a 'cruel kestrel':

> Houses by three, and seven, and ten he raseth,
> To make the common glebe his private land:
> Our country cities cruel he defaceth;
> The grass grows green where little Troy did stand.

Writers have pointed out that much of the outcry must be discountenanced as a gross exaggeration, and that although hardship was undoubtedly inflicted on individuals when the commons were taken away, the total acreage involved in the inclosure movement, for the parts most affected, the midlands, was only 6·23 per cent. Moreover, it should be borne in mind that inclosure for arable was proceeding side by side with inclosure for pasture. Thomas Tusser and others who studied the problem at first hand are emphatic on the point that this form of inclosure, while it destroyed the old open-field system, greatly increased the productiveness of the soil—a fact which probably counterbalanced any possible shrinkage of the corn supply owing to the action of the graziers. So far as can be ascertained there was no visible diminution of the food supplies

---

[1] *The Anatomie of Abuses*, ed. F. J. Furnivall, pp. 116, 117.

of the kingdom as a whole during the period.[1] Nor can it be said that the chief economic problem of the age arose out of scarcity. In spite of recurrent bad harvests, by all accounts England was a land of plenty. 'Thanks be to God,' wrote Bishop Jewel in 1570, 'never was it better in worldly peace, in health of body, and in abundance of victuals.' Nine years later Thomas Churchyard rhapsodized in verse over the 'blessed state of England':

> Here things are cheap and easily had,
> No soil the like can show,
> No state nor kingdom at this day
> Doth in such plenty flow.

What troubled the economists was not a falling off of the food supply, but the existence of a 'dearth' (or dearness) in the midst of apparent plenty. Why did the price of all commodities, food included, continue to rise, as rise it did, even when harvests were satisfactory? 'All this dearth notwithstanding', said John Stow (1575), 'there was no want of anything to him that wanted not money.' The craftsman blamed the husbandman, and the husbandman blamed the merchant and the landlord. The merchant passed on the blame to the foreigner, who charged more for his wares, and the landlord complained that he could not maintain his position in society unless he increased rents and fines, and inclosed more land for grazing. All classes, in fact, were in the grip of an upward movement of prices, for which no one in particular was responsible, and which had for its efficient cause the influx of American silver and the consequent cheapening of the purchasing power of money.

Public opinion, however, laid the onus for the dearth on the individual who added field to field for grazing purposes; and the government, always afraid of a popular rising, took the same line. Thus the history of Elizabeth's reign, so far as inclosures are concerned, is largely a continuation of her predecessors' policy. While no action was taken against the incloser for tillage, who was really an asset to the community, legislative war was proclaimed against the other type, whose activity caused the destruction of buildings, unemployment, and depopulation. In 1563 an act was passed confirming the agrarian legislation of

[1] According to J. U. Nef (*The Rise of the British Coal Industry*, i. 190), it was the supply of timber rather than the supply of grain that suffered through the extension of inclosures.

Henry VII and Henry VIII: all land that had been under the plough for the space of four years since 1528 was to remain in tillage for ever, and no land now under the plough must be converted to pasture. Thirty years later (1593) the 'great plenty and cheapness of grain' tempted the government to repeal this statute, which, it would appear, was 'obscure and imperfect'. But in 1598, under pressure of renewed high prices, famine, and popular outcry against inclosures, there was a reversion to the former policy. Two statutes were passed with a view to increasing the area under tillage and arresting the depopulation of the countryside. In the first it was stated that 'of late years more than in times past there have sundry towns, parishes, and houses of husbandry been destroyed and become desolate, by means whereof a great number of poor people are become wanderers, idle and loose'. It was therefore ordered that the destroyed houses should be rebuilt and suitable land assigned to them for cultivation. The second, in which it was pointed out that since the repeal of the 1563 statute 'there have grown many more depopulations, by turning tillage into pasture, than at any time for the like number of years heretofore', it was enacted that all land that had been under the plough for twelve consecutive years and had been converted to pasture since the first year of the reign should be restored to tillage. Clearly, from the government standpoint, the agrarian question was by no means settled. But the next century was not far advanced when all anti-inclosure legislation was definitely abandoned as unnecessary, showing that the Elizabethan government was perhaps unduly nervous as to the dangers of the movement.

Of the sufferings of the poorer classes there can, unfortunately, be no doubt. Those who were torn away from their moorings in the country by the agrarian changes, and sought an entrance into industry, found their way barred by stringent gild restrictions or exorbitant fees; while the town corporations, resenting the influx of casual labour from the outside, which added to their liabilities in respect of poor relief, legislated against the settlement of aliens in their midst. Begging was of little avail, for private charity had grown cold since the dissolution of the monasteries. 'A poor man', says Robert Greene,[1] 'shall as soon break his neck as his fast at a rich man's door.' 'They lie in the

---

[1] Pamphleteer, dramatist, and novelist (1560?–1592).

streets', writes Stubbs, 'in the dirt as commonly is seen . . . and are permitted to die like dogs or beasts without any mercy or compassion showed them at all.' Verily it must have seemed a heartless world to those whom the relentless pressure of economic fact drove to beggary. Caught in the maelstrom of forces that swept them hither and thither with a violence as blind as it was merciless, they could do nothing to avert their fate.

But if some classes were sinking in the social scale, others were rising. The freeholders, who had broken away from the manor, and, unlike the 'customary' tenants, enjoyed the protection of the law in the payment of fixed rents, benefited by the cheapening price of money and the higher price of victuals. These men were the yeoman class, the backbone of England, of whom Harrison speaks so enthusiastically, 'who come to great wealth insomuch that many of them are able and do buy the lands of unthrifty gentlemen, and often sending their sons to the schools, to the universities, and the inns of court, or otherwise leaving them sufficient lands whereupon they may live without labour, do make them by these means to become gentlemen'. The prosperous yeoman was a typical product of rural society in Elizabeth's day. Often he was more than a wool grower and supplier of markets with butter, cheese, and beef: he dabbled in the new industrial undertakings of the time, sinking coal pits, setting up mills, digging for alum, opening up salt springs, or supplying the leather-goods manufacturers with material from his own tan-yard.

The expansion of industry, however, was the dominant feature of the period. New crafts were springing up, often as the result of the incoming refugee protestants from the Netherlands and France, and the older crafts received an impetus from the same source. The felt industry, thread-making, lace-manufacture, silk-weaving, engraving, the manufacture of parchment, needles, and glass were introduced. Manchester benefited by the development of 'cottons',[1] Birmingham by the addition of brass to her iron industry; and in both cases much was due to the skill of the foreign workman. Even the staple industry, cloth, profited by the immigration, especially in the finer branches of the craft, i.e. 'bays'. Not less noteworthy than the acclimatization of new industrial pursuits was the extension of industry to parts hitherto unaffected. Thus the cloth industry, originally centred

[1] 'Cottons' were not cottons in the modern sense, merely a species of woollens.

in the eastern counties, at Norwich, Bury St. Edmunds, the valleys of the Colne and Blackwater, in Essex, and the Kentish Weald, had spread to the north and west, to Yorkshire and Lancashire, Wiltshire, Gloucester, Somerset, Devon, the Welsh border districts, and the southern coastal counties from Yarmouth to Plymouth. Nor was cloth-weaving confined to urban centres: it took root in the cottages of rural workers, who procured their supplies of wool from some neighbouring market, and returned later to the same market to sell the woven cloth. This practice was quite common in the west riding of Yorkshire and in Devon, and may have been more prevalent than we suspect. Moreover, although the circumstances of the time favoured the capitalist employer, like John Winchcombe, who paid his workers a mere pittance of fourpence or sixpence a day and amassed a great fortune, they also contributed to the rise of the 'small master', the urban counterpart of the prosperous yeoman in the country, who worked not for wages but for profit, borrowed capital when necessary, and, if successful, established himself as a capitalist on a small scale. He, too, was a typical product of the later sixteenth century, and his existence materially helped to maintain self-respect among the working classes at a time when effective combination against rapacious employers was impossible.

Although the development of the cloth industry and its allied trades was the outstanding feature of the time, there were other directions in which the ebullient energy of the nation found an outlet. The extractive industries were not neglected. Companies were formed on the joint stock principle for the exploitation of natural resources and the manufacture of metals, e.g. the Company of the Mines Royal and the Mineral and Battery Company, both of which were incorporated in 1568. The 'hostmen' of Newcastle developed the coal industry of Durham and Northumberland, and practically controlled the entire traffic in this commodity, although as yet the consumption was small, owing to the continued use of wood in iron-smelting. So considerable was the wastage in timber at the iron-works in Sussex, Nottinghamshire, and in the neighbourhood of Birmingham that it became a matter of grave concern to the state. 'These iron times', sang the poet Drayton, 'breed none that mind posterity.'

It was, in many ways, a chaotic world. Company promoters

and speculators preyed on the credulity of a penny-come-quick public; and patentees and monopolists, many of them courtiers or friends of courtiers, sought and obtained from the queen the exclusive privilege of dealing in certain articles of commerce, or industrial processes, to the detriment of the common weal and their own enrichment. Everywhere private interests predominated over public: every one 'gaped' for gain. Money was power —it was more than this, it was the national divinity, whose virtues even poets could hymn:

> Money, the minion, the spring of all joy;
> Money, the medicine that heals each annoy;
> Money, the jewel that man keeps in store;
> Money, the idol that women adore!

Nor was the worship of Mammon confined to the laity. Richard Cox, bishop of Ely, a man of scholarly attainments, was also noted for his meanness and avarice. He began his episcopal career with a lawsuit against his popish predecessor, Thomas Thirlby, for alleged conversion of funds belonging to the see to his own private use, and then proceeded to lease the episcopal lands to graziers and sell the timber, in order to enrich himself and his heirs. Edwin Sandys of Worcester and John Scory of Hereford are similar examples of clerical cupidity. On the whole, however, the church was not a highway to wealth: there were too many greedy cormorants among the laity hovering round ecclesiastical endowments to permit of it. On the other hand, there were hundreds of openings for the lay speculator to establish a fortune, provided he was willing to 'chance his arm' in questionable transactions. Contractors for supplies to the army, treasurers, paymasters, and all who handled public funds or their disbursement were favourably circumstanced for giving reign to their cupidity; for public money was regarded as fair game by the embezzler, and embezzlement was one of the curses of the age. Sir George Carey, the bearer of an honourable name, whose loyalty to the state was above suspicion, made a fortune for himself as treasurer of the war in Ireland during the later years of the reign, by passing falsified clothing accounts, by manipulating the exchange, and by purchasing provisions and paying the troops in the base Irish currency, while he himself received sterling from England. Even Sir Thomas Gresham, the oracle of the city, merchant prince and trusted financial

agent of the government, whose munificent benefactions were
a feature of the time, finished his career, in 1575, with a substantial sum unaccounted for in his transactions on behalf of
the Crown. The plain fact is that public morality was low, and
the great officials who died poor, like Edward Baeshe, surveyor-
general of the navy victualling department, were the exception
rather than the rule.

Yet, although the desire for monetary wealth was great, it
was not a hoarding age. Men made money in order to enjoy it,
and the surest way to enjoy it, and at the same time to gratify
one's social ambitions, was to lay it out in land. The successful
burgess, lawyer, industrialist, or merchant who had savings to
invest, itched for real estate in the hope of founding a county
family. Moreover, land could be acquired in various ways: by
the purchase of derelict estates, by lending to thriftless landowners on the security of their estates, and selling them up when
they failed to redeem their bonds, or by buying reversions and
mortgages. All that was needed for this kind of speculation
was shrewdness, prudence, and patience—qualities possessed in
ample measure by men like Mr. George Stoddart of London,
who in a few years transformed himself from a grocer and
usurer into a landed proprietor with broad acres.

Undoubtedly there was a land-hunger in Elizabethan England. Not only were capitalists dabbling in real estate: the law
courts were busy from one end of the country to the other with
claims arising out of land, or disputed successions to manors.
Men flew to the law on the slightest provocation, if they thought
they could thereby establish an advantage to themselves at the
expense of their neighbours. The succession of a minor to a
great estate was sometimes the occasion for a general scramble
after plunder by all who deemed they had a claim, bogus or
real, to a share in the inheritance; and a sober, industrious landlord like William Darrell of Littlecote, Wiltshire, might find
himself, through the greed or malice of his neighbours, condemned to a whole lifetime of litigation.

England, to be sure, was a quieter place than it had been
in the days of Sir John Paston. The statute of retainers had
done its work, and the 'overmighty subject' was now a thing of
the past. But respect for the law was still only skin deep. Suits
could be won by bribing judges, suborning witnesses, or by
the help of a powerful patron; and legal decisions were not

infrequently followed by reprisals. Thus a defeated claimant to an estate might avenge himself on his successful rival by leading a band of rioters into his grounds, laying waste his fields, plundering his manor house, and, generally speaking, indulging in a miniature private war, although such a procedure inevitably led to a further action for battery and assault or forcible entry. When the mood was on him, a gentleman would divert himself by hunting deer or felling timber on his neighbour's estate, with singular disregard for the rights of private property. In short, behind the imposing façade of the law, society was at war with itself; and the root of this turbulence and factiousness was covetousness.

The main tendencies of the period are easy enough to detect. It is not so easy, however, to determine what part the government played in shaping or controlling the course of events. From one point of view the forces at work appear to have been entirely economic and beyond the scope of government action; but from another the government figures as the *deus ex machina* of the revolution, guiding with its paternal hand the giant impulses at work, so as to minimize their potentiality for evil and maximize their usefulness to society as a whole. It was, we must remember, the beginning of the mercantilist age, when governments were endeavouring to concentrate under their direction all the sources of national power; and England led the way in this new development. Yet it may be doubted whether the state could do more than remove obstacles, encourage enterprise, and seek by remedial measures to keep the swollen stream from overflowing its banks. Within these limits, the Elizabethan government was exceedingly active. By the reform of the coinage (1561) the confidence of the trading community was strengthened. The statute of apprentices (1563) helped to stabilize industry and encourage agriculture, by insisting on long contracts, and by ensuring a plentiful supply of labour in the fields when labour was most needed. It also instituted a system of wage assessment, based upon the conditions obtaining in different parts of the country, so as to 'yield to the hired person, both in time of scarcity and in time of plenty, a convenient proportion of wages'. This elastic principle was an improvement on the older statutes of labourers, which attempted to fix a flat rate for the entire country. Similarly the legalization of interest at a maximum of 10 per cent.

(1571) probably gave a stimulus to the employment of capital in increasing amounts. It is noticeable, too, that the privy council kept a wary eye on the food supplies, in order to mitigate the sufferings of the poor in time of dearth. Markets were watched. The export of corn was checked when prices rose higher than 10*s.* (1563) and 20*s.* (1593) per quarter, and totally prohibited to any part of the Spanish dominions after the quarrel with Spain developed. Efforts were also made, not so successfully, perhaps, to check engrossing, forestalling, and regrating at the markets, and to check fraud by standardizing weights and measures. In both cases the government acted through the clerks of the markets. Encouragement was given to enterprises that made for national security, such as the working of metals, the production of sulphur, saltpetre, flax, and hemp, and the manufacture of canvas. The same minute attention was paid to the construction or enlargement of ports and harbours, the preservation of timber, and the development of the fisheries by the enforcement of compulsory fish days. The institution of a 'political lent' was definitely intended to strengthen the maritime population and increase the supply of seamen for the navy.

On the other hand nothing was done to improve communications. The roads or common ways of England were allowed to remain, as they had been for centuries, ill conditioned and neglected. Their upkeep fell upon the parishes, who treated their obligation lightly, fulfilled it reluctantly, or totally ignored it when circumstances permitted. Only the great arterial roads connecting London with Dover, Plymouth, Bristol, Chester, and Berwick, which were necessary for government communications, seem to have been suitable for constant traffic, although they, too, had their 'noisome' corners, 'sloughs', and 'holes'. The smaller roads were little better than mere tracks, deeply rutted for most of the year, and practically impassable for foot passengers in winter. All highways were, of course, plentifully provided with commodious and well-equipped inns—the best in the world, said travellers who had sampled those of other countries. But once the traveller left their friendly shelter, he ran the risk not only of the physical hardships of the road but also of falling a victim to the 'close-booted gentlemen' who lay in wait for fat 'booties' by the wayside. Every exposed place had its gang of desperadoes. Gadshill near Rochester, Shooter's hill by Blackheath, Salisbury plain, Newmarket Heath, &c.,

were notorious haunts of thieves. Frequently, if we may believe Harrison, these land pirates received their information as to travellers' movements and the nature of their business from the ostlers and tapsters of the inns. Falstaff's disgraceful exploit at Gadshill was probably a not uncommon occurrence.

Nevertheless, in spite of the dangers and difficulties, the main roads were serviceable enough for the needs of the time. Rapid movement was rendered possible by the provision of post-horses at fixed stages, which could be hired at 3d. per mile (for private persons) and 2½d. (for persons on government business), with 6d. for the post-boy, who returned with the 'mount' after completion of the stage. In addition to post-horses there were wheeled vehicles on the roads—carts, coaches, and 'caroches'. But the day of the vehicle was only just beginning: the lack of springs made coaching a sorely trying experience. In London the old-fashioned horse litter was still in vogue for ladies, and some indulged in the newfangled sedan chair, which became popular in the reign of Charles I. In the far north, again, where the roads were few and mountainous, the saddle-horse and the pack-horse remained the sole means of communication and transport.

If the government neglected the roads, the great floating population of vagabonds who used them presented a problem which could not be ignored. Here the need for action on a nation-wide scale was now more than ever apparent, for in spite of all previous attempts to control the plague of beggars their numbers had increased so greatly as to constitute a grave menace to public order. According to Harrison,[1] the vagabonds or sturdy beggars alone numbered 10,000. Harman,[2] the contemporary anatomizer of roguery, asserts that there were no fewer than twenty-three different categories of thieves and swindlers—'rufflers', who pretended they had been at the wars and lived on charity, extorted by violence 'when they may be bold', or by pity when violence would not pay: 'upright men', who were 'of so much authority that, meeting with any of their profession, they may command a share and snap unto themselves': 'hookers', who pilfered linen and clothes from windows with a six-foot pole: 'palliards' or 'clappendogens', who went about in patched clothes, carrying a testimonial, and blistered their

[1] W. Harrison, *Description of England.*
[2] T. Harman, *A Caveat or Warning.*

limbs with spearwort, ratsbane, or other corrosive: 'priggers of prancers' or horse-thieves: 'Abraham men', who 'feign themselves to have been mad': 'fresh-water mariners' or bogus sailors, 'whose ship was wrecked on Salisbury plain': 'counterfeit cranks', whose specialty was to dissemble the falling sickness by working their mouths into a lather with soap: 'dummerars', who pretended to be dumb by 'holding down their tongues doubled' and 'holding up their hands piteously': 'demanders for glimmer', who said their goods were consumed by fire; and so on to the 'bawdy baskets', 'morts', 'doxies', 'dells', and common harlots. Such was the composition of this 'merry England' that slept in haylofts, sheepcotes, or on doorsteps, spreading terror in the country and disease in the towns.

The official attitude to the whole fraternity of vagabonds had always been, and still was, one of fear-ridden ferocity: they were the true 'caterpillars of the commonwealth', who 'lick the sweat from the labourers' brows'. But the impotent poor, the poor by casualty, who were poor 'in very deed', were acknowledged to be a charge on public benevolence. The vital question was what form this public maintenance should take. Slowly and painfully the state was being driven by the colossal dimensions of the problem to the conviction that responsibility in the matter could not be left to the conscience of the individual, but must be enforced by law upon every one. Thus while the first pauper enactment of Elizabeth's reign (1563) simply revived the statutes of Henry VIII and Edward VI with regard to the sturdy beggar, it recognized for the first time, in principle at least, the necessity for a compulsory levy for the maintenance of 'impotent, aged, and needy persons'. Men of 'froward and wilful mind', who refused to yield to the gentle suasion of the bishop and contribute their quota to the parish poor fund, were to be handed over to the justices of the peace and coerced into payment. More precise instructions, however, were necessary before the new weapon of compulsion could be used effectively by the local authorities. Consequently in 1572 a further enactment instructed all justices, mayors, sheriffs, bailiffs, &c., to levy a reasonable rate on all inhabitants dwelling within their respective jurisdictions and to settle the impotent in permanent abiding-places. Pauper children were also to be bound out to service. At the same time the treatment of the rogue was made harsher: he was to be whipped and bored through the gristle of the right ear, and if he continued in his roguery he was to suffer, in the last resort, death for felony. The term 'vagabond' was defined so as to include a remarkable variety of flotsam and jetsam: procurators who went about with counterfeit licences, idle persons who indulged in unlawful games, palmists, physiognomists, fencers, bearwards, players of interludes and minstrels 'not belonging to any baron of the realm', jugglers, pedlars, tinkers, petty chapmen, loiterers, scholars of Oxford and Cambridge who begged without the authority of their university, shipmen 'pretending losses by sea', and all who 'being mighty of body' refused to work for wages. In 1576 this statute was amended by another, which compelled every city and town corporate to provide a store of wool, hemp, flax, iron or other stuff, so that the honest poor, if able of body, might be set to work and paid 'according to the desert of the work'. Rogues, on the other hand, were to be sent to 'houses of correction', two of which were to be erected in each county, and there disciplined to labour. Finally came the comprehensive statute of 1598, which corroborated and exploited the same remedies still further. The contribution for poor relief was fixed at a maximum of 2d. per parish. Provision was made for the compulsory apprenticing of pauper children, until they reached the age of twenty-four (for males) and twenty-one (for females). Begging was prohibited. Justices of the peace in each county were to appoint parish 'overseers' of the poor, whose duty it was to raise 'by taxation of every inhabitant' stocks of material for the workhouses, and be responsible for their management. Parents and children of the impotent, who had the necessary wherewithal, were to relieve them 'of their own charges'—a shrewd concession to the principle of self-help. By the same act the statute of 1576 against vagabonds was repealed in favour of more humane methods. Boring of the ear was dropped, but the sturdy beggar, after being whipped 'until his or her body be bloody', was to be sent back to his birth-place, or, if that was unknown, to the place where he last dwelt, and there to be thrown into a house of correction or the common jail until he could be placed in service. If he was not able to work, his destination was an almshouse. Dangerous rogues were to be banished the realm or sent to the galleys, and, if they returned, were to be treated as felons. Private endowment of hospitals, houses of correction, *maisons de Dieu*, or abiding-places for the poor was to be encouraged, provided the endowment amounted to the 'clear yearly value of £10'. This law, which was at first a temporary one, was published with slight alterations in 1601, and became the final expression of Elizabethan statesmanship in the sphere of poor-law administration.

It will be evident that the act completely secularized the whole question of relieving poverty, and at the same time established the principle of corporate responsibility as fully as the time permitted. If it showed no mercy to the 'work-shy', it did everything that was humanly speaking possible for those who were willing to work, and acknowledged a very significant obligation to the children of paupers.

Between the economic and the social condition of a country a close connexion always subsists. Alterations in the distribution of wealth bring new classes to the forefront of society, and they, in turn, create a new social environment. A hundred different ways of spending are discovered and practised, old fashions give place to new, and the whole outlook of society is transformed. Such a change occurred in Elizabeth's reign side by side with the industrial and commercial expansion.

The *foyer* of this social revolution was London, but it spread with the rapidity of an epidemic through the length and breadth of England. Camden noted the phenomenon for the first time in 1574 when, he observes, 'our apish nation' displayed 'a certain deformity and insolency of mind', 'jetted up and down in silks glittering with gold and silver', and began to indulge in a 'riot of banquetting' and 'bravery in building', to the 'great ornament of the kingdom, but to the decay of the glory of hospitality'. Most of the novelties demanded by society for the adornment of the person came from abroad, and the foreign merchant drove a thriving trade. 'One trifling toy not worth the carriage, coming from beyond the sea,' says a writer of the time, 'is more worth with us than a right good jewel easy to be had at home.' In a play entitled *The Three Ladies of London* (1584), one of the characters says: 'Then, signor Mercator, I am forthwith to send ye from hence to search for some new toys in Barbary and in Turkey.' The economist might point out that the purchase of this 'foreign trash' cost England annually £100,000. The puritan satirist might castigate the garish tastes of society with the fury of a Juvenal. Neither could make any impression on the widespread craving for luxury in every shape and form. 'They be desirous of new-fangles,' said Stubbs, 'praising things past, condemning things present, and coveting things to come; ambitious, proud, light-hearted, unstable, ready to be carried away by every blast of wind.' 'Nothing is so constant', wrote Harrison, 'as inconstancy in attire.' Fynes Moryson,[1] who had travelled more widely than most men of his time, thought his countrymen 'more sumptuous than the Persians', because 'they affect all extremes'.

It was not only the rich who flaunted their 'silks, velvets, and chains of gold': all who could come by the necessary means— farmers, peasants, artisans—followed in the train of fashion. Many were the gibes levelled by writers of the period at the vanity and presumption of the lower classes. 'Now', says Greene, 'every lout must have his son a court noll, and these dunghill drudges wax so proud that they will presume to wear on their feet what kings have worn on their heads. A clown's son must be clapped in a velvet pantoufle, though the presumptuous ass be drowned in the mercer's book, and make a convey of all his lands to the usurer for commodities.' 'The ploughman', comments Thomas Lodge, 'must nowadays have his doublet of the fashion with wide cuts, his garters of fine silk of Granada, to meet his Sis on Sundays.' To become a fine gentleman, according to Ben Jonson, it was necessary to hie to the city, 'where at your first appearance 'twere good you turned four or five hundred acres of your best land into two or three trunks of apparel'.

Naturally a 'babylonian confusion' of classes ensued, and the social world turned itself topsy-turvy. The 'mingle-mangle' of dress made it impossible to say who was 'noble, worshipful, gentle, or even a yeoman'. 'To such outrage is it grown now-a-days,' writes Stubbs, 'every butcher, shoemaker, tailor, cobbler, husbandman, and other; yea every tinker, pedlar, and swine-herd, and every artificer and other *gregarii ordinis*, of the vilest sort of men must be called by the vain name of master at every word.'

The highly artificial composition of men's dress is portrayed humorously in Thomas Dekker's *Seven Deadly Sins of London*.

'The Englishman's dress', says the dramatist, 'is like a traitor's body that hath been hanged, drawn, and quartered, and is set up in various places; his cod-piece is in Denmark, the collar of his doublet

---

[1] Author of *An Itinerary: containing his ten years' travell*, &c.

and the belly in France; the wing and narrow sleeve in Italy; the
short waist hangs over a Dutch butcher's stall in Utrecht; his huge
slops speak Spanishly. . . . And thus we that mock every nation for
keeping of one fashion, yet steal patches from every one of them to
piece out our pride.'

So whimsical was the national taste that even the enterprising tailor had to acknowledge himself beaten. Beneath an
old wood-cut depicting a naked Englishman carrying a roll of
cloth over his arm and a pair of tailor's shears in his hand,
stands the following doggerel:

> I am an Englishman, and naked I stand here,
> Musing in my mind what raiment I shall wear;
> For now I will wear this and now I will wear that;
> And now I will wear I cannot tell what.

The permanent elements of male costume, on which all the
elaborate changes were rung, were not really complicated. A
doublet tightly fitting the figure, carrying a small puff at the
shoulder, and a high collar at the neck edged with a frill of
lawn, cambric, or holland: a pair of round breeches, generally
showing the knees: long stockings reaching to the thighs: small
pointed shoes or long wide-mouthed boots: a short cloak hanging
loosely from the shoulders: a flat or elongated hat bearing a
band with a feather; and, finally, a rapier and dagger—such
were the essentials. But foreign 'cuts', individual fancies, and
the enterprise of the sartorial artist embellished and enriched
each item almost beyond recognition. Thus the velvet doublet
and breeches were filled out with 'bumbast' of hair or wool
(5 or 6 lb.), 'slashed' with silk or other costly material, embroidered with gold lacing, and fitted with crystal buttons, or
buttons of gold or silver. The cloak, reaching to the girdle,
knee, or even to the heel, was of the most elaborate workman-
ship, and made of the costliest material, within as well as without. As a rule it was of silk, taffeta, and velvet. The stockings
were seldom of cheaper material than silk, and were frequently
'cross-gartered' above and below the knee. Footwear, whether
boot or shoe, was high-heeled, so as to raise the individual
'a finger or two' above the mire of the streets, and like the other
parts of the dress was of silk, velvet, or costly leather work
'slashed' or 'pinked' (perforated), and variegated with embroid-
ery and silver or gilt buckles. Hats, decorated with bands and
feathers, were sometimes drawn out 'like the battlements of a

house' or elongated 'like a spear or shaft of a steeple'. They were made of silk, taffeta, velvet, wool, sarcenet, or fine hair. Even the underwear was of the finest cambric, lawn, or holland, treated with silk needlework, and flounced or frilled at the neck and hands. Finally the dagger, sword, and rapier bore gilded or silvered hilts, 'damasked' engraven blades, and were sheathed in velvet scabbards. All this fantastic dress, it should be noted, was a glow of colour of the most bizarre description— popinjay blue, pease-porridge tawny, goose-turd green, lusty-gallant, Judas colour, &c. The cost was, of course, heavy. Ornamental breeches ran from £7 to £100, stockings from 15s. upwards, shirts from 10s. to £5, and boots from £4 to £10.

Women's dress was no less revolutionary than that of the men. 'A ship is sooner rigged', said Stubbs, 'than a gentle-woman made ready.' Elizabethan London showed a strange re-semblance to renaissance Venice in regard to feminine fashions. In both places it was the custom to expose the breasts freely, to paint the face and dye the hair, 'almost changing the accidents into the substance'. Apparently, too, the stylish women of England followed their Italian sisters in the quest for the blond, and were not above inveigling children with golden locks to part with them for a penny. On the edge of their 'bolstered' hair they hung 'bugles', 'ouches', 'rings', and other ornaments of silver or gold; and across their foreheads were laid, from temple to temple, a gilded or silvered fillet. The inordinate lust for these 'gee-gaws' of the foreign merchant—for they came from abroad—was the cause of many a sly hit in the drama of the time.

All classes and ranks of feminine society affected similar expensive tastes. The merchant's wife and the gentlewoman had their French hoods, artificers' wives had their hats of velvet, while the humble cottager's daughter had her taffeta or woollen headgear, well lined with silk or velvet. Likewise corked shoes, pantoufles, slippers, 'some of black velvet, some of white, some green, and others yellow', embroidered with gold or silver work, were the prerogative of no special class. But the chief fashionable addictions of the period were the ruff and the farthingale. The ruff was a pleated collar, worn round the neck, made of linen, lawn, or cambric, and stiffened with wire ('supportasses') and starch until it stood out from the person by about a foot. The effect, according to one critic, was of

John the Baptist's head on a platter. Sometimes several ruffs were placed *gradatim* one on the top of another, and 'all under the master devil ruff'. As a rule feminine taste demanded that the material should be crusted over with gold, silver, or silk lace—'speckled and sparkled here and there with the sun, moon, and stars'. The other special feature of female attire, the far-thingale or *verdugado*, was a Spanish invention. Introduced into England about the same time as the ruff, it supplanted the old-fashioned gothic skirt which clung to the figure. Essentially it was a petticoat of canvas, distended with whalebone, cane, or steel hoops, and covered with rich cloth of taffeta, velvet, or silk. The effect was greatly to enlarge the hips; and this was still further accentuated by the habit of squeezing the upper part of the figure into a stiff, pointed doublet—the so-called 'spagnolized' waist. 'A trussed chicken set upon a bell' is how one critic described the general effect.

In so artificial a society the barber drove a thriving trade. His 'cuts', cosmetics, and perfumes were in great demand. Setting aside the old 'Christ's cut' which gave a man's head the appearance of a 'Holland cheese', as if it had been 'cut round by a dish', the tonsorial artist purveyed the latest styles of Italy, France, and Spain. The Italian manner was 'short and round and frounced with curling irons so as to look like a half moon in a mist': the Spanish cut was 'long at the ears and curled'; while the Frenchman affected a love-lock drooping to the shoulders, 'where you may wear your mistress' favour'. The beard was similarly treated. The 'peak' might be 'short and sharp amiable like an inamorato, or broad pendant like a spade, to be terrible like a warrior and a soldado'. But the physio-gnomy of the client had to be considered. 'And therefore', says Harrison, 'if a man have a lean and straight face, a Marquess Otto's cut will make it broad and large; if it be platter-like, a long slender beard will make it seem the narrower; if he be weasel-beaked, then much hair left on the cheeks will make the owner look big like a bowdled hen, and grim as a goose, if Cornelius of Chelmsford say true.' The moustaches were separately treated, being 'fostered about the ears like the branches of a vine', or cut down to the lip in the Italian manner. Of the unguents and perfumes in common use, perhaps the most popular were civet, musk, oil of tartar, lac virginis, and camphor dissolved in verjuice.

It was not only a richly dressed and highly perfumed, but also a 'militant civility' that paraded the London streets. Everyone went armed from eighteen years of age and upwards, bearing at least a dagger, a broadsword, or a rapier—sometimes both dagger and rapier. Even the labouring man in the country had his weapons, which he laid down in a corner of the field when working. Apprentices had their knives, and women their bodkins. The tendency of the more lethal weapons was to grow longer; and in 1580 the government was compelled to issue a proclamation limiting the length of swords to three feet and daggers to twelve inches, inclusive of the handle. Obviously street fighting was frequent, for the Englishman at this time aped the Italian: he had a 'factious heart, a discoursing head, and a mind to meddle in all men's matters'. Mercutio's taunt to Benvolio shows him in action: 'Thou wilt quarrel with a man for cracking nuts, having no other reason but because thou hast hazel eyes. . . . Thou hast quarrelled with a man for coughing because he hath wakened thy dog that hath been asleep in the sun.'

If the dress was sumptuous, the Elizabethan house and household appurtenances were no less an expression of the general desire for luxury. The old, rude, timber-and-clay dwelling, with its smoky interior, its rush-strewn floor, and its straw pallet, was no fit habitation for 'velvet breeches', however suitable it might have been for 'cloth breeches'. 'Oak, which, in times past', writes Harrison, 'was dedicated wholly to churches, religious houses, noblemen's lodgings, and navigation', now became the rage. Hornbeam, plum-tree, and sallow willow—the old materials of construction—were discarded, and the 'royal wood' appeared in every pretentious dwelling. Timber, clay, and wattles gave place to stone, lath, and plaster: wainscoting appeared on the inner walls and plaster of Paris on the ceilings: the feather bed replaced the straw pallet: carpets took the place of rushes; and stoves and chimneys confined the wandering smoke of the medieval house within manageable limits, while the use of glass dispelled its darkness and draughts. As tapestry and arras work beautified the walls of the rooms, so goblets of silver, glass, or pewter, metal spoons, chairs, and table linen helped to make the act of dining a more elaborate and less gluttonous performance. In noblemen's establish-

ments the value of the plate amounted to £1,000 or even
£2,000: in smaller houses, such as the merchant class occupied,
to £500; and in artisans' or farmers' cottages it was not
altogether unknown. Some idea of the length to which the
wealthiest class was prepared to go in the worship of luxury may
be gathered from Laneham's flamboyant description of Kenilworth, Leicester's country seat. This stately pile, one of the
'wonder houses' of the time, was 'all of hard stone, every room
so spacious, so well lighted, and so high-roofed within: so seemly
to sight, so glittering of glass a-nights by continual brightness of
candle-fire and torchlight, transparent through the lightsome
winds, as if it were the Egyptian Pharos relucent unto all the
Alexandrian coast'. When the queen visited it in 1575 she
was treated to a banquet that must have been the *chef d'œuvre*
of culinary art in the sixteenth century. It was served in a
thousand dishes of glass or silver by two hundred gentlemen.

Apparently all classes fed liberally, save the poor, who
confined themselves to 'white meats', bread made of rye or
barley, when they were obtainable, and a coarser bread of beans,
oats, or acorns in time of dearth. Of the fashionable dinner-
table Stubbs writes: 'Nowadays, if the table be not covered
from the one end to the other as thick as a dish can stand by
another, with delicate meats of sundry sorts, one clean different
from another, and to every dish a several sauce appropriate to
his kind, it is thought unworthy of the name of a dinner.' The
nobleman never wanted, in season, his beef, mutton, veal, lamb,
kid, pork, coney, capon, red and fallow deer, fish, and wildfowl,
together with sundry 'delicates', 'wherein the sweet hand of the
seafaring Portugal is not wanting'. The gentlefolk and the
merchants used ordinarily four, five, or six dishes, but on special
occasions aped the fuller tables of those above them, showing a
keen relish for rare and outlandish delicacies. Beef formed the
main constituent in the diet of the husbandman and the artificer;
but if they happened to 'stumble upon' a haunch of venison and
a cup of very strong beer or wine, 'they think their cheer so
great and themselves to have fared so well as the lord mayor of
London'.

The money expended on spices and condiments was out of all
proportion to that expended on food; but there is no reason to
suppose that excessive eating was carried on to any marked
degree. Excess in drinking, on the other hand, was probably one

of the commonest vices of the time. 'Large garaussing', says Moryson, came over with the soldiers and captains returning from the Low Countries, and was evidently due to contact with the Germans. The wealthy 'garaussed' in their fifty-six different kinds of French wine, and their thirty brands of Italian, Grecian, Spanish, and Canarian; while the poorer folk frequented the public inn or alehouse, and became 'cup shotten' on common ale. The chief sinners were not the winebibbers, who drank from Murano glass and silver goblets, but the 'vulgar sort' who copied the vices of the swaggering soldier. 'It is incredible', says Harrison, 'how our malt bugs lug at their liquor even as pigs lie in a row lugging at their dam's teats till they lie still again and be not able to wag.' Drunkenness, however, was regarded by the better classes as a 'clownish' and 'reproachful' vice; and by a very simple device they prevented over-indulgence. The drinks were kept on a sideboard, not on the table, and the guests drank only when necessity urged, thereby avoiding 'the note of great drinking'.

'Tobacco taking' was another luxury of the age. It was probably introduced by Hawkins from America in 1566, in the form of pipe-smoking. Harrison noted its prevalence some eight years later. By 1598 smoking seems to have become general, for Hentzner, who visited England in this year, observed that the English were constantly puffing 'the Nicotian weed' in public places. It was a costly indulgence, for tobacco was retailed at 3s. per ounce; but the practice of providing a common pipe at the inns, taverns, and alehouses placed it within the reach of everybody. By the end of the reign the novelty had become a social necessity, and was practised equally 'in the courts of princes, the chambers of nobles, the bowers of sweet ladies, the cabins of soldiers'.

The social revolution affected also the gardens of the time, which Harrison thought more wonderful than the gardens of the Hesperides, and Fynes Moryson rated above those of any other nation. 'So curious and cunning are our gardeners now in these days', says Harrison, 'that they presume to do in manner what they list with nature.' The old-fashioned garden of 'humble roots' was, by comparison with the new, no better than a 'dung-hill'. The artistic sense of the age demanded that the kitchen garden, with its rank odours, 'which are scarce well-pleasing to perfume the lodgings of any house', should be removed to a distant part of the grounds, and that precedence should be given to the vineyard, the orchard, and the flower-bed. Like the Italian, the English gardening expert made the garden the setting for the house. Surrounded with a wall of stone or a thickset hedge of hornbeam or holly, and formed on the pattern of a square, it was laid out with the utmost stiffness and formality, in path and flower-bed, with arbours and vases of lead and stone to diversify the somewhat monotonous repetition. Strange plants, herbs, and fruits adorned it—grapes, plums, apples, pears, walnuts, apricots, almonds, peaches, figs, oranges, lemons, and wild olives. In short, the gardens of Elizabethan England were symbolic of the luxurious and far-flung life of the period.

But of all the sights of the time none could compare with the city of London. The hundred and twenty thousand inhabitants who thronged its congested and plague-infested streets were an epitome of the energy, enterprise, and indomitable spirit of a great people—a people, says Stubbs, 'audacious, bold, puissant, and heroical . . . in all humanity inferior to none under the sun'. The Thames, which formed the principal highway between east and west, was the centre of the Londoner's world, and a source of endless surprise to strangers. Its crowded wharves, 'shaded with masts and sails', gave it the appearance, remarks Camden, of a 'wooded grove'. No less wonderful were the palaces and riverside houses, stretching from Temple stairs to King's bridge, each with its garden and water-gate. In the city itself were many attractions to hold the eye of the curious: West Cheap with its far-famed Goldsmiths' Row, noted for its glittering tower and fountain that played continuously: Fleet street with its puppet-shows, where visitors could view a dead Indian for ten doits, and many other marvels: Holborn, the sixteenth-century Kensington, famous for its gardens and the freshness of its air: the Royal Exchange in Cornhill, the haunt of merchants and loungers: the Tower with its armoury and primitive zoological gardens: the monuments at Westminster; and, if a bird's-eye view of the entire panorama were desired, it could be had for a penny by mounting St. Paul's tower.

The principal places of fashionable resort, where all classes sunned themselves, were St. Paul's, the theatre, and the bear garden. Each, in its own way, represents a facet of the variegated life of the period. St. Paul's, intimately associated with all the epoch-making pronouncements of the reformation from the days of the Henrician settlement, had now become one of the hubs of the social universe. In vain the Homilies supplied for the use of the pulpit inveighed against its desecration: the cathedral, in Elizabeth's day, was a 'house of talking, of walking, of brawling, of minstrelsy, of hawks, and of dogs'. The middle aisle of the nave was the scene of a daily pandemonium. Here the gallants and the idle gentlewomen of the capital met to exchange favours. Here, too, came the hawkers and riff-raff of the streets. In one corner were to be found the lawyers and scriveners; in another, shopkeepers exposed their wares for sale, and indulged in their various cries, using the tombs and the font as counters for the payment and receipt of accounts. 'Lordless' men paraded up and down offering their services for hire. And, to increase the din and confusion, horses and mules were led through the cathedral as a short cut, profaning the sacred precincts with the filth and litter of the streets. Outside, in the churchyard, was the book-market of London, where the 'merry books' of Italy were bought and sold; and, in all probability, it was in this very market that Shakespeare purchased the Italian romances on which many of his plays are based.

It was in the theatre, however, that the mundane spirit of the age found its most acute expression. What drew the Elizabethan populace to the playhouse? We are sometimes invited to consider the high poetic value of a Shakespearian drama, and to marvel at the dizzy heights of appreciation required of audiences at the Globe, the Swan, and Blackfriars, when witnessing masterpieces like *Romeo and Juliet*, *Hamlet*, or *A Midsummer-Night's Dream*. We should remember that plays of this nature rose far above the habitual level of contemporary dramatic art. Normally the Elizabethan play staged scenes that would revolt a modern audience—scenes portraying human nature in the raw, that sometimes appear to us to be no fit subject for art at all. Yet this must have been a main source of attraction to the apprentices, mechanics, and artisans who filled the pit. The plain fact is that the popularity of the theatre lay in its direct appeal to the senses. Just as it was the rough and tumble at the bear garden, the 'biting, clawing, roaring, tugging, grasping, tossing, and tumbling' of the bear, and the nimbleness of the dogs, that delighted the spectators, so it was the vehemence of the action, the licence of expression and sentiment, the

resounding declamation, the pageantry, and the fine dresses of
the actors that charmed the playgoers. To the puritan, the
theatre was an eyesore, to the city fathers a plague-spot, but
it was, nevertheless, the glory of Elizabethan London, and a
microcosm of the age.

It cannot be said that the Elizabethan invented any new form
of sport. He indulged in bear and bull baiting, cock-fighting,
hunting the deer and the hare, falconry, fowling, and angling—
all of old standing; but fox-hunting had not yet reached the
dignity of a true sport. The 'sweet and comfortable recreation'
of baiting the chained bear with mastiffs was the national sport
par excellence, being patronized by the queen when official enter-
tainments had to be given in honour of distinguished foreigners.
But cock-fighting was probably more popular with the masses,
since 'cocks of game' were more easily obtained than bears, and
there was more excitement in the sport by reason of the betting
that accompanied it. Partisans of both sports were to be found,
on Sunday afternoons or holidays, at Paris garden, in Southwark, close by the theatres, or at Lewin street, Shoe lane, and
St. Giles-in-the-fields, where the most famous cockpits were
situated.

In the country the premier sport was the deer hunt. The
Elizabethan loved his fat venison, but he was not over-nice as
to the manner of killing it. Sometimes the deer were driven into
prepared ambushes and shot down with the cross-bow; sometimes they were chased to bay with hounds. 'Under this thick
grown brake we'll shroud ourselves,' says the keeper in *Henry VI*,
'for through this laund the deer will come.' Leicester, who was
a keen huntsman, preferred the more active pursuit, 'riding
about from bush to bush with a cross-bow on his back'. The
queen, too, was an ardent follower of the chase, riding so hard
on one occasion that she had to keep her bed for several days,
and could not interview the French ambassador. But when the
Elizabethan sportsman wanted a real thrill, it was the hare he
hunted, not the deer. The hare was the 'king of venerie', the
'most marvellous beast there is'. Shakespeare's stanzas on 'poor
Wat', the hare, in *Venus and Adonis*, contain, without doubt,
the most memorable literary portrait of animal life in the
period.

Next to the chase, falconry was probably the most popular

pursuit of the country gentry. The very terms of the sport were household words. 'An a man have not skill in the hawking and hunting language', says Master Stephen in *Every Man in his Humour*, 'I'll not give a rush for him.' Training of the hawk was a science in itself. Sometimes he was unhooded and flown off the fist when the quarry rose, sometimes liberated first, to hover over the falconer while the game was flushed. The long-winged species were used for the killing of heron, bittern, wild duck, when long-distance flights were necessary; but goshawks and sparrowhawks, whose range was shorter, sufficed for partridges, pheasants, and woodcock in woodland or enclosed country. Dogs, of course, were necessary for retrieving the birds or driving them out of cover. Other methods of bird-taking were practised by the fowler, who used the bow and the bird-bolt, springes, gins, bird-lime, and, more rarely, the caliver and shot.

The use of the caliver as a fowling-piece shows that fire-arms were now beginning to invade the realm of sport. They had already established their supremacy in war—a disturbing fact to all who remembered the glory of the long bow, with which so many battles had been won over the French. But so it was. The man trained in the deadlier arquebus was now more prized than the archer, despite Roger Ascham's spirited defence of the bow in *Toxophilus*. Harrison tells us that Frenchmen and German 'rutters', clad in corslets, derided the English archers, crying 'Shoot, English!' Archery, in fact, was fast becoming a pastime practised for skill rather than a serious preparation for war. Similarly the old-fashioned sword and buckler were giving way to the more stylish and deadly rapier. Justice Shallow remembered the time when 'with a long sword I would have made you four fellows skip like rats'; but every gallant now studied to become an expert in the 'immortal passada' of the Italian and Spanish schools of rapier-fencing. Tybalt's style, as described by Mercutio, was probably copied from the Spaniard, who fought on the 'geometric' system: 'He fights as you sing prick-song, keeps time, distance, and proportion; rests me his minim rest, one, two, and the third in your bosom.' It was a ceremonious age, when men stood on points of honour, and duelling was a common occurrence. A quarrel 'by the book', according to Touchstone, consisted of seven stages: first, the 'retort courteous', then the 'quip modest', the 'reply churlish', the 'reproof valiant', the 'countercheck quarrelsome', the 'lie with circumstance', and, finally, the 'lie direct'.

Doubtless the England of Elizabeth had its saints and martyrs; and in the quiet backwaters of the country, far beneath the troubled surface of the age, the main current of religious life ran its course, as it had done for centuries before. Nevertheless it was a time when, as we have seen, the bulk of men paid little more than lip service to the self-denying virtues of the christian religion. As in politics, so also in society, the prevailing tone was markedly secular and worldly. The day of the puritan ascetics had not yet dawned—they were still voices crying in the wilderness, whose lamentations and denunciations hardly penetrated to the ears of those for whom they were intended. 'They speak', wrote Thomas Nash, 'as though they had been brought up all their days on bread and water . . . as though they had been eunuchs from their cradle or blind from the hour of their conception.' 'Dost think, because thou art virtuous,' says Toby Belch to Malvolio, in *Twelfth Night*, 'there shall be no more cakes and ale? . . . A stoup of wine, Maria!' The confusing doctrinal changes of the previous generation, the extensive spoliation of the church and the consequent discredit of religion, coupled with the rising tide of material prosperity and the unscrupulous pursuit of gain, had contributed not a little to the atmosphere of egoism, paganism, and epicureanism in which the Elizabethan lived and moved and had his being. But it was the infiltration of Italian vices, brought back by those who had made the 'grand tour' of that over-civilized land, that gave the age its peculiar flavour of naughtiness. The Englishman italianate, said a contemporary Italian proverb, is a devil incarnate. 'Thou comest not alone', wrote Greene—himself no paragon of virtue—'but accompanied with a multitude of abominable vices hanging on thy bombast, nothing but infectious abuses and vainglory, self-love, sodomy, and strange poisonings, wherewith thou hast infected this glorious isle.' Among other things, the *italianato* was the patron of the 'bawdy books' that streamed from the press in ever increasing volume, to the alarm and disgust of sober men like Roger Ascham. 'Ten *Morte Arthures* do not the tenth part so much harm', complained the schoolmaster, 'as one of these books, made in Italy and translated in England. They open, not fond and common ways to vice, but subtle, cunning, new, and diverse shifts, to carry young wills to vanity, and young wits to mischief, to teach old bawds new school points.' No wonder the pulpit spoke of 'godly preaching heard without remorse', of 'fasting kept without affliction', of 'almsgiving without compassion', and of 'Lent holden without discipline'. It was an age that claimed and exercised the right to unlimited self-expression—'robustious', ostentatious, licentious: when he that could 'lash out the bloodiest oaths' was accounted the bravest fellow. 'Canst thunder common oaths like the rattling of a huge, double, full-charged culverin?' says a character in Marston's *Scourge of Villainie*: 'Then, Jack, troop among our gallants, kiss my fist, and call them brothers.'

VIII

LITERATURE, ART AND THOUGHT

As a race the Elizabethans set a supreme value on the active life. 'We cannot deny', wrote Lawrence Keymis, Raleigh's companion in the ill-fated Guiana expedition, 'that the true commendation of virtue consists in action: we truly say that *otium* is *animae vivae sepultura*.' Nevertheless a description of the age that concerned itself only with the men of action would rob it of half its glory and many of its greatest names. In the realm of literature the England of Elizabeth stands on a pinnacle by itself, and the writers who contributed most to its brilliance were as distinctively national as the anglican church or the poor law. Some connexion doubtless exists between the two aspects of the period, between the heroic deeds of the seamen and soldiers and the wonderful sunburst of poetry, drama, and speculative thought with which the names of Spenser, of Marlowe, of Shakespeare, and of Bacon are associated; for an age rich in exploit is seldom without its poets and dreamers. The out-goings of the spirit are conditioned by its in-takings, and the imaginative reach of the man of letters keeps pace with the widening horizon of his experience. The countless adventures on land and sea that made England respected and feared, and opened up the world to English enterprise, produced an exaltation of soul which the poet and the philosopher by the alchemy of their genius transmuted into the unforgettable beauties and enduring strength of a great literature. Like Drake and Cavendish, Shakespeare and Bacon circumnavigated the earth and grew rich on its spoils.

Let us beware, however, of supposing that the literature of an age is necessarily a mirror of its history. It is infinitely more than this. The literature of the later sixteenth century, as we have seen, has much to say of contemporary life in its social aspect; but the struggle for power in the political arena, the institutions, laws, policy, and administrative machinery that regulated the political life of society, held as little fascination for the writers of the time as for the multitude in the streets. Politics and literature had not yet come together, but ran their courses in entirely different channels. As for the truly great writers, they are less parochial and more universal in their appeal than at any other period. Those who search for 'political allusions' in their works do not find much to reward them for their pains, and the little they do find is of comparatively small value. The only outstanding exception to this general rule is Richard Hooker, whose *Ecclesiastical Polity* ranks with Shakespeare's *Plays*, Bacon's *Essays*, and Hakluyt's *Voyages*, as one of the representative books of the time. But Hooker's work, for all its massive learning and intellectual power, seldom rises to the height of great literature, and only does so when the author abandons the logomachy of the scholastic divine, and launches out into disquisitions on the nature of law, the structure of society, or the duty of man. The main body of his work belongs to the polemical theology of the day.

If the Elizabethans were non-political in their writings, so too they cannot (Spenser alone excepted) be called romantics —if by romance is meant an imaginative longing for a golden age in the past. Ages of romance are essentially decadent ages, when the present loses its flavour and men turn to the past for their inspiration; but the age of Elizabeth was an age of optimism, of experiment, of constructive achievement. The present was too full of interest for men of letters to fall back on the storehouse of memory, or to indulge in vain regrets for a vanished splendour. They looked forward not backward. They blazed fresh trails and opened up new channels of literary expression which subsequent generations turned into broad, beaten highways. Like pioneers, too, they fumbled and blundered; but their irrepressible exuberance and fertility of mind carried them through to amazing success. In the end they created a great literature worthy of a great people.

Finally it may be observed that the term 'Elizabethan' as applied to literature is not coincident in scope with the beginning and ending of Elizabeth's reign. It connotes a period of time stretching from about the end of the second decade of that reign to almost the closing years of her successor's. The first authentic sign that a great movement was under way in the literary world was the publication of Lyly's *Euphues* and Spenser's *Shepheards Calender* in 1578. Twenty years later, when the queen was already passing into the sear and yellow leaf, the giants of literature began to reach their maturity. When she died the greatest of Shakespeare's plays were still unwritten, and the dramatists who stand nearest to him in the quality of their work—Middleton, Webster, Beaumont and Fletcher, Ford, and Rowley—were still in their 'teens. Ben Jonson had some thirty years of life and work before him, and Bacon was only forty-two. Of the better-known writers only five ended their lives within the confines of the reign, viz. Sidney, Marlowe, Spenser, Nash, and Hooker, and two of them died prematurely—Sidney of a wound received on the battle-field of Zutphen, and Marlowe in a tavern brawl. Clearly, then, we must claim a poetic licence when we speak of Elizabethan literature, and pursue our study to its natural conclusion even if it takes us into Jacobean times. To chop and divide a literary movement in accordance with the political calendar would be neither reasonable nor just, nor would it enable us to appreciate the extent of the debt we owe to the Elizabethans.

It is difficult to describe the manifold literary activities of such a prolific period: a mere catalogue of the authors and their works would almost fill the space at our disposal. Let us begin by noting the salient features of the more ephemeral literature. In the first place, there was a colossal amount of printed matter emanating from, or in some way connected with, the prevailing patriotism and national pride. The self-satisfaction of the individual cast a glamour over his whole environment—his family, his ancestors, the soil of his native land, its history and antiquities. It was as if a search-light had suddenly been turned on history and geography. In consequence of this a continual stream of biographical, historical, and geographical works poured from the press. The labours of Leland, Hall, and Fabyan, which had met the needs of the previous generation in regard to such matters, had to be wrought over again on a grander scale; and a new race of antiquaries, topographers, and chroniclers made their appearance. William Camden followed Leland with a description of England, county by county, borough by borough, noting every natural feature and every point of local antiquarian lore. His *Britannia* and *Remains Concerning Britain* are monuments of patient research and physical toil. Similarly William Lambarde and Richard Carew devoted themselves to descriptive surveys of Kent and Cornwall. Other learned works—Norden's *Speculum Britanniae*, Saxton's *Atlas of England*, Nowell's *Maps of the English Countries*, and Speed's *Theatre of the Empire of Great Britain*—show the awakening interest in geography; while Richard Grafton, John Stow, and

Raphael Holinshed enriched historical knowledge by their chronicles. The last-mentioned work was prefaced by William Harrison's *Description of Britain*, probably the most valuable contemporary account of Elizabethan England, and certainly the best known. Sir John Fortescue's *De Laudibus Angliae*, a notable book of the late fifteenth century, was reprinted several times in order to satisfy the demand of those who desired to be told why the laws of England 'are plainly proved far to excel as well the civil laws of the Empire, as all other laws of the world'. Interest in recent history was carried to a high pitch by Fox's *Book of Martyrs*, Camden's *Annales . . . Regnante Elizabetha*, and Hayward's *Annals of the First Four Years of Queen Elizabeth*; while John Speed's *History of Great Britain* and Samuel Daniel's *History of England* attempted historical surveys on a more extended scale. The best of these historians, Camden and Daniel, were eclipsed in fame by Bacon, whose *History of Henry VII* is generally held to be the most literary work of its kind during the period. But Camden, for all his shortcomings in form and finish, still holds the higher place in the affections of those who value historical research, both for his indefatigable efforts to establish the truth, and his industry in collecting material. Lastly we must note the great 'prose epic' of English exploit—Hakluyt's *Principal Navigations, Voyages, Traffics, and Discoveries of the English Nation*: a work which, in sheer massiveness, stands without a rival in our literature.

Both Camden and Hakluyt believed in what critics describe as an *ouvrage de grande haleine*. Both were confronted with a task which only a profound patriotism and self-effacement could carry through, and both were gifted, or cursed, with the same desire for meticulous accuracy, the same scholarly probity of mind, and the same engaging humility. Camden's story of the struggle with the mass of manuscripts which Lord Burghley placed at his disposal, 'in the rigging and searching whereof I laboured till I sweat, being covered over with dust', is almost an epic in itself. Twice he threw down his pen in dismay at the magnitude of the task before him, and only at the third attempt did he succeed in bringing it to a successful conclusion. It was perhaps natural, after so exhausting travail, that he should dedicate his *Annales* 'to God, my Country, and Posterity, at the Altar of Truth'. Hakluyt, that 'great zealot of the map and the flag', encountered difficulties no less great; but in his case it was the search for documents that wearied him rather than the 'rigging and searching thereof'. 'What restless nights, what painful days, what heat, what cold I have endured; how many long and chargeful journeys I have travelled; how many famous libraries I have searched into; what variety of ancient and modern writers I have pursued . . . albeit thyself canst hardly imagine, yet I by daily experience do find and feel, and some of my entire friends can sufficiently testify.' Only his glorious theme and ardent love of his country sustained him, and 'as it were with a sharp goad provoked me and thrust me forward into this most troublesome and painful action'. By redeeming the exploits of his countrymen from oblivion he wrote all unwittingly a paean to imperial England, by far the most impressive that has ever been written.

But history was written in verse as well as prose, for the popularity of verse at this period was greater. The *Mirror for Magistrates*, a composite work by many hands and based upon Lydgate's *Fall of Princes*, became the acknowledged depositary of England's tragic history. Although of very meagre poetic value and crude in compilation, the *Mirror* was enriched by the pen of Sir Thomas Sackville (afterwards Lord Buckhurst), who wrote the Induction, or Prologue, and set the fashion for a new type of plaintive history. Contrasting with this semi-medieval pageantry of lamentation over the downfall of the great, there appeared the heroic poem, which glorified the achievements of the English race in the present as well as the past. 'Heroical works', writes William Webbe, 'are the princely part of poetry.' Sidney was no less enthusiastic: 'There rests', he remarks in his *Defence of Poesy*, 'the heroical, whose name (I think) should daunt all back-biters.' Thus we have Warner's *Albion's England*, Samuel Daniel's *Wars of the Roses*, and Michael Drayton's *Barons' Wars* and *Poly-Olbion*. The only importance of these works is their contemporary popularity: they reflect the spirit of the age and pander to its desires; but their pompous garrulity had little enduring poetic worth.

Side by side with the awakened interest in history went a movement for the study of the national literature and language. This, too, is understandable as an off-shoot of the self-complacency and national pride of the age, its love of things English, and its desire to glorify the past in the light of the present. Manifestoes, treatises, epistles on orthography and pronunciation, on the merits of rhyme as against classical metres, on the need for purity in the use of the native tongue, on the superiority of English over all other languages, followed one another from the press, as writer after writer lent the weight of his authority to the cause. 'I cannot but wonder', wrote Daniel, 'at the strange presumption of some men that dare so audaciously adventure to introduce any whatsoever foreign words, be they never so strange, and of themselves as it were, without a parliament, without any consent or allowance establish themselves as free denizens in our language.' 'The most English words', said George Gascoign, 'are of one syllable: so that the more monosyllables that you use, the truer Englishman you shall seem.' There was also a widespread desire that the glories of English literature, up to this time treated with sublime indifference by continental peoples, should be made patent to the whole world. 'O that Ocean did not bound our style', sang Drayton,

> Within these strict and narrow limits so:
> But that the melody of our sweet isle
> Might now be heard to Tiber, Arne, and Po:
> That they might know how far Thames doth outgo
> The music of declined Italy.

National pride did not, however, turn the eyes of the Elizabethans entirely inward. If they gloried in their own literature and history, they appropriated to themselves the history and literature of other peoples, both ancient and modern. Translators looted the classics and presented their compatriots with Homer, Herodotus, Aesop, Aristotle, Demosthenes, Plutarch, Theocritus, Ovid, Caesar, Cicero, Terence, Seneca, Lucan, Quintus Curtius, Sallust, Pliny, and many others in English dress. Some of the translations came through the medium of French, for instance North's *Plutarch*; but there were many editions of classical writers in the original. Modern French literature was well represented by John Florio's *Montaigne* (usually regarded along with North's *Plutarch*, Chapman's *Homer*, and Sylvester's *Du Bartas* as the most successful of contemporary translations), and English versions of the writings of La Noue, Estienne, La Primaudaye, Commynes; while Ronsard and the poets of the Pléiade were plundered wholesale by the sonneteers. Spanish literature also found its translators, who introduced their countrymen to *Lazarillo de Tormes*, Guevara's *Dial of Princes*, the

Diana of Montemayor, and medieval romances such as *Palme-rin* and *Amadis*. But the foreign literature which commanded the greatest popularity in English garb was that of Italy. Boc-caccio, Tasso, Ariosto, Castiglione, Machiavelli[1] were trans-lated either wholly or in part, and there was a constant demand for the *novellieri*—Straparola, Cinthio, Bandello—whose tales furnished the dramatists with matter for their comedies. Alto-gether the influx of foreign literature and its consumption by the Elizabethans must have been on a truly vast scale, and the ramifications of its influence on the native literature is a chap-ter which still remains to be completed.

So far we have only touched the fringe of Elizabethan litera-ture. When we enter into the heart of the subject, the luxuriance and variety make the task of selection difficult. The muse of poetry tried her hand with every species of metrical composi-tion except the epic, whose place was no doubt taken by the drama. The lyric, the ode, the madrigal, the ballad, and the sonnet were all brought to a high pitch of perfection which has never been surpassed. Erotic verse, of course, predominated, but not to the exclusion of other types; and even within the eroticism of the age it is possible to mark a distinction between the sensuousness and sensuality of poems like Marlowe's *Hero and Leander*, Shakespeare's *Venus and Adonis*, or Marston's *Pygmalion*, and Spenser's chaste imagery in his *Amoretti* or Sid-ney's in the sonnet sequence *Astrophel and Stella*. Hymns to heavenly love and spiritual sonnets were not wanting, but religious poetry in the strict sense of the word was conspicu-ously absent from the front rank of literature. The only poem that gained subsequent renown in this direction was Robert Southwell's *Burning Babe*, the creation of a martyred jesuit who saw his 'vision splendid' behind the bars of a state prison.

On the whole it is probably true that much of the poetry of the period sprang from the fancy rather than from the depths of the soul's experience. It is certainly true of the sonnet, the vogue of which was second only to that of the drama. Every one who aspired to poetic honours experimented with this easily mastered and more or less stereotyped form of erotic verse. The Dianas, Delias, Phyllises, Corinnas, Auroras, and other goddesses of the sonneteers were more often than not mere figments of the fancy; and the feigned emotion which

[1] Only the *Arte della Guerra*.

they gave rise to palls by its monotony and conventionality. Indeed it may be said that only Sidney, Spenser, and Shakespeare embodied a real experience in their sonnets, and in the case of Shakespeare it is still impossible to determine with any degree of accuracy the identity of the celebrated 'dark lady' on whom the poet lavishes his affection. Nevertheless, in sonnet literature as in dramatic, Shakespeare undoubtedly stands out as the greatest genius of his time.

It is perhaps permissible to speak with more warmth of the ballad; for the ballad, being less artificial, less ornate, and more popular in its topics and composition, became the vehicle for a simpler and truer emotion that appealed directly to all who could not appreciate the intricacies and subtleties of erotic verse, but loved to hear of the exploits and sufferings of familiar heroes—Robin Hood, Clymme of the Clough, Sir Patrick Spens, Edom o' Gordon, William of Cloudesley, Percy and Douglas, the Earl of Essex, &c.[1] Everything calculated to move the heart of the multitude was grist to the mill of the ballad-writer, and his range of subject was practically infinite. Even men of letters and scholars could avow themselves lovers of this lowly literature. 'I never heard the old song of Percy and Douglas', wrote Sidney in an oft-quoted passage, 'that I found not my heart moved more than with a trumpet; and yet it is sung but by some blind crowder, with no rougher voice than rude style.'

If we think only of pure poetry, and exclude the drama, the greatest exponent of the art which the age produced was the somewhat sombre and melancholy Edmund Spenser (1552?-99). He stands apart from all his contemporaries, not merely because he was the first to reveal poetic beauty to his generation in *The Shepheards Calender* (1579), but also because of the unique character of his genius, and his authorship of the *Faerie Queene*, the most remarkable poem of the age, and 'the only long poem in the language that a lover of poetry can sincerely wish longer'. The source of Spenser's poetry may be found in the impact of his idealism on the crude realities of his material environment. He was a disillusioned man, who consistently aimed at political power and just as consistently failed to attain it. All through life he hovered between satire and pessimism, between veiled attacks on the corruptions and backslidings of

---

[1] For a full account of the subject, which is voluminous, see Sir Charles Firth's chapter on Ballads and Broadsides in *Shakespeare's England*, vol. ii.

the age and lamentations on its degeneracy. Probably no great writer ever found himself less in sympathy with the world in which he lived. In spite of the fact that his lot was cast in the most expansive period in English history, Spenser was essentially a *laudator temporis acti*. His philosophy is largely summed up in the derivation he gives of 'world', namely 'warre-old', or that which gets worse as it grows old. Obsessed by the evils, the misery, and the apparent confusion of contemporary society, he seems to have turned his back on its immense potentialities for good, and to have taken refuge more and more in a platonic world of his own creation, where the love of beauty is equivalent with the love of God, where sensuous desire and spiritual aspiration are one, and where men may 'follow their instincts and call their joys by the same name as their duties'.

Spenser's more important works can be grouped as follows: his satires—*The Shepheards Calender*, *Mother Hubberd's Tale*, the *Complaints*, and *Colin Clout's Come Home Again*: his erotic verse —the *Amoretti*, the *Epithalamion*, and the four platonic *Hymns* to love and beauty: the *Faerie Queene*; and his prose pamphlet entitled *A View of the Present State of Ireland*.

Of all these works, as well as of others which we have not mentioned, the literary critic has much to say, both in regard to the matter and the form. Here, however, we must limit ourselves to a very brief consideration of the *Faerie Queene*, the poem into which Spenser distilled more of his personality, art, and technique than may be found in all his other poems put together. Critics will always be in doubt as to the meaning or purpose of this strangely exotic romance, with its medieval setting, its allegorical figures, and its classical mythology. The author himself conceived of it as a didactic poem, designed 'to fashion a gentleman or noble person in virtuous and gentle discipline'. But there is no trace in the poem of this didactic element, nor is the allegory obtrusive enough to spoil the story. What we get, in effect, is a renaissance version of the old Arthurian romance, shot through with veiled allusions to contemporary personalities in the political world, and transacted against the background of an idealized Irish landscape. All these disparate factors are woven together by the magic of the poet's genius into verse of superb beauty which seems to fit the theme like a glove. 'We hear a slow music', says M. Legouis, 'whose perpetual return rocks the mind and sways it from out the real world into a world of harmony and order, of which it seems to be the natural rhythm.' The Spenserian stanza, in fact, is the 'regulative clock' of fairyland.

It may seem an abrupt transition to pass at one bound from the ethereal heights of Spenser's verse to the sober levels of Elizabethan prose. But the development of prose as an instrument of literary expression must be regarded as one of the minor triumphs of the age, and the writers who took part in it, although they erected no 'star-ypointing pyramid', and were, for the most part, overshadowed by the genius of those who indulged in the more popular art, are nevertheless worthy to be remembered. Conventionally the study of Elizabethan prose begins with the publication of John Lyly's *Euphues* (1579) and *Euphues His England* (1580). These books, crammed with far-fetched similes derived from medieval bestiaries, herbaries, and lapidaries, with their tricks of alliteration, prolixities, and pedantries, were merely academic exercises in the ornate—notable only for their novelty of form and elegant moral tone. Lyly was a moralist who set himself the task of teaching the young; but the moral of his tale was of less importance than the studied graces of the style he affected. Much the same might be said of Sir Philip Sidney's pastoral story of love and chivalry, the *Arcadia*, which was written about the same time, although there is more spontaneity in his heavily jewelled prose than in the ponderous conceits of *Euphues*. So infectious, however, was euphuism as a cult that it spread like an epidemic, becoming for many years a kind of universal language of the literary world. In this vein Robert Greene wrote his romantic novels *Mamilia*, *Menaphon*, and *Pandosto*, and Thomas Lodge his *Rosalinde*. Both authors interspersed their prose with exquisite lyrics, and Lodge displayed a remarkable talent for witty dialogue; but the main interest attaching to them is that Greene's *Pandosto* supplied Shakespeare with his plot for *A Winter's Tale*, and Lodge gave him both plot and chief character for *As You Like It*. Clearly writers had not yet realized that the function of prose is not to follow in the eccentric footsteps of poetry or to charm the imagination, but to instruct the reason, and to describe real things with precision, simplicity, and restraint. Lodge probably had some distant glimmering of the truth when he tried to lend verisimilitude to his narrative by com-

posing it, not in the study but at sea, 'where every line was
wet with a surge, and every humorous passion counter-checked
with a storm'. But it was Greene who first broke new ground
in the right direction with his coney-catching tracts (1591–2).
Here at least was reality of a sort—the reality of London's
underworld, with its thieves, swindlers, and loose women—of
which the dissolute, tavern-haunting author could speak with
all the vigour and authority of an eye-witness. To the same
class belong the writings of Thomas Dekker, the author of the
Gul's Hornbook, the *Seven Deadly Sins of London*, and many other
pictures and sketches of London life, both grave and gay,
macabre and humorous. With Thomas Deloney we enter yet
another field of descriptive prose; for Deloney's novels, *Jack of
Newbury* and *The Gentle Craft*, take us into the craftsman's
world, among the weavers and cordwainers, and give us intimate pictures of industrial life in the capital. Greene, Dekker,
and Deloney led English prose along the path that culminated
a century later in the writings of Swift, Addison, Steele, and
Defoe. Meanwhile Thomas Nash, a disciple of Greene, had
turned his mordant wit to effective use in the acrimonious disputes of the period, particularly the Martin Marprelate controversy; and pamphlet after pamphlet flowed from his sardonic
pen with the force and velocity of an avalanche. In spite of his
grotesque style and frequent lapses in taste, no one did so much
for the development of Elizabethan prose as the 'frantic' Nash.
He was more than a controversialist. He castigated the superstitions and absurdities of his generation with the satire of a
Juvenal, or mocked at them with the humour of a Rabelais. He
scarified the super-pedant of the age, Gabriel Harvey, tilted at
Marlowe's 'spacious volubility', wrote a picaresque novel (*The
Unfortunate Traveller, or The Life of Jacke Wilton*) to counteract
the vogue of euphuism, and produced perhaps the best burlesque
of the period—*Lenten Stuff*, in which he glorified the red herring,
the source of Yarmouth's wealth.

Although prose found its descriptive powers and potentialities in the hands of realists and satirists, there were writers who
applied it with telling effect in other directions. Sidney's *Defence
of Poesy*, Daniel's *Defence of Rime*, and the writings of William
Webbe are illustrations of its use in literary criticism. Roger
Ascham wrote it brilliantly in his *Scholemaster*, Spenser even
more brilliantly in his *View of the Present State of Ireland*, the most

valuable contemporary analysis of Irish society in the sixteenth century. Sir Thomas Overbury, John Stephens, and John Earle found it the best medium for their 'character' studies. Dramatists adopted it for many of their comic scenes and less exalted personages. Shakespeare wrote it like a master when occasion served, and Ben Jonson employed it entirely in two of his plays. Finally it reached its highest pitch of perfection in Hooker's *Ecclesiastical Polity*, Hakluyt's *Voyages*, the *Essays* of Bacon, and the *Authorized Version* of the Bible. It is a notable fact, however, that when Bacon wrote his great philosophical work, the *Novum Organum*, he reverted to Latin, as being the more suitable medium for the expression of exact ideas. 'These modern languages', he remarked, 'will at one time or other play the bankrupts with books.'

There remains the drama, the crowning achievement of Elizabethan England in the field of literature. More so than any other literary creation the drama gathered up and expressed the emotional and intellectual life of the age in all its length, breadth, height, and depth. The stage play, in fact, became a truly national cult. The speed with which the movement progressed is one of its minor surprises. At the beginning of Elizabeth's reign the drama was still struggling to free itself from the worn-out morality and miracle plays of the middle ages: there were no public theatres, and actors were treated as vagabonds by the authorities. Until 1580 dramatic literature could be comprised in a single slender volume, and consisted of academic and scholastic productions like Nicholas Udall's *Ralph Roister Doister*, a comedy based on Plautus; Sackville and Norton's *Gorboduc*, a tragedy in the manner of Seneca; episodical plays with classical themes like *Damon and Pythias*, *Appius and Virginia*, and *Cambyses*; and a roaring farce, *Gammer Gurton's Needle*, which resembled in content the interlude of the old miracle play. All these plays—*Gammer Gurton's Needle* alone excepted—were designed for select audiences at school, university, or inns of court. But the foundation of the first actors' company under the patronage of the earl of Leicester, in 1574, and the erection of the first public theatre, in 1576, indicated that a new age was dawning. Then came the court drama of John Lyly (1581–90)—*Campaspe, Sapho and Phao, Endymion,* and *Midas*—followed by George Peele's *Arraignment of Paris* and *David and Bethsabe*. But these writers with their dainty rhymes,

conceits, and flatteries of the queen, which formed the substance of their plays, were quickly swept aside by Thomas Kyd,
Christopher Marlowe, and the unknown author of *Arden of
Feversham*, who made plain for the first time the immense possibilities of the stage play as an instrument of popular amusement,
and a sounding-board for the deeper emotions of the period.
Kyd's *Spanish Tragedy* (1586) is a drama of the horrific type, a
crude study in realism photographed from life, which makes
the flesh creep with its passion and barbarity. In *Arden of
Feversham* the crime *motif* again dominates the play; but in this
case the author is interested in the psychological accompaniments of the crime rather than the crime itself. Prior to the
appearance of the Shakespearian plays, *Arden* is the best psychological drama of the period. But it was Marlowe who invented
the true medium of the Elizabethan drama. His 'drumming
decasyllabon' and 'bragging blank verse', coupled with the spirit
of revolt and defiance which animated all his plays, took the
public by storm, and created a tradition which his successors,
including Shakespeare, adopted and perfected. More than any
of his contemporaries Marlowe displays the *terribilità* of the
Italian renaissance. There is a massive simplicity and grandeur
in the structure of his dramas, a singleness of aim and concentration of purpose that lift them far above the crude realism
of Kyd. If at times he blunders into the grotesque in his effort
to obtain effect, he more than redeems himself by passages of
exquisite beauty and pathos. His heroes and villains are slaves
to the 'will to power'—the power of military might, in *Tamburlaine* (1586), the power of knowledge, in *Doctor Faustus* (1588),
the power of wealth, in the *Jew of Malta* (1589), and the power
of cunning, in Mortimer (*Edward II*, 1592). It is highly probable
that when Marlowe died prematurely, in 1592, he had already
revealed the full scope of his genius: he certainly could have
added nothing to the quality of his work, though he would
undoubtedly have increased its quantity. He was a man of
intense but limited vision, a fore-runner with a message preceding the true incarnation of dramatic power—Shakespeare.

All the playwrights before Shakespeare were university men,
as, indeed, were the poets and other writers of the period. But
Shakespeare was a discovery of the despised actor community,
whose phenomenal rise to the dignity of authorship caused no
small jealousy among the professed masters of the dramatic

art. Greene alluded to him as 'an upstart crow beautified with our feathers', who, with his 'tiger's heart wrapped in a player's hide', imagined himself to be 'the only Shake-scene in the country'. At the time Greene penned his indictment the new playwright had nothing to his credit except *Love's Labour's Lost*, the *Comedy of Errors*, the *Two Gentlemen of Verona*, and three parts of *Henry VI*—none of which displayed much genius or originality, and certainly gave little inkling of the robust vitality that was to follow. If Shakespeare had died, like Marlowe, at the age of twenty-nine his fame would have been entirely overshadowed by that of the author of *Tamburlaine*. But by a curious coincidence chance and circumstance combined to eliminate all rivals from his path, when most he needed a free field for the development of his genius. Marlowe and Greene died in 1592, Kyd in 1594; Lodge gave up letters for medicine; Lyly ceased work; and Peele plunged into dissipation. In fact, from 1592 to practically the close of the century Shakespeare had the stage to himself, and established an unassailable reputation as the greatest living playwright. It was during this period that he wrote the bulk of his historical plays, *King John, Richard II, Richard III, Henry IV*, and *Henry V*; the fairy play, *A Midsummer-Night's Dream*; the lyrical tragedy, *Romeo and Juliet*; and all the romantic comedies, the *Merchant of Venice, Much Ado about Nothing, As You Like It, Twelfth Night*, the *Merry Wives of Windsor*, and the *Taming of the Shrew*. Then came the second and greatest phase of his activity (1601–8), when a pessimistic strain mingled in his work, and the note of tragedy prevailed. It was the period of the gloomy comedies, *Measure for Measure, All's Well that Ends Well, Troilus and Cressida*, of the profound tragedies, *Hamlet, Othello, Lear, Macbeth*, and *Timon of Athens*, and of the Roman plays, *Julius Caesar, Antony and Cleopatra*, and *Coriolanus*. The final phase (1608–16) saw the tragic vein mellow into the romance of *Pericles, Cymbeline, The Winter's Tale*, and *The Tempest*. Thus Shakespeare may be said to have completed the cycle of human experience, passing from the fanciful plays of his youth to the tragedy of his mature years, and ending on a note of disillusionment, tolerance, and romance. But it will always be a disputable point how far this evolution was conditioned by his environment, and how far it was the natural development of his mind; for the life of Shakespeare apart from his art is wrapped in much mystery,

and the attempted reconstructions of the interplay of environment and genius are usually as futile as they are fanciful.

The greatness of Shakespeare rests upon the extraordinary
range of his gifts. His power of characterization, objectivity,
inventiveness, and universality of appeal are all unique. Unlike Lyly, Kyd, Marlowe, or even Ben Jonson, he had no
mannerisms that could be copied. He wrote lyrical, historical,
tragical, comic, romantic, and fairy plays with equal facility
and equal distinction; and a reference to the chronology of his
works will show that he could turn from one genre to another,
even within the same period, without any apparent effort. It
has been said with truth that there are more immortal characters in one of his greatest plays than in all the plays of his
contemporaries taken together. When he took up the role of
historiographer he not only revived the past, he recreated it;
and the lineaments of the historical personages that throng his
gallery stand out with greater permanence than reality itself.
So strong was the impress of his genius on national history that
palpably false creations like his Richard III persist in defiance
of historical research; and a great critic, Froude, could aver:
'The most perfect English history which exists is to be found, in
my opinion, in the historical plays of Shakespeare.' In addition
to all this the felicity of his diction, imagery, and thought has
made his plays the mightiest store-house of aphorisms in the
English language. In short, we might apply to him the words,
slightly modified, spoken by Enobarbus of Cleopatra: 'Age
cannot wither nor custom stale *his* infinite variety.'

Strictly speaking, Ben Jonson belongs to the generation later
than Shakespeare, and ought to be considered in conjunction
with Dekker, Marston, Middleton, Fletcher, Tourneur, Webster, and Heywood. But his marked individuality and versatility bring him into close juxtaposition with Shakespeare, from
whom, however, he differed profoundly in his whole conception and practice of the dramatic art. Without doubt Jonson
was the most learned of the Elizabethan playwrights. His
knowledge of the classics was profound, and his reverence for
the dramatic proprieties, as understood by the Greeks, amounted
almost to fanaticism. In the technique of plot-construction as
well as in accuracy of detail he far surpassed Shakespeare,
whose carelessness in such matters is proverbial. This is particularly noticeable in Jonson's Roman plays, *Sejanus* and

Catiline, where he indulges in a thousand touches, derived from extensive reading in Latin literature, which Shakespeare could never aspire to. But in spite of his wider knowledge of literature and greater facilities, Jonson failed just where Shakespeare achieved his greatest success, namely in characterization. All his characters are caricatures rather than living persons. He was interested in human eccentricities, or, as he would put it, in men's 'humours'—a feature of his work which is strikingly displayed in two of his best-known comedies, *Every Man in His Humour* and *Every Man out of His Humour.* It is but a short step from exposing men's eccentricities to satirizing their follies; and Jonson, as his powers matured, rapidly became the greatest master of satirical comedy in his time. To this category belong *Volpone, or the Fox* (1605), *Epicene, or the Silent Woman* (1609), *The Alchemist* (1610), and *Bartholomew Fair* (1614), all of which, in plot, character, and incident, are entirely creations of the author's own brain, and results of his observation of contemporary society. Never again did he reach the heights attained in these four plays: they stand, says one critic, among the most remarkable dramatic works of the English renaissance.

But Jonson was more than a great playwright. Some of the finest lyrics of the period are from his pen; and as a producer of masques he was without a rival. His worst fault was his egotism and aggressive self-assertion, which drew him into controversy with several of his contemporaries, and filled his life with bitterness and contention. This aspect of his character and career is startlingly reflected in his play *The Poetaster,* in which he apotheosizes himself as the Horace of the age.

Of the plays written by other authors during the great period of the Shakespearian drama and subsequently, special mention may be made of the following: George Chapman's *Tragedy of Biron* (1608), Thomas Dekker's *Shoemaker's Holiday* (1599), *Old Fortunatus* (1600), and *Honest Whore* (1604), Thomas Heywood's *A Woman Killed with Kindness* (1603), *Eastward Ho!* by Marston, Chapman, and Jonson (1605), Thomas Middleton's *Witch* (1622), Cyril Tourneur's *Revenger's Tragedy* and *Atheist's Tragedy* (1607 and 1611), John Webster's *White Devil* (1611) and *Duchess of Malfi* (1614), Beaumont and Fletcher's *Knight of the Burning Pestle* (1609) and *Maid's Tragedy* (1611). Few of these plays failed to reach a high level of dramatic art, and some are worthy to be compared with Shakespeare's; but none of the authors mentioned was capable of the sustained brilliance and profound thought that characterized the great dramatist at his best. It was as if Shakespeare's genius had been split into 'fragments and distributed. Dekker could create living characters, Heywood was a master in pathos, Webster was a specialist in the morbid and the *macabre*, and Beaumont and Fletcher were experts in stage-craft and scenic effects. But the combination of qualities that resulted in the production of great literature as well as drama was the prerogative of Shakespeare alone.

It is not necessary to carry the study of Elizabethan literature farther. The age of great poetry was drawing to its close with the 'metaphysical' verse of John Donne and the Spenserian imitations of George Wither, William Browne, Giles and Phineas Fletcher, and Drummond of Hawthornden. The drama, though still robust, was entering upon its silver age with the plays of Philip Massinger, John Forde, and William Rowley. In every direction save prose, literature was losing touch with the lively, ebullient spirit of Elizabethan England. It was perhaps a sign of the times; for the shadow of puritanism was already cast over the land.

If Elizabethan literature developed a truly national tradition, the same thing cannot be said, except with considerable reservations, of the fine arts. There were plenty of collectors of pictures among the nobility, like the earl of Leicester and Lord Lumley, whose galleries were famous; and Holbein had created a vogue for portraits. It was also the fashion, according to Harrison, for the walls of houses to be hung with tapestry, arras work, and 'painted cloths', 'wherein divers histories, herbs, beasts, knots, and such like are stained'; and there were art shops in Blackfriars, Fleet street, the Strand, and especially Robert Peake's in Holborn, where pictures could be purchased. But the artists of the day were mostly foreigners trained in the great schools of Flanders and Holland, who had migrated to England to escape the severities of Alva's rule. Among these foreign artists three stand out as particularly noteworthy for their excellence in portraiture, viz. Marcus Gheeraerts of Bruges, Paul van Somer of Antwerp, and Daniel Mytens of The Hague. The only English-born painter of note was Cornelius Johnson (or Jansen), and he, too, was of Netherlandish descent, though born in London. Except perhaps in miniature work, with which the names of Nicholas Hilliard and Isaac Oliver are associated, there was no native school of painting. The humble art of copper-plate engraving, either for book illustration or map-making, was in a similar position, largely dependent upon the stimulus from abroad; but there were some English engravers, like Augustine Ryther, the map engraver, and William Rogers and Thomas Cockson, portrait engravers, who made a name for themselves. Probably the best known, if not the most famous, of the engravers who 'devilled' for the bookseller were the Droeshout brothers, Martin and John, members of a refugee family from Brussels, who gained notoriety from the fact that Martin was responsible for the portrait of Shakespeare which adorned the frontispiece of the First Folio.

In regard to sculpture and architecture the same general criticism applies. It is more than doubtful whether England possessed a single creative genius of outstanding merit fit to be compared with the great masters of the Continent. The sculptors and tomb-makers of Southwark were in the main either Flemings or Germans, or of Flemish or German extraction, and the only vast public building erected during the reign, viz. Gresham's Royal Exchange, begun in 1566 and completed in 1570, was the work of a Flemish architect, Henri de Pas (or Paschen), who was responsible for the Hôtel des villes hanséatiques at Antwerp. The fact that Gresham employed not only a foreign designer but also foreign workmen, and imported the marble for paving the piazza of the Exchange —an action for which he was severely criticized by contemporaries—would seem to suggest that English craftsmen had not yet mastered the higher technique of the renaissance. As a matter of fact it merely shows that the 'classical' or 'italianate' architecture, that was all the rage abroad, was not fully 'received' into England. Gresham's building was intended to be an imitation of the italianate Bourse at Antwerp, and it was natural that he should make use of Flemings in its construction.

When we turn, however, from public buildings to private, and consider the domestic architecture of the period, a different picture presents itself. The glory of Elizabethan England lies depicted for us in the magnificent dwelling-houses, the 'cloud-capp'd towers' and 'gorgeous palaces' of the rich, and the humbler but no less stately manor houses of the gentry, which were now rising in great profusion all over the country. Here something like a blend was effected between classical models and traditional English design, and the architects or 'surveyors', as they styled themselves, were altogether English, albeit we know very little about their careers and individual achievements. Two of the most famous were John Thorpe, who constructed Kirby Hall, Northamptonshire, and Robert Smithson, the designer of Wollaton Hall, near Nottingham. The general trend of their work has been described as 'a gothic framework with a classical overlay'. It may not have been a very happy combination from a purist point of view. There was a touch of the exotic, not to say bizarre, in the balustraded and ornamental parapets, the fantastic gables, the pilastered, columned, and entablatured fronts, and many other features which the more pretentious houses affected. Like the dress of the period, architecture of this sort must have struck observers as outlandish. This, however, was not the opinion of the enthusiastic Harrison, who writes with obvious pride: 'If ever curious building did flourish in England, it is in these our days, wherein our workmen excel and are in manner comparable with old Vitruvius, Leo Baptista, and Serlo'.

It was perhaps natural that an age when men were rapidly emancipating themselves from medieval habits of living and had not yet adapted themselves to the new conditions should indulge in extravagances. The old militarized architecture of feudal England, when a man's home was his castle, was out of date in a society whose peace and security were daily becoming more firmly established. The 'halls', 'courts', and 'places' of Elizabethan England—the very names are significant —had no use for moats, drawbridges, portcullises, machicolations, stanchioned windows, and the other relics of a barbarous past, except for decorative purposes. The cry now was for comfort, symmetry, and light—especially light. Rooms were multiplied, galleries introduced for recreational purposes, windows enlarged into bays and oriels, and piled one on the top of another so as to form glittering façades of glass. So great, indeed, was the craze for illumination that Bacon remarked: 'You shall sometimes have fair houses so full of glass that one cannot tell where to be come to be out of the sun or cold.' As an antidote to over-illumination he advocated the building of 'double houses' (i.e. rooms back to back), so 'that you may have rooms *from* the sun, both for forenoon and afternoon'.

In spite, however, of the demand for light, the Elizabethans do not appear to have favoured a southern exposure, as we moderns do. On the contrary there is evidence to show that they preferred to site their houses so as to avoid it. The south wind was the English mistral of the sixteenth century. 'A south west blow on ye and blister you all o'er!' was Caliban's curse on Prospero. Thomas Tusser, the gardening expert, wrote:

> The south as unkind draweth sickness near:
> The north as a friend maketh all again clear.

And the well-known hygienist of the age, Andrew Boorde, in his advice to builders, urged them to 'order and edify' the house so that the main prospect might be east and west, preferably north-east, 'for the south wind doth corrupt and doth make evil vapours'. The east, on the other hand, is 'temperate, fresh, and fragrant', and the north 'purgeth all vapours'. Wherefore, he concluded, it were better that windows should 'open plain north than plain south'.

Opinion seems to be unanimous that the interpenetration of English architecture by foreign influences during Elizabeth's reign did not fundamentally affect its character. The typical dwelling-house of the period was neither Italian, nor French, nor Flemish, nor German: it was distinctively English. We know it by certain well-marked traits—its high-pitched gables, wreathed chimneys, mullioned and transomed windows with leaded lights, heraldic devices over the doorways, oak-panelled rooms, and central hall with open timbered roof. Sometimes it was constructed of stone, sometimes of timber, sometimes of half-timber with brick filling between the wooden uprights. Local materials were necessarily used in construction because of the difficulty of transport. In the Gloucestershire Cotswolds, in the valley of the Nene in Northamptonshire, and in Yorkshire and Derbyshire, where stone was plentiful, this was the material chiefly used; but elsewhere the preference was for the timbered or half-timbered house, with a slight 'overhang', the upper story projecting beyond the lower. Even in stony Derbyshire this type was to be found side by side with the more fashionable stone structure. Finally we may note that the dependence upon materials locally available made possible the perfect harmonizing

of the Elizabethan dwelling with the surrounding landscape—
one of the first principles of artistic architecture—and at the
same time contributed greatly to the maintenance of conser-
vatism and the national tradition in construction.

The position of music was slightly different. Here England
could boast a galaxy of talent, indisputably national in character, and of a high order. This is all the more remarkable in
that the Elizabethan composers had no longer the elaborate
ritual of the Roman church to stimulate them to the noblest
flights of musical genius, but were compelled to cultivate their
art mainly for secular purposes. Even so, however, there were
several, notably William Byrd, Thomas Tallis, Christopher
Tye, and Orlando Gibbons, whose work in the field of sacred
music compares not unfavourably with that of the more famous
masters abroad. The first three were catholics, the last an
anglican. Byrd enjoyed, even in his own day, the reputation
of being the *homo mirabilis* of the musical world. He towered
above his contemporaries, both in the quality and quantity of
his compositions, much in the same way as Shakespeare out-
distanced all rivals in the realm of letters. Depth of feeling,
spontaneity, the perfect blending of musical form with the
sense it is intended to convey, constitute the essence of his
title to greatness. 'There is', he remarks, in the first book of
the *Gradualia*, 'a certain hidden power in the thoughts underlying the words themselves, so that as one meditates upon the
sacred words and constantly and seriously considers them, the
right notes, in some inexplicable fashion, suggest themselves
quite spontaneously.' The secret could not be better expressed;
but it remained a secret, even from himself. Byrd was a voluminous composer, and all his works have not yet been published;
but it is now generally admitted that the *Gradualia*, *Cantiones
Sacrae*, and the four *Masses* place him in a category by himself
as the equal of Palestrina.

If, however, we would see the true spirit of the time reflected
in its music, we must turn from the sacred to the secular, from
the composers of solemn organ music to the song-writers; for
it is in the lyric, the ballad and the madrigal, that the soul of
'Merrie England' resides. Byrd could write in both styles, but
the strongly religious bias of his temperament restricted his
success in lighter music. Here the honours lie with Thomas
Morley, John Wilbye, Thomas Weelkes, John Dowland, Thomas

Campion, and John Danyel. Morley, Wilbye, and Weelkes excelled in the madrigal, a species of composition resembling the modern part-song, generally set for three or more voices, polyphonic in character, and unaccompanied by any instrument. Dowland, Campion, and Danyel, on the other hand, were writers of solo songs to the accompaniment of the lute. Dowland's 'ayres' are said to represent the high-water mark of English song-writing. On the one side they hark back to the days of the troubadour; on the other they forecast the song of to-day. Nothing more beautiful in the elegiac vein has ever been written than his 'In darkness let me dwell'; and his contemporaries did no more than justice to his genius when they referred to him as one

> . . . whose heavenly touch
> Upon the lute doth ravish human sense.

In the realm of physical science the age of Elizabeth coincides with the beginnings of a great revolution in human thought, traceable, in the first instance, to the Polish astronomer Copernicus, who introduced the scientific world to a new cosmic system based upon observation, calculation, and deduction. Copernicus was, in fact, the Columbus of the heavens. His investigations into the movements of celestial bodies, which culminated in the publication, in 1543, of his *De Revolutionibus Orbium Celestium*, exercised an influence on astronomical thought comparable to that which the unveiling of the Atlantic and the discovery of America exercised on geography. For over fourteen hundred years the *Almagest* of Ptolemy, with its conception of a geocentric universe, in which the sun, moon, planets, and stars revolved in their separate concentric 'spheres' (or crystalline envelopes) round the earth, had been the canonical book of astronomers. Copernicus replaced this erroneous thesis with a new solar system of astronomy, wherein the planets, including the earth, revolved about the sun; and he coupled with this an equally revolutionary theory of the rotation of the earth on its own axis, and of the planets, like the earth, in their orbits. He regarded his system as merely an hypothesis; and an hypothesis it remained until the appearance of Kepler's epoch-making treatise *De Motibus Stellae Martis* (1609) and Galileo's *Sidereus Nuncius* (1610). But it spread with great rapidity over Europe, and gained a number of supporters in Tudor England. Robert

Recorde, the author of *The Castle of Knowledge* (1556), was familiar with it, and another English astronomical scholar, John Field, avowed himself a convinced Copernican in his *Ephemeris anni 1557 currentis juxta Copernici et Reinhaldi Canones*, published in the following year. In Elizabeth's reign the vogue of the new astronomy increased, and a coterie of enthusiastic Copernicans made its appearance in Dr. John Dee, Thomas Digges, William Gilbert, and Thomas Harriot, who not only corresponded with foreign astronomers and sometimes visited them, but contributed not a little by their researches to the advancement of the science. Harriot was the friend and collaborator of the great Kepler himself; Digges was in touch with Tycho Brahe, Kepler's immediate precursor and the learned author of an important work on comets; and Gilbert published a remarkable book 'on the magnet and magnetic bodies and especially the great magnet of the earth, &c.' (1600),[1] which has been classed as one of the major books on physical science of the century. It would therefore appear that English scientists were fully abreast of the continental development. In one particular direction, indeed, they may have been actually in advance of it; for they possessed an instrument, referred to at the time as 'perspective glasses', which, in principle at least, anticipated the invention of the telescope.

It is important to note, however, that knowledge of the scientific progress of the period was confined to a very small group of enthusiasts. The lay mind was slow to disabuse itself of the Ptolemaic cosmogony, the very nomenclature of which had become embedded in the language of poetry, philosophy, and everyday life. 'Come, Mephistophiles, let us dispute again, and argue of divine Astrology,' says Dr. Faustus in Marlowe's play, and Mephistophiles responds with a disquisition couched in the best Ptolemaic manner. Shakespeare, again, echoes Pythagoras when he speaks of the 'music of the spheres', and pays homage to the old order in the line, 'And certain stars shot madly from their spheres.' Hooker likewise accepts it when he writes: 'If celestial spheres should forget their wonted motions, and by irregular volubility turn themselves any way ... what would become of man himself, whom all these things do now serve?' In short, the conventional heaven of the layman remained unaffected by science, not only during the reign

[1] *De Magnete magneticisque corporibus, et de Magno Magnete Tellure, Physiologia nova.*

of Elizabeth but for a long time afterwards. It was not until the 'age of reason' dawned that the Ptolemaic imagery was finally eradicated.

Side by side with this persistent belief in the old cosmic order there existed in Elizabethan England an inveterate conviction—fostered by the pseudo-science of astrology—that the heavenly bodies exercised a profound influence on human affairs. Eclipses were harbingers of disaster: comets were associated with the deaths of illustrious persons. 'When beggars die', says Calpurnia in *Julius Caesar*, 'are no comets seen: the heavens themselves blaze forth the deaths of princes.' The twelve 'signs' of the zodiac and the seven planets fixed men's natures and fates, and each part of the human body was regarded as governed by some particular 'sign' or planet. Aries governed the head and face, Taurus the neck and throat, Gemini the shoulders, arms, and hands, Leo the back and heart, Cancer the breast, stomach, and lungs, Libra the reins and loins, Sagittarius the thighs, Capricornus the knees, Aquarius the legs, Pisces the feet, &c. The prognostication of events by the movement of heavenly bodies, the decision as to favourable moments for great public functions, and the drawing of individual horoscopes were, of course, the province of the astrologer or astrologaster, who, accordingly, occupied a far more important place in society than the bona fide observer of celestial phenomena. He was the 'applied scientist' of the age, who linked up the occult influences of the heavens with the daily life of humanity.

Kepler regarded the whole subject of astrology with a tolerant contempt as the 'foolish daughter of a wise mother', but admitted the obvious truth that without the help of her foolish daughter the mother could not have lived. On the other hand it is important to realize that although astrology was deeply rooted in the popular mind, the common sense of the age was beginning to revolt against its absurdities. 'It standeth upon nothing else', wrote Philip Stubbs, 'but mere conjectures, supposals, likelihoods, guesses, observations of times and seasons, conjunctions of signs stars, and planets, with their aspects and occurrents, and the like, and not upon any certain ground, knowledge, or truth, either of God, or of natural reason'; and again: 'It is the malice of the devil, the corruption of our own nature, and the wickedness of our own hearts that draweth us to evil, and so to shameful destinies, and infamous ends, and not the stars or planets.' Cassius made short work of the pseudo-science in his well-known remark to Brutus:

> The fault, dear Brutus, is not in our stars,
> But in ourselves, that we are underlings.

And Edmund, in *King Lear*, reduces it all to absurdity with the statement: 'When we are sick in fortune—often the surfeit of our own behaviour—we make guilty of our disasters the sun, the moon, and the stars; as if we were villains by necessity, fools by heavenly compulsion, knaves, thieves, and treachers by spherical predominance, drunkards, liars, and adulterers by an enforced obedience of planetary influence; and all that we are evil in, by a divine thrusting on.' By the close of the Elizabethan period astrology was clearly doomed, and its practitioners were falling into the category of charlatans, cozeners, and common swindlers, who hung on the skirts of society.

Alchemy, or the *ars chemica* of the middle ages, was also moribund. Unlike modern chemistry, which was still unborn or struggling to be born, and which rests upon the hypothesis of the indestructibility of matter—an hypothesis supported by the evidence of the balance—alchemy was built on a false theory of the unity of matter, which had no basis except in the imagination of the schoolmen. To describe it as a bastard science would, however, be unjust and unhistorical. Actually the alchemist was not attempting anything miraculous: he was merely endeavouring to carry out artificially in his laboratory what he conceived to be the processes of nature. What was wrong was not the method, but the presuppositions on which the method was founded. Since it was assumed that all substances were composed of one primitive matter—the *prima materia*, which is metamorphosed into different forms by the imposition of different qualities upon it, it seemed reasonable to suppose that if by certain chemical processes these qualities could be abstracted, and the primitive matter laid bare, then by further treatment this primitive matter could be transformed into any particular substance required. The transmutation of metals was thus one of the basic assumptions of the alchemical art. But it had another branch, which allied it with medicine. Besides being a search for the philosophers' stone, contact with which could transform the basest metal into gold, it was also a search for the elixir of life or *aurum potabile*, 'the efficacy of which is so certain and so wonderful that by it all infirmities whatsoever are easily curable, human life is prolonged to its natural limit, and man wonderfully preserved in health and manly strength both of body and mind'. It will be clear that the two aspects of alchemy were analogous to each other. Just as the philosophers' stone could banish baseness or imperfection from metals and create the perfect metal, gold, so the elixir of life could eliminate corruption from the body and create perfect health.

Needless to say, it was the multiplying of the precious metals rather than the prolongation of life that provided the driving force in the pursuit of alchemy, and gave it its great popularity. Although forbidden by the law in England, encouragement was given to it by the queen and by many important personages. In 1564 Dr. Dee was appointed the royal adviser in mystic secrets, including alchemy, and a year later Cornelius Alvetanus (or De Lannoy) dedicated a treatise entitled *De Conficiendo Divino Elixire sive Lapide Philosophorum* to Elizabeth, which secured for him a contract to produce for her 50,000 marks of pure gold each year at a moderate price. He was given quarters in Somerset House and set to work. Evidently, however, his experiments ended in complete failure, for he was committed to the Tower, in 1567, for 'abusing the queen's majesty by promising to make the elixir'. Dee does not appear to have been any more successful in his studies. The fact is, as Reginald Scot[1] pointed out, that the end of the alchemist was generally poverty. 'They have nothing else left them in lieu of lucre', he remarks, 'but only some few burned bricks of a ruinous furnace, a peck or two of ashes, and such light stuff, which they are forced peradventure in fine to sell, when beggary hath arrested and laid his mace on their shoulders.'

Doubtless there were some who practised alchemy with honest intent; but the great majority were sharks who preyed on the credulity and greed of an acquisitive age. Scot describes them as 'rank cozeners and consuming cankers', 'to be rejected and excommunicated from the fellowship of honest men'. Dramatists poked fun at their pseudo-scientific jargon and deceptions. Thomas Nash was of the opinion that alchemy 'hath wrought

[1] Author of *The Discoverie of Witchcraft* (1584).

such a purgation . . . of men's purses . . . that it hath clean fired them out of all they have'. But the supreme exposure comes from the pen of Ben Jonson in the *Alchemist*, where he depicts the swindling of Sir Epicure Mammon by Subtle and his gang. After spending a considerable part of his fortune in the vain hope of turning Devonshire and Cornwall into 'perfect Indies' and attaining the longevity of the patriarchs, Mammon sees the experiments come to an end with the blowing up of the furnace at the crucial moment, and is advised to go home and repent, while Subtle has made away with his ill-gotten gains.

Another department of scientific inquiry that came under the revolutionizing influence of renaissance thought was medicine. During the generation immediately preceding Elizabeth's accession the most striking feature of the history of European medicine had been the weakening authority of Galen, the second-century Greek physician, whose encyclopaedic writings on every aspect of the healing art had been accepted without question by everyone. Step by step the doctrines of this great master had been challenged, amended, and re-written by investigators who based their work on first-hand observation and experiment. Paracelsus (1493–1541), the Swiss physiologist of Basel, began the revolt by casting the writings of Avicenna, Galen's medieval Arabic interpreter, into the flames; and he followed up this symbolic act by inaugurating a new study of pharmacy allied with chemistry. The next shock to Galenic ideas came from Vesalius (1504–64), the Belgian professor of anatomy at Padua, who began the dissection of human subjects and created a new school of anatomical research. His book entitled *De Humani Corporis Fabrica*, which was published in the same year as Copernicus's revolutionary treatise on the heavenly bodies, created a mighty ferment in Europe no less far-reaching than that caused by the astronomical revelations. Then followed the startling discovery by Servetus, a Spanish physiologist, of the pulmonary circulation of the blood, which he published to the world in his *Restitutio Christianismi* (1553). These epoch-making innovations were only the more important of a series of extensive inroads into Galenism, the cumulative effect of which was like that of a battering-ram of irresistible momentum.

It is to be noted, however, that so far as the movement had

gone, England had taken no active part in it. Until the appearance of William Harvey's *De Motu Cordis* (1628), indeed, this country remained largely in the background—willing enough to absorb the new lore, but unable to take an independent share in the higher creative work. It is a significant fact that although Henry VIII had established professorships of medicine at Oxford and Cambridge (1547), together with the College of Physicians (1518) and the Company of Barber-Surgeons (1540), all of which were in full swing in Elizabeth's day, the pick of English youth who desired a sound education in medical science chose to seek it at the universities of Padua, Montpellier, Basel, Heidelberg, or Leyden rather than in their own native land. Padua was especially favoured because it was the headquarters of the Vesalian school of anatomy. Thither went Harvey, in 1598, to study the action of the venous valves, under Fabricius ab Acquapendente, that led directly to his own wonderful discovery in the early seventeenth century. It is perhaps more significant that when Vicary, surgeon to St. Bartholomew's Hospital, wrote his *Treatise of the Anatomy of a Man's Body* (1577), he made no mention whatever of Vesalius or any other continental scientist of the century, but contented himself with reproducing a fourteenth-century manuscript!

Nevertheless the study of the new anatomy and physiology was not neglected in England. In 1565 the queen granted the College of Physicians the right to carry out human dissections on its premises, and fellows of the college were compelled, under penalty of fine, to take part in public demonstrations of anatomy. There were also similar demonstrations at Barbers' Hall, where an articulated skeleton was kept for instructional purposes, and at Gonville and Caius College, Cambridge, where the Vesalian surgeon Dr. Caius lectured and dissected. The most notable surgeons of the time were John Gale (1507–87), William Clowes (1540–1604), and John Woodall (1569–?), all of whom gained their technical experience in war. Gale served in the English army in Henry VIII's reign, and later in the Spanish army in the reign of Philip II. He wrote a notable book on the treatment of gunshot wounds (1563). Clowes served with Warwick in France during the English occupation of Havre, and later under Leicester in the Netherlands. He, too, was the author of several works on wounds. And Woodall was sent to France with the troops placed at Henry IV's disposal in the last decade of the reign. He became an expert in amputations. There is no doubt that these men were abreast of their time in regard to anatomical science, but their fame was essentially local.

The progress of medicine, as distinct from surgery, was greatly hindered by the Galenic theory of 'humours', which persisted intact until well into the following century. According to this theory four 'humours' or fluids entered into the composition of the body, viz. blood, phlegm, choler (yellow bile), and melancholy (black bile); and the predominance of one or other of them determined the temperament of the individual. Thus we have the sanguine, the choleric, the phlegmatic, and the melancholic temperaments. Excess of any 'humour' constituted a morbid condition, which could only be cured by 'cupping' or 'purging', i.e. by lessening the quantity of blood or by reducing the bile by drugs. Nor was there any English pharmacopoeia in Elizabeth's time, the first being published in 1618; and the sale and purchase of drugs was in the hands of apothecaries, who were still without a charter and under the supervision of the College of Physicians. If Shakespeare is to be trusted, they were a poor and starveling set of men, whose poverty made them susceptible to bribes.

> And in his needy shop a tortoise hung,
> An alligator stuff'd, and other skins
> Of ill-shaped fishes; and about his shelves
> A beggarly account of empty boxes,
> Green earthen pots, bladders, and musty seeds,
> Remnants of packthread, and old cakes of roses,
> Were thinly scatter'd to make up a show.

It was from such a one that Romeo obtained his poison. Moreover, clinical study of disease was only beginning. Dr. Caius was the first to introduce it with his account of the sweating sickness, which he published in his *Liber de Ephemera Britannica* (1555). Clowes followed in his footsteps with observations on the condition of his patients at St. Bartholomew's. But the first to devote himself wholeheartedly to minute study of the subject was Sir Theodore Mayerne, who became physician to James I. We might also note that the first treatise on tropical medicine to be published in England was issued from the press in 1598. It bore the title *The cure of the diseased, in remote regions: preventing mortality, incident in foreign attempts of the English nation.* The author was G. W. (George Whetstone?), who had suffered from yellow fever while a prisoner in Spain, and was anxious to give his countrymen the benefits of his experience.

Philip Stubbs's account of the physicians and surgeons of his day is far from flattering. He describes them as avaricious and frequently unlearned. 'For now-a-days every man, rag and tag, of what sufficiency soever, is suffered to exercise the mystery of physic, and surgery, and to minister both the one, and the other, to the diseased, and infirm persons.' He also accuses some of gross malpractices, of being in league with the apothecaries, who sell 'druggy baggage' to the ignorant, and of deliberately causing the death of their patients. He is especially bitter on the subject of the itinerant quacks 'who run straggling (I will not say rogueing) over the countries, and bear men in hand of their great knowledge', and 'rake in great sums of money, which when they have got, they leave their cures in the dust, I warrant you, and betake them to their heels as to their best refuge'. Whatever truth there may be in this criticism—and it seems to be grossly exaggerated—it is reasonable to suppose that although the granting of licences to practise by the two qualifying bodies of the time, viz. the College of Physicians and the Company of Barber-Surgeons, was strict enough so far as the London area was concerned, there must have been men who practised illicitly in the country and escaped prosecution. But let us remember, on the other hand, the sympathetic and humane picture of the sixteenth-century doctor drawn by Shakespeare in *Macbeth*. After observing Lady Macbeth's symptoms in the famous sleep-walking scene, he remarks:

> This disease is beyond my practice.
>
> . . . .
>
>   . . . Unnatural deeds
> Do breed unnatural troubles: infected minds
> To their deaf pillows will discharge their secrets:
> More needs she the divine than the physician.
> God, God forgive us all! Look after her;
> Remove from her the means of all annoyance,
> And still keep eyes upon her. So, good night.

Two other branches of knowledge, allied with medicine, were also making progress during the Elizabethan period, viz. zoology and botany. In 1552 Edward Wotton had published his *De Differentiis Animalium*, the first printed book by an English-

man on zoology. It was not an original work, nor was it,
strictly speaking, scientific: it was simply a refurbishing of
Greek and Latin animal lore, the fruits of a wide and scholarly
reading of Aristotle, Pliny, and the great authors of antiquity.
A more valuable work was Dr. Caius's treatise on British dogs,
which was published in English in 1576, and became the standard book on that subject for the century. Meanwhile John
Maplet had issued his comprehensive study of plants, animals,
and stones or metals, under the title of *A Green Forest or a
Natural History* (1567), which introduced into England for the
first time the term 'natural history'. Once again, however,
the matter of the book was derivative, being drawn from
medieval sources which, in their turn, were largely dependent
upon Aristotle and Pliny. Maplet's method of classification was
purely alphabetical, not zoological. His description of the cat
is typical both of his matter and style. 'The Cat', he says, 'in
Latin is called *Catus*, as if you would say *cautus*, wary or wise.
In Greek she is named *Galiootes*, with the Germans *Katz*. She
is to the mouse a continual enemy: very like to the lion in tooth
and claw; and useth to pastime or play with the Mouse ere she
devoureth her. . . .' Mixed up with these attempts to handle
the subject in a scholarly way were unedifying repetitions of
mythical lore about dragons, griffins, cockatrices, basilisks,
phoenixes, pelicans, salamanders, unicorns, and other fabulous
beasts, in the manner of the medieval bestiaries. Quite clearly
zoology in England was only in its infancy, and hardly on the
level to which it was raised by Conrad Gessner, the great
German-Swiss naturalist, and other continental scientists of the
period. It was passing from the credulous to the collecting
stage, from mere repetition of obsolete authorities to attempts
at scientific classification; but it had still to move on to the
anatomical and microscopic stage before a proper method of
classification could be achieved. Towards the close of Elizabeth's reign a beginning was made in this direction by the
application of dissection to animal subjects.

Botany was farther advanced. There was a passion for collecting strange plants in England. Burghley was a great
collector, and had a garden of medicinal herbs at Holborn. So,
too, were Lord Zouche, Lord Hunsdon, and Nicholas Lete, a
London merchant. 'It is a world also to see', said Harrison, 'how
many strange herbs, plants, and annual fruits are daily brought

unto us from the Indies, Americans, Taprobane, Canary Isles, and all parts of the world.' Not only so: there was a real renaissance of botanical studies, due to the labours of William Turner, the 'father' of English botany. Turner had published, in 1548, a treatise entitled *The Names of Herbs*, which, as its title implies, was an attempt—the first attempt—to bring the popular plant vocabulary in England into line with the nomenclature adopted by scientists abroad. He followed this up with another work, *The New Herball* (1568), in which he tried to define and fix the names of familiar and cultivated plants. Turner's work supplied the connecting link between the old and the new botany so far as England was concerned. A number of other books appeared shortly afterwards, notably Matthias Lobel's *Adversaria* (1571), which attempted to classify plants according to their natural affinities. This was perhaps the greatest book on scientific botany that appeared in England during the reign; and it was written by a Dutchman, who, strangely enough, published it in London. Later on, Lobel was to become botanist to James I. Meanwhile Henry Lyte superseded Turner's herbal by his *Niewe Herball* of 1578, a copy of a similar Dutch compilation of 1554; and John Frampton, a merchant of Seville, added to the widening stream of botanical knowledge with his *Joyful News out of the New World* (1577)—a translation of Nicholas Monardes's *De las drogas de las Indias* (1565)—which introduced his countrymen to the vegetable products of America. Finally, John Gerard summed up the progress of the times in his comprehensive account of the subject entitled *The Herball, or Generall Historie of Plantes* (1597). It was not an original contribution to learning, but a sound journalistic effort, accompanied by copious illustrations, and it remained for a long time a standard work. When all is said and done, however, it must be admitted that botanical studies in England were behind those of the Continent. The fact that Cesalpino's *De Plantis* (1583), the most significant investigation into plant physiology of the age, was not known in this country until the next century speaks volumes in itself.

It was probably in the pursuit of geographical studies, rather than in physical science, medicine, or biology, that Elizabethan England came most closely into touch with European scientific thought. Until 1550 or thereabout, as we have seen, the country had lagged behind the rest of the world in maritime activity, and consequently in the knowledge of geography. But as soon as the forward movement began under Elizabeth, and the practical problems of navigation called for solution, astronomers, mathematicians, and men of science became interested in this new department of inquiry. Side by side with the old-fashioned 'rutters' of the sea, or sailing-charts, there appeared tracts and treatises on the subject of navigation, almanacs and sea manuals, discourses on compass variation, on the method of determining the position of ships at sea, on the use of nautical instruments, on surveying, &c. At the same time books of travel, voyages, and discoveries, ancient and modern, some of them translations from Spanish, French, Italian, and Portuguese works, others original contributions by Englishmen who had either taken part themselves in exploration or were in personal touch with the pioneers, added yearly to the mass of information available about the maritime achievements of the chief sea-faring nations. And what appeared in print was probably a mere fraction of what circulated in manuscript. The consequence was that before the third decade of the reign drew to a close there was a school of scientists in England whose knowledge of geography and allied subjects was not so very far behind that of the great continental teachers and researchers. True, there were no cosmographers of the calibre of Gerard Mercator of Duisburg, or Abraham Ortelius of Antwerp; but in Dr. Dee, whose astronomical and alchemical studies we have already referred to, England had a scientist and speculator hardly less famous. Up to 1580 Dee was behind all the leading sea enterprises conducted by Englishmen, and corresponded as an equal with his confrères abroad. There were also surveyors and cartographers, like William Bourne, William Borough, and above all Christopher Saxton, who stood in the forefront of their profession. Saxton's great atlas of England and Wales was issued from the press in 1579. Nor must we forget Richard Hakluyt of the Middle Temple, whose knowledge of geography and interest in the expansion of English trade became the inspiration of his namesake and still more famous cousin. Nevertheless English geography was still in a measure derivative: there was no specifically English geographical literature of sufficient weight and importance to evoke the admiration of continental scholars.

Then came with surprising suddenness Camden's *Britannia*
(1586) and the younger Hakluyt's *Voyages and Discoveries* (1589).
These two works, very dissimilar in character but animated by
the same spirit, placed England in the forefront of geographically minded nations. As monuments of industry and learning,
and as historical milestones, they have an imperishable interest.
The enthusiasm with which they were received is shown by the
fact that Camden's work was republished and enlarged four
times before the end of the reign, and Hakluyt's, which was
originally a single volume, grew by successive additions to the
three-volume edition of 1600.

While full credit should be given to these achievements, both
from a literary and historical point of view and as contributions to the progress of geographical knowledge, it is necessary
also to realize their limitations. Neither Hakluyt nor Camden
was, in the strict sense, a scientific geographer, that is, interested primarily in acquiring information about purely geographical phenomena. Hakluyt interlarded his work with a
considerable amount of matter—letters of merchants, reports
on trade, political correspondence—that belongs rather to state
papers or the conduct of business than to a geographical
compilation. His interest lay in the extension of markets, in
the development of trade, in the utilizing of products of new
territories for the benefit of home industries: questions of climate, vegetation, winds and currents, and earth-knowledge
generally were secondary. In short, he was utilitarian in his
pursuit of geography, not a disinterested inquirer bent on
obtaining information for its own sake. Camden, on the other
hand, was essentially an antiquarian. His interest in topography
was subordinated to an overpowering love of historical lore of
all kinds. The contours of the land, the description of scenery
or climate, the accurate positioning of roads, and other matters
of first-rate geographical importance find a very limited space,
if any, in the *Britannia*.

'In my description of each country', he says in the Preface, 'I will
show with as much plainness and brevity as I can, who were the
ancient inhabitants, what was the reason for the name, what are the
bounds of the country, the nature of the soil, the places of greatest
antiquity, and of greatest eminence at present, and lastly, who have
been dukes or earls of each, since the Norman Conquest.'

Quite clearly the word 'chorographical' (or 'regional') bore a very different meaning in the sixteenth century from what it does to-day. To Camden a 'region' was interesting mainly because of the people who inhabited it, their manners and customs, religion, origin, and history. Much the same criticism applies to other works of a similar character written during this period. Giles Fletcher's *Russe Commonwealth* (1591) devotes only three chapters to purely geographical detail, and all the rest of the book concerns policy, wars, religion, customs, and other 'human' interests. The truth is that geography, for all the advance it was making, was only beginning to realize its real vocation. Like the other branches of inquiry which we have already dealt with, it was in process of transition to the scientific stage proper, but had not yet clearly grasped its objective. Observers were not observing the right things: instruments were inaccurate: technique was undeveloped. Nevertheless, despite its limitations, the spirit of investigation was steadily and surely moving towards that clearer definition on which the science of geography would one day be built.

Lastly, in the sphere of education the age was not inactive. No fewer than five schools, which became famous at a later day, were founded during the reign—Repton (1559), Merchant Taylors' (1561), Rugby (1567), Uppingham (1584), and Harrow (1590); and Harrison asserts that almost every corporate town now had at least one grammar school 'with a sufficient living for a master and usher'. The universities were also expanding. At Cambridge Trinity College chapel was rebuilt (1555–64) and the great court completed (1597), and Sir Walter Mildmay founded Emmanuel College (1584). At Oxford Jesus College was founded (1571) and Sir Thomas Bodley built a notable addition to the ancient library (1587–1602) which was henceforth called by his name. Outside the old universities the only other academic erection was that of Gresham College, London (1596), built by the executors of the great financier from a bequest left by him for that purpose. The distinctive feature of this college lay in the fact that it was intended to provide education in the 'seven liberal sciences' in contradistinction to the traditional curriculum at Oxford and Cambridge.

Internally both universities made progress during the reign. Ten years of constitutional unrest at Cambridge came to an end with the appointment of Sir William Cecil to the chancellorship;

T

and in 1570 the university received its new statutes. Leicester, as chancellor of Oxford, was instrumental in securing for it an act of permanent incorporation (1571), which made it unnecessary to seek a fresh charter from each successive ruler. He also provided it with a new printing-press. Great enthusiasm was evoked by the visit of the queen to Cambridge in 1564 and to Oxford in 1566 and 1592. But in spite of the fact that she held out hopes of doing something, like her forebears, for the advancement of learning at both institutions, nothing appears to have eventuated beyond a small contribution to the foundation of Jesus College. So far as the curriculum was concerned, there was no material change introduced at either university. The subjects taught were virtually the same as those prescribed in the time of Edward VI, viz. theology, civil law (expanded to include the ecclesiastical laws of the realm), the philosophy of Aristotle and Plato (with the addition of Pliny), medicine as expounded by Galen or Hippocrates, mathematics (including arithmetic, geometry, astronomy, with the addition of cosmography), dialectic, rhetoric, Greek, and Hebrew. History was not a part of the *studium*, the pursuit of this subject being the work of the Society of Antiquaries, founded by Parker in 1572.

Yet if we are to believe Harrison, neither university was in a very healthy condition. In the election of scholars gross corruption prevailed. It was difficult, he says, owing to the 'bribage' used, for a poor man's son to obtain a scholarship; and the sons of the rich, when they were placed, 'often bring the universities into much slander'. They 'ruffle and roist it out', says Harrison, 'exceeding in apparel and ranting riotous company,' and when they were charged with breach of order, 'say they be gentlemen'. Packing was also used in elections to fellowships, with the same discriminating effect against the poor: so that 'not he which best deserveth, but he that hath most friends, though he be the worst scholar, is always surest to speed'. This indictment was probably just, for it is borne out by Sir Humphrey Gilbert, who was in hope of doing something whereby the universities 'shall better suffice to relieve poor scholars, where now the youth of nobility and gentlemen, taking up their scholarships and fellowships do disappoint the poor of their livings and advancements'.

School education was, of course, conditioned by the higher

studies at the university. In all schools the subject-matter of instruction and the method of teaching was much the same: they were laid down by the fathers of English renaissance education, Colet and Lily. Latin predominated—was, in fact, the only *pabulum* in the ordinary schools, Greek being taught only at Eton, Harrow, Westminster, Shrewsbury, and a few others. Lily's Latin grammar was the recognized primer, enlarged by successive editions; but there was a wide variety in the prescribed texts. Works by renaissance scholars, such as the *Bucolica* of Baptista Spagnolo, the *Zodiacus Vitae* of Palingenius, Erasmus's *Colloquia*, an old favourite, and Vives' *Linguae Latinae Exercitatio*, were studied side by side with the ancient classical authors. A new note was also struck by the compulsory inclusion in all school curricula of Christopher Ocland's *Anglorum Praelia*, a long rhymed account in hexameters of England's wars from Edward III's reign to the close of Mary's, and the same author's *Elizabetha*, a panegyric of the queen's peaceful rule. Another interesting innovation, if it could have been carried out, was the proposal made by Richard Mulcaster, head of the Merchant Taylors' and St. Paul's schools, that English should be accorded a place in the curriculum. But, speaking generally, the supremacy of classical literature was not seriously challenged. Modern languages played no part in the Elizabethan system of school education; nor did history, although the great Camden was head of Westminster for many years. This exclusive insistence upon the ancient languages did not, however, meet with the entire approval of intelligent critics who, like Humphrey Gilbert, would have liked to see learning brought into closer relation with public life. In or about 1564 Gilbert published a treatise entitled *Queene Elizabeth's Achademy*, in which he outlined a scheme for the training of youth in 'matters of action meet for present practice, both of peace and war'. English was to be studied: there were to be lectures on 'civil policy'; and boys were to be taught shooting, riding, marching, navigation, with the elements of medicine, surgery, and natural science. The academy was to be provided with a library to which 'all printers in England shall be for ever charged to deliver at their own charges one copy, well bound, of every book . . . that they shall print'. The idea was attractive, but like many other reforms mooted during the reign it remained merely a *ballon d'essai*.

More interest attaches to Roger Ascham (1515–68), the greatest educationist of his time and the author of *The Schole-master*, which was published two years after his death. Ascham was no mere pedagogue versed in the traditional school culture of the renaissance: he was a philosopher, moralist, and reformer, whose supreme aim was to see the youth of England 'so grounded in judgement of learning, so founded in love of honesty, as, when they should be called forth to the execution of great affairs, in service of their prince and country, they might be able to use and order all experiences, were they good were they bad, and that according to the square, rule, and line of wisdom, learning, and virtue'. Like Vittorino da Feltre, Vives, and other schoolmasters of the renaissance, he believed that the study of classical literature, organized on rigorous lines, was the only possible basis of a liberal education; but he combined with this insistence on linguistic training the wider ideal set forth by Castiglione, the great Italian master of courtesy, whose *Corte-giano*, 'advisedly read, and diligently followed but one year at home in England, would do a young gentleman more good, I wis, than three years' travel abroad spent in Italy'. Like Casti-glione, he favoured all manner of 'courtly exercises and gentle-manlike pastimes'—riding, vaulting, leaping, running, tilting, shooting at a mark with bow or gun, wrestling, dancing, singing and playing of instruments—provided they 'be joined with labour, used in open place, and on the day light, containing either some fit exercise for war, or pleasant pastime for peace'. 'I was never either Stoic in doctrine or anabaptist in religion,' he says, 'to mislike a merry, pleasant, and playful nature, if no outrage be committed against law, measure, and good order.' Verily the author of *The Scholemaster* and *Toxophilus* was no believer in a fugitive and cloistered virtue.

Three great principles are announced by Ascham in relation to education, namely the necessity of gentleness in instruction, especially at the initial stage: the importance of cultivating the 'hard wits' rather than the 'quick wits'; and the superiority of learning to experience as a sure and safe way to wisdom. In expounding these principles he criticizes, sometimes bitterly, sometimes even savagely, the prevailing practice of his time. The 'butcherly fear of making latins', which characterized grammar-school instruction, the foolish discrimination against the slow-witted child, the brutality of masters, and the habit of sending young men to Italy to acquire experience—all come under his lash.

'I remember', he says, 'when I was young, in the north, they went to the grammar school little children: they came from thence great lubbers: always learning, and little profiting: learning without book everything, understanding within the book little or nothing.' In order to remedy this defect of learning 'without the book', i.e. of studying accidence apart from syntax—which he describes as 'tedious for the master, hard for the scholar, cold and uncomfortable for them both'—Ascham advocates the combination of both accidence and syntax in one organic system. Once the child is familiar with the parts of speech, the joining of substantives with adjectives, of nouns with verbs, of relatives with antecedents, let the master take some simple passage from one of Cicero's epistles, explain the occasion and purpose of it, construe it into English, parse it over perfectly, and when he is sure that the child has understood it all, ask him to construe and parse in the same manner. The next stage is for the child to make a translation of the passage in his exercise-book, unaided; after which, an hour or so having elapsed, he must retranslate his own English version into Latin, without recourse to Cicero. The pupil's effort is now compared with the original, errors are pointed out, reasons given for this or that turn of expression, case, gender, &c., as against that, the grammar book is consulted for the rules, and questions are invited. 'This', says Ascham, 'is a lively and perfect way of teaching of rules.' But the teacher must have patience: he must not chide his pupil for faults, for 'that shall both dull his wit, and discourage his diligence'. Rather let him praise him for what he has done well, for 'there is no such whetstone to sharpen a good wit and encourage a will to learning as is praise'.

In pointing out the need for gentleness on the part of the master Ascham exposes the fallacy of supposing that, by nature, children dislike learning. It is not a question, he says, of the disposition of the child, but of the method employed by the teacher. Beat a child for dancing ill, and cherish him even if he be backward in learning, and he will shun dancing and fly to his books. This was the secret of Lady Jane Grey's love of books: God had blessed her, she said, with a gentle schoolmaster and sharp and severe parents. Moreover the master must study his pupils, not favouring the quick-witted and punishing the slow-witted, but 'discreetly consider the right disposition of both their natures, and not so much weigh what either of them is able to do now, as what either of them is likely to do hereafter'. Quick wits, he says, are like over-sharp tools, whose edges are very soon turned: they are apt to take, unapt to keep, for 'this I know, not only by reading of books in my study, but also by experience of life abroad in the world, that those which be commonly the wisest, the best learned, and best men also, when they be old, were never commonly the quickest of wit when they were young'. 'Hard wits' on the other hand, provided they are not over-dull, heavy, and lumpish, and are not thwarted and wrought 'against the wood' by schoolmasters, are 'both for learning, and whole course of living, always the best', and in the end 'do come to that perfection of learning . . . that quick wits seem in hope, but do not in deed, or else very seldom, ever attain unto'.

Finally in the attainment of wisdom, the be-all and end-all of education, learning is a better preceptor than experience, for learning is the eye of the mind, 'to look wisely before a man, which way to go right and which not'. It is possible to learn more in one year from books than from experience in twenty, and it is safer and infinitely less costly. Experience often makes one more miserable than wise, and he is an unhappy shipmaster who learns his craft by many shipwrecks. 'We know by experience itself', says Ascham, 'that it is a marvellous pain to find out a short way by long wandering.' Hence the folly of those who search for wisdom by travelling in Italy. Italy was the Circe's court of Europe, where vice was cultivated to a fine art, and whence those who succumbed to her charms returned swine and asses, carrying in one body 'the belly of a swine, the head of an ass, and the womb of a wolf'. 'I was once in Italy myself', says Ascham, 'but I thank God that my abode there was but nine days. And yet I saw in that little time, in one city, more liberty to sin than ever I heard tell of in our noble city of London in nine years.' To scorn both protestant and papist, to ignore the Scripture, to mock the pope, to rail at Luther, to be atheists in doctrine and epicures in living, to aim at their own present pleasure and private profit, to expound all the mysteries of religion with the formula *Credat Judaeus Apella!*[1]

[1] i.e. Let the Jew Apella believe that (if he likes)!

—these, according to Ascham, were the fruits of copying the Italians. No wonder he did not think Italy a safe place for wholesome doctrine, or godly manners, or a fit school for the young gentlemen of England to be brought up in.

The greatness of Ascham as an educationist has never been questioned. *The Scholemaster* will remain one of the most instructive books in the English, or, indeed, in any language on the subject of linguistic training, but it will be remembered even more for its wisdom than for its technique.

We may conclude this chapter with a glance at some of the superstitious beliefs and practices of the time, for in spite of its learning, culture, and realism Elizabethan England was permeated with superstition. The supernatural world pressed in on the natural and became mingled with it to a remarkable degree. Apart from the occult influences which the stars were supposed to rain on human life, there was a widespread belief in elves and fairies, Robin Goodfellows, lubber fiends, firedrakes, hobgoblins, incubuses and succubuses, Tom Thumbs, and many other incarnations of a supramundane sphere, in so much that on a dark night 'a polled sheep is a perilous beast, and many times is taken for our father's soul, especially in a churchyard, where a right hardy man heretofore scant durst pass by night, but his hairs would stand on end'. If a raven croaked from a neighbouring roof, a man made his will: if a sea voyage was intended, he anxiously consulted the almanac and chose a Sunday or some other lucky day. The efficacy of charms, spells, incantations, divinations, and amulets was accepted as part and parcel of everyday existence. 'How superstitiously we mind our evils,' says Delio in the *Duchess of Malfi*:

> The throwing down of salt, or crossing of a hare,
> Bleeding at the nose, the stumbling of a horse
> Or singing of a cricket, are of power
> To daunt whole man in us.

Dying men were credited with miraculous insight into the future. 'Methinks', says Shakespeare's John of Gaunt on his death-bed, 'I am a prophet new inspired, and thus expiring do foretell of him—' (i.e. King Richard II). Ghosts of the dead walked the earth between midnight and cock-crow, and communed with the living, because they had either some secret to impart, or wrong to set right, or reparation to make. 'My father's spirit in arms!' exclaims Hamlet when informed of the apparition on the battlements at Elsinore. 'All is not well: I doubt some foul play.' Subsequently he meets the ghost and hears a tale of incest and fratricide that starts him off on his career as avenger. Similarly Macbeth is haunted, tormented, and mocked by the ghost of 'blood-boltered' Banquo, and Brutus is visited by the spectre of the murdered Caesar in his tent at Sardis—the harbinger of disaster at Philippi. There were charms for every evil and grief. To be cured of the falling sickness it was advisable to 'drink in the night at a spring water out of the skull of one that had been slain', or 'eat a pig killed with a knife that slew a man'. Headaches could be conjured away by tying 'a halter about your head, wherewith one hath been hanged'; a scorpion's bite could be cured by 'saying to an ass secretly, and as it were whispering in his ear—I am bitten with a scorpion'; a woman in travail could be released by throwing over the house 'a stone or any other thing that hath killed three living creatures, viz. a man, a wild boar, and a she bear'; and the prophylactic against the quotidian ague was to 'cut an apple in three pieces, and write upon the one "The Father is uncreated", upon the other, "The Father is incomprehensible", and upon the third, "The Father is eternal"'. Fairies, again, were friends of 'human mortals', rewarders of their minor virtues and punishers of their shortcomings. Like other supernatural beings, they came and went with the hours of darkness. They had great powers, for Puck, 'that merry wanderer of the night', could 'cover the starry welkin with a fog as black as Acheron' and 'overcast the night'. But since they were sympathetic to mortals they did not provoke nature into war against them. It was only when the fairies were unhappy that things went wrong with the world.

> And this same progeny of evil comes
> From our debates, from our dissensions,

says Titania in *A Midsummer-Night's Dream*. Shakespeare, apparently, has two kinds of fairy queen—the Titania of the moonlit glades and Mab the mischief-maker, who sometimes gallops her chariot

> over a courtier's nose,
> And then dreams he of smelling out a suit;
> And sometimes comes she with a tithe-pig's tail
> Tickling a parson's nose as a' lies asleep,

Then dreams he of another benefice;
Sometimes she driveth o'er a soldier's neck,
And then dreams he of cutting foreign throats,
Of breaches, ambuscadoes, Spanish blades.

A very different creature was the witch. If the fairy was a denizen of the world of folk-lore, the witch was as grim a reality as nature herself—or so the age believed. Her powers were malignant and destructive. Popular belief credited her with being able to raise and suppress lightning, thunder, hail, rain, winds, tempests, and earthquakes, pull down the moon and stars, send needles into the livers of her victims, transfer corn in the blade from one place to another, cause disease or sterility in man and beast, fly in the air, dance with devils, transubstantiate herself and others into animals, keep devils (i.e. familiars) in the shape of toads or cats, dry up springs, change the course of nature, turn day into night and night into day, pass through auger-holes, sail in an egg-shell, cockle-shell, or mussel-shell through or under tempestuous seas, bring souls out of graves, slay lambs with a look, deprive a cow of its milk, &c. The essence of the business was a supposed secret compact with the Devil. 'A witch', said Coke, 'is a person who hath conference with the Devil, to consult with him or to do some act.' Reginald Scot put it ironically thus: 'In the estimation of the vulgar people it [i.e. witchcraft] is a super-natural work contrived between a corporal old woman and a spiritual devil. The manner thereof is so secret, mystical, and strange, that to this day there hath never been any credible witness thereof.' The most famous case recorded in British history of the sixteenth century both of the power and the fate of those who indulged in the craft is that of the four witches of North Berwick, who were accused of conspiring to destroy James VI and his queen, Anne of Denmark, in 1590. According to the evidence submitted by themselves—under torture—they raised a storm by their incantations and by casting a 'conjured' cat into the sea at Leith, hoping thereby to wreck the royal ship on its way from Denmark to Scotland. Failing to achieve their end by this means, they prepared a waxen image, which they melted, in accordance with their cult, saying, 'This is King James the sixth, ordained to be consumed at the instance of a noble man Francis earl of Bothwell'. Sentence of death was pronounced against them by a court of justice, and they were burnt at the stake on Castle hill, Edinburgh, in 1591.

The universality of belief in witchcraft rendered it almost impossible for the person accused of it to escape. Even the enlightened Bishop Jewel could aver, in a sermon delivered before the queen in 1572: 'These eyes have seen most evident and manifest marks of their wickedness. Your Grace's subjects pine away even unto death, their colour fadeth, their flesh rotteth, their speech is benumbed, their senses are bereft. Wherefore your poor subjects' most humble petition unto your Highness, is, that the laws touching such malefactors may be put in due execution.' Proof of witchcraft was comparatively easy. 'She was at my house,' says the plaintiff: 'she would have a pot of milk: she departed in a chafe, because she had it not: she railed, she cursed, she mumbled and whispered, and finally she said she would be even with me; and soon after my child, my cow, my sow, or my pullet died, or was strangely taken.' Little more than this was sufficient, in ordinary circumstances, to decide the fate of the accused. There were some, however, who refused to believe in such trumpery charges, notably Scot, whose sturdy common sense penetrated the absurdity of the witch's supposed powers, and found naturalistic explanations for the evils attributed to them. Butter would not churn, he says, not because the witch has cast her spells over it, but 'if either the maids have eaten up the cream, or the goodwife have sold the butter before in the market'.

'My question', he says in his preface,[1] 'is not whether there be witches or nay; but whether they can do such marvellous works as are imputed to them. Good Master Dean, is it possible for a man to break his fast with you at Rochester, and to dine that day at Durham with Master Doctor Matthew; or can your enemy maim you, when the Ocean sea is betwixt you? May a spiritual body become temporal at his pleasure? Or may a carnal body become invisible? . . . Alas, I am sorry and ashamed to see how many die, that being said to be bewitched, only seek for magical cures, whom wholesome diet and good medicines would have recovered.'

But it was a long time before the illuminating ideas of Scot could penetrate into the ignorance even of the educated, or influence practice. He was severely criticized by his contemporaries, and taken to task by one no less than James I.

[1] *The Discoverie of Witchcraft.*

There remains the 'impious and damnable' magic of the sorcerer, the fruit of an illicit desire for forbidden knowledge—in a word, the so-called 'black art'. It was a superior craft, and might be defined as the attainment of power by intellectual means. As such it appealed to the learned rather than to the ignorant, to men of science like Paracelsus or Dr. Dee rather than to the humble people who practised the mystery of witchcraft. Sorcerers, says Scot, 'pass the degree of witches, and entitle themselves to the name of conjurors'. Unlike the witch, who went to work with toads and cats, the sorcerer could by his adjurations summon devils from hell 'quicker than the pope could release souls from Purgatory'. The seventy and nine principal and princely orders of infernal spirits, with their massed legions of petty devils, were at his disposal. Each had his special qualities and powers. 'Marbas' could reveal all secrets and cure all diseases: 'Furcas' could impart wisdom and cunning in all mechanical arts: 'Beroth' was master of things present, past, and to come: 'Asmodai' could render men invisible and reveal hidden treasure: 'Allocer' could procure the love of any woman: 'Orias' could give dignities and prelacies, and reconcile enemies: 'Vepar' had the secret of killing men by putrefying their wounds: 'Amduscias' could bring it to pass that trumpets and musical instruments might be heard and not seen: 'Buer' and 'Bifrons' had the power to make men live long: 'Caym' could render intelligible the voice of birds and beasts; and so on through the long list. Popular belief had it that in return for the enjoyment of his power over demons the sorcerer yielded to the Devil the reversion of his soul; but there was a different type, who acquired his magic not by devilish compact but by philosophical and cabalistic studies. The contrast is brought out by Marlowe's Faustus, who sold his soul in order to gratify his senses, and Shakespeare's Prospero, whose 'rough magic' depended upon book, wand, and mantle, and was used simply to deliver himself from evil men who had dispossessed him of his dukedom. Faustus's art was diabolic, Prospero's harmless. The former had as his 'familiar' the satanic Mephistophelis: the latter was served by the dainty sprite Ariel. In the end Faustus, his time accomplished, is carried off by demons; while Prospero, having achieved his purpose, breaks his staff, drowns his book 'deeper than did ever plummet sound', and reverts to his former life as the learned prince of Milan.

Notwithstanding the strong hold which sorcery had on the imagination of the Elizabethan age, there were indisputable indications that its foundations were beginning to crumble. The spread of the true scientific spirit, the gradual demolition of the impressive edifice of medieval 'science' by investigation into natural causes, and, lastly, the application of common sense to psychic phenomena—all contributed to its defeat. Scot asserted that he could see no difference between the practices of the sorcerer and the conjurations of the popish exorcist, except that the 'papists do it without shame openly, the other do it in hugger mugger secretly'.

IX

ELIZABETH AND THE NETHERLANDS
1575–86

IN the field of foreign policy Elizabeth was essentially a machiavellian, although there is no reason to suppose that she actually borrowed her maxims of statecraft from *The Prince*.[1] The chief object of her diplomacy, from which she never consciously wavered, was to establish her throne and kingdom in a position of unassailable security and power. To attain this end she was prepared to use every instrument that gave promise of being serviceable, every ally that chance or necessity threw in her way, while at the same time avoiding commitments that might jeopardize her own freedom of action, or lessen the advantage she possessed as the ruler of an insular state. Both a realist and an opportunist, she made interest the determining factor in all her political manœuvres and combinations, and reason of state a sufficient justification for every act. She was loyal to treaties only so long as they served her purpose, and broke them unscrupulously, in spirit if not in the letter, when they threatened to hamper her freedom of movement. A master of prevarication and deceit, when occasion drove her to it, she was an adept at finding subterfuges for actions of doubtful legality, and always had emergency exits at hand when strategic retreats became necessary. Watchfulness and flexibility were the very essence of her system; for each situation, as it arose, had to be examined afresh in the light of England's vital needs and policy modified or redirected accordingly. Thus while the ultimate goal remained as fixed as the stars the lines of approach to it might be as tortuous as the path of a comet. A more self-centred or less doctrinaire statecraft it would be difficult to imagine: it was the very incarnation of *sacro egoismo*.

Continental protestants, oppressed by tyrannical governments or struggling for their lives against the superior might of catholic armies, beckoned her in vain to a crusade on behalf of the reformed faith. The fervent appeals of her own ministers,

---

[1] The influence of Machiavelli on Tudor political thought and policy was practically nil. For a discussion of the subject see M. Praz, 'Machiavelli and the Elizabethans', in *Proc. Brit. Acad.* xiii, 1928; also L. A. Weissenberg, 'Machiavelli and Tudor England', in *Pol. Sci. Quar.* xlii, 1927.

who believed that the protestant cause was also England's cause, struck no responsive chord in her cold and calculating brain. Her one thought was how she could use their respective idealisms for the furtherance of her own secular aims. Experience had taught her—and she was an apt learner—that to confuse state policy with religion, or to put religion first and the state second, could only lead to chaos, bloodshed, and disaster. Her foreign policy, like her home policy, was entirely free from fanaticism. When the German protestants, sunk in their interminable credal dispute concerning the 'ubiquity and omnipresence of the body of Christ', resisted her efforts to make them think as political beings, she made no secret of her contempt for their 'unprofitable dissensions'. But let any one, be he protestant prince or catholic monarch, infringe upon what she deemed to be the permanent interest of her crown and country, she acted with a swiftness and truculence that disconcerted her friends as well as her foes.

Yet in spite of her fulminations she was no firebrand. Peace with security was her avowed aim.[1] 'Elle est une femme nourrie à la paix et repos,' wrote La Mothe Fénelon, summing up his impressions after seven years' residence at her court: '. . . veut jouir son état tant qu'elle vivra sans guerre ni trouble.' Substantially he was right in his diagnosis. Although Elizabeth had fought two wars against France in the earlier part of her reign, she quickly realized that diplomacy, not war, was her 'long suit'. War, in fact, she abominated: it was the destroyer of peoples, the enemy of that material well-being which she made it the object of her rule to promote. If she could, she would have eliminated bloodshed from international relations, beaten the swords of the nations into ploughshares, abolished confessional strife, and created a world in which all disputes would be settled by the peaceful method of arbitration. For men of blood she had an unconquerable dislike. It is a notable fact that, for the greater part of her reign, her voice was heard amid the din of a Europe given over to violence in every shape and form, counselling moderation and offering her services as a

[1] 'It may be thought simplicity in me that all this time of my reign I have not sought to advance my territories, and enlarge my dominions; for opportunity hath served me to do it. . . . And I must say, my mind was never to invade my neighbours, or to usurp over any; I am contented to reign over mine own, and to rule as a just prince.' (Queen's speech to parliament at its dissolution in 1593: Townshend, p. 48.)

mediator. Possibly she under-estimated the strength of the
forces arrayed against her—William of Orange and Henry of
Navarre were shrewder judges of mankind than she—but she
laboured on in the conviction that even fanatics could be persuaded, cajoled, or coerced into a reasonable frame of mind.
It is only fair to add, however, that the peace she envisaged
was an English peace; that is, a peace designed to safeguard
English interests abroad.   And in the second place, if she
deprecated war, it was not merely because of her humanita-
rian impulses: she had neither the means nor the inclination
to embark upon the costliest of all gambles. Here again her
realism probably came into play, for her parsimony kept her
solvent at a time when the continental monarchies, with their
much greater resources, were either bankrupt or staggering on
the verge of bankruptcy.[1] In a world cursed by prodigal expenditure she stood forth a shining example of the power that
comes from thrift. English gold, carefully husbanded and carefully doled out, was more potent, in the long run, than all the
precious metals of the Indies; and Europe moved more at
Elizabeth's bidding than at the impressive military gestures of
King Philip.

In 1575 England's position in Europe, though far from
secure, was relatively safe. The foreign debt, which had hung
like a millstone round her neck since the commencement of the
reign, was practically extinguished. Trade was beginning to
boom, and for the next eleven years wealth poured into the
country. The queen's credit stood high in Europe—much higher
than King Philip's, who could not borrow money at less than
from 12 to 18 per cent., while Elizabeth could have it at 8 or 9
per cent. The renewal of the treaty of Blois with Henry III
removed any immediate danger from France: the crisis with
Spain had been disposed of by the treaty of Bristol and the
reopening of traffic between England and Antwerp; and Scotland was quiescent under the anglophil government of Morton,
the king being a minor of nine years. Apart from the unalterable enmity of Rome and of the exiles who flitted about the
Continent like uneasy ghosts, living meagrely on Spanish
pensions and hoping for the 'enterprise' that would restore them

---

[1] Philip II was bankrupt in 1575 and again in 1596: France was in a chronic
state of insolvency.

in triumph to their native land, there was little to trouble the queen except the continued turmoil in the Netherlands. If only peace could be re-established there, without involving a complete Spanish victory, she might at last breathe freely.

Such a pacification was not only desirable: it was imperative for a variety of reasons—first, because Antwerp was still the principal market for English cloth on the Continent; secondly, because so long as the struggle continued, the bugbear of a French intervention could not be effectually dispelled; and thirdly, because if the Spaniards mastered the country they might use the 'sea gate' of Flanders as the starting-point for a conquest of England. But how was it to be accomplished? To Elizabeth, with her intensely English outlook, the only feasible solution of the problem was a return to the *status quo* before the outbreak of the troubles: that is, the country must remain under Spanish sovereignty, but its political and civil liberties must be restored. If this could be secured, she was ready to let Spain settle the religious question in her own way, in accordance with the general formula that rulers had the sole right to prescribe the religion of their subjects. Indeed she could hardly do less, since she claimed and exercised the same right with regard to her own dominions.

But there was little likelihood of either William of Orange or King Philip accepting such a solution. Philip was determined to master the rebellion even if it meant making a desert of the country. He would rather lose the provinces altogether than suffer any diminution of his sovereign rights, or infringement of 'our holy catholic religion'. Unconditional submission, he said, must precede any peace negotiations. Orange was no less uncompromising in his attitude. Eight years of struggle had taught him the futility of seeking a compromise with a monarch who was dominated by a fixed idea, who notoriously kept no faith with heretics, and whose hatred of himself was implacable. So distrustful was he of the Spaniard that he likened peace gestures from Madrid to the piping of children in the market place when they try to ensnare little birds. Hitherto he had masked his intentions with the convenient fiction that his quarrel was with the king's ministers, not the king himself; but the mask was wearing thin. Already the doctrines popularized later by Duplessis-Mornay[1] in his *Vindiciae contra Tyrannos* were

[1] The enlightened huguenot adviser of Henry of Navarre.

beginning to circulate among the huguenots in the prince's entourage, and many held the view that oppressed peoples had the
right to dispose of the sovereign power in their own interest.

The strength of the stadholder, now as always, was concentrated in Holland and Zeeland, whose geographical position
gave them control of all approaches to the Netherlands by sea,
and rendered them practically invulnerable to attack from
Spain. He also possessed a formidable navy, increasing in
power each year. But two things were necessary for a rapid
triumph over his enemy, viz. a close blockade of the river
Scheldt, the only waterway that gave access to the heart of the
'Belgian' parts, and a reliable ally to share the burden of the
war on land. Orange apparently attached great importance to
the blockade, for he is reported to have said that 'if he could but
hinder that in a year's space there come no salt into Flanders,
he could win such a peace as he wished for'. On this point,
however, he encountered the inflexible opposition of Elizabeth,
who refused to allow any interference with the free navigation
of the Scheldt: the newly revived trade with Antwerp was
just as necessary for England's prosperity as the starvation of
Antwerp was, in Orange's opinion, necessary for the success
of the rebellion. There was considerable bickering over this
question. Similarly when the prince, in his quest for help,
turned to France, whose age-long enmity to the house of
Habsburg seemed to him the one sure fact on which to build
amid the shifting sands of European politics, Elizabeth again
interposed her veto, giving him to understand—a favourite
expression of hers—that if he brought in the French she would
immediately 'bend all her forces to the assistance of the king
of Spain'. Public opinion in Holland and Zeeland was pardonably annoyed at this callous indifference to their fate, and
Englishmen were severely criticized as 'those that do put on
religion, piety, and justice for a cloak, to serve humours withal
and please the time, while policy only is made both justice,
religion, and God'. Even in England there was some honest
heart-burning at the treatment meted out to the prince. 'There
is no honest man nor of good religion', said Sir Thomas Smith,
'that doth not pity his case and wish it were better.' But
Elizabeth held on her course, impervious to the calls of sentiment and religion. She would neither help the Dutch herself
not permit them to help themselves; and if they flouted her

efforts to mediate a peace, and by their stubbornness prolonged the war, she would teach them a lesson. Nay, if they maltreated her merchants, she would exterminate them all!

Nor did the Spaniard escape his share of blame for the unquiet state of the Netherlands. Philip was informed that unless he took steps to remedy the grievances of his subjects the French would march in, and Elizabeth would be compelled 'for our own safety to put in execution that remedy for their relief that we would not willingly yield unto otherwise than constrained thereto'. Again and again—'en une infinité de manières et de termes aigres', said Champagny, Requesens's envoy—the queen came back to her fundamental principle that the Spaniards must not 'impatronize' themselves in the Low Countries. They would be bad neighbours, she said, and would try to encircle (cerner) her dominions. From an economic standpoint, too, they were a danger, for, as Burghley remarked to Guaras, 'You people are of such a sort that wherever you set foot no grass grows.' And Elizabeth herself corroborated this with the sage, if somewhat trite, saying that 'where countries be governed and replenished with men of war, the haunt of merchandise will cease'. If the king really desired peace, let him arrange an armistice and let diplomacy get to work. Should his terms be reasonable, and his subjects refuse to accept them, he might rest assured that she would give him her full support in reducing them to obedience.

Meanwhile the struggle in the Netherlands went on. In September 1575 Requesens began his great attack on the islands of Duiveland and Schouwen, and laid siege to Zerickzee, thereby cutting the communications between Holland and Zeeland. By October both islands were in his hands, and an English eyewitness wrote to Burghley that 'all Holland is like to be lost, and not without great danger to the prince his own person'. Captain Edward Chester was of the same opinion. 'Without present relief', he wrote, 'this state will small while stand.' So serious was the situation in December that Orange dispatched a mission, headed by St. Aldegonde and Paul Buys, with the request that Elizabeth should take the two provinces into her protection, lend them a sum of money for military purposes, and accept as security for its repayment the towns of Flushing, Brill, Dordrecht, and Enckhuizen. For a brief space it looked as if the queen's hand would be forced, for the council was strongly in favour of intervention. But the 'business seemed not to ripen'; and in March 1576 Elizabeth reaffirmed her intention to seek a peace by mediation—'an honourable and advantageous settlement' for both parties. Thus the Orange deputation was turned away empty-handed, much to its chagrin.

Worse still: in the same month an incident in the narrow seas nearly provoked war between the two governments. An English ship returning from Antwerp was attacked within a few miles of Dover by cruisers in the service of Zeeland and taken as a prize to Arnemuiden, her crew and passengers being spoiled of their goods. Apart from the gross insult to the flag 'within her Majesty's stream', which was sufficient in itself to warrant sharp reprisals, the treatment of a distinguished foreigner—the bride of the Portuguese ambassador in London—who happened to be travelling in the ship under a safe-conduct from Elizabeth so incensed the queen that she ordered the immediate seizure of Dutch shipping at Falmouth. The admiralty of Zeeland replied with a similar arrest of vessels belonging to the Merchant Adventurers at Flushing. During the heat of the controversy, which lasted all summer, Elizabeth treated Orange to some pretty plain speaking. He was told, in effect, that he was entirely mistaken if he imagined that England's safety depended upon him, and that 'in the judgement of the world, the aptest mean for her Majesty to withstand or prevent the peril that he conceived might grow to her by his overthrow were to join with the king of Spain against him'.

By this time, however, the tension in Holland and Zeeland had relaxed, owing to the death of Requesens (3 March 1576) and the mutiny of the Spanish troops for arrears of pay. Since there was no hope of financial relief from Spain, for Philip had been unable to meet his creditors during the previous year, and had suspended all payments in September 1575, the mutiny rapidly became a reign of terror in the Flemish provinces. Towns and villages were plundered or held to ransom by a licentious soldiery, who treated the civil population with all manner of barbaric cruelty. In self-defence the peasantry and burghers flew to arms, resistance was organized by the provincial estates, and a cry arose for the expulsion of the Spaniards bag and baggage. This was the moment for which Orange had long hoped and prayed, and for which he had worked assiduously since the beginning of the troubles. His dream of a united Netherlands seemed at last on the verge of realization. Accordingly, soon after the 'terror' broke loose, he entered into negotiations with the nobles of Brabant and Flanders for a general agreement. Time was precious, for at any moment the new governor designate, Don John of Austria, might arrive and frustrate the chance of concerted action. But for once fate smiled upon the efforts of the prince. Stimulated by the atrocities of the Spaniards, which culminated in the 'Fury' of Antwerp (Sept.),[1] the states-general concluded the Pacification of Ghent (8 Nov.), and for the first time for nine years the whole seventeen provinces drew together in a common resolve to rid themselves of military rule and to re-establish their political liberties as in the days of Charles V. Although the question of religious toleration was still too dangerous to be handled, and was left in suspense, it was agreed to abolish the placards against heresy, to declare the illegalities of Alva null and void, to confirm Orange in his stadholderate of Holland and Zeeland, and to insist upon the immediate dismissal of all the Spanish garrisons.

Thus when Don John arrived he was confronted, not with a riot to put down and an aggrieved people to be pacified, but with a population in arms and determined not to receive him until he accepted the terms of the 'Pacification' and removed every Spanish *tercio* from the country. It was hardly the sort of reception that the most renowned military commander in christendom expected at the hands of subjects. He had come with a message of peace, entrusted by the king with power to satisfy the provinces on all points compatible with the maintenance of obedience to the Crown and the catholic church. But it was understood[2] that once this was accomplished he would turn his forces against England. This latter project was veiled in the deepest secrecy and was to be executed without warning; but it was nevertheless the *raison d'être* of Don John's mission to the Netherlands.

---

[1] For details see Motley, *The Rise of the Dutch Republic*, iii. 43–56.

[2] i.e. by Don John and his secretary Escovedo and also in Rome, where the *empresa* was the dominant theme in diplomatic circles. There is no ground, however, for believing that Philip II was at this time in favour of a blow at England. He was anxious not to compromise the pacification of the Netherlands by associating it with a crusade against Elizabeth (see P. O. de Törne, *Don Juan d'Autriche et les projets de conquête de l'Angleterre*, 1928).

The new governor, however, soon found himself in a dilemma. If he agreed to the withdrawal of the Spaniards, how could he obtain the military force necessary for the invasion of England? On the other hand, if he refused to send away the Spaniards, he would not be received as governor, and would have to fight for his position against a united Netherlands. Whichever policy he adopted, his mission was doomed. In the circumstances he tried dissimulation, and offered to evacuate the troops by sea, hoping to divert them, once they were safely embarked, to some convenient port in England. But the ruse was defeated by the states-general, backed by the vigilance of Elizabeth, whose fears were justifiably roused at the prospect of a Spanish force coasting along her shores, when rumours were afoot that some hostile stroke was intended against herself. Consequently, in February 1577, Don John was compelled to come to terms with the states-general at Marche-en-Famine. By this treaty, which was the work of the catholic party in the Low Countries, and was generally known as the 'Paix des Prêtres', or by the inappropriate title of the 'Perpetual Edict', he undertook to observe the Pacification of Ghent, to respect the ancient liberties of the provinces, and to remove the Spanish troops within twenty days by land to Italy. Having now disarmed himself—for the evacuation was carried out to the letter—Don John was received as governor in Brussels, on 1 May, with the cry 'There was a man sent from God whose name was John'! The only one who refused to join in the general rejoicing was Orange, who had taken no part in the peace negotiations, and remained sullen, mistrustful, implacable.

Elizabeth had little cause to be dissatisfied with the settlement. She had laboured all winter to bring it about, urging the states-general 'in God's name not to lose any occasion for obtaining the said peace', and plying Don John with threats that, if he sought to conquer the provinces, she would 'aid them with all the might and power we can'. But Orange was an insurmountable obstacle to the complete pacification on which she had set her heart. Instead of listening to the pleadings of Don John, who 'surpassed Circe' in his efforts to charm, the prince grumbled his suspicions to Elizabeth, renewed his appeals for an alliance, and pointed out that with the havens, ships, and mariners of Holland and Zeeland on her side she could defy her enemies with impunity.

What was the queen to do? She could not coerce Orange, because she might need his help if he were right about the ulterior designs of the Spaniard; she could not abandon him altogether, for that would throw him into the arms of the French; nor, again, could she support him without involving her whole Netherlands policy in ruin. Her dilemma was the more acute because some of her council, notably Leicester, Walsingham, and Wilson—the leaders of the so-called war party—were openly on the side of the prince, and gravely concerned about the position into which England had drifted as a result of the queen's attachment to the cause of peace. Even Burghley was not happy. 'There is no trusting suspected friends at this time,' wrote Wilson, 'but plain dealing with a protestation to make peace will be the best assurance'; and again, 'Better not to deal at all than not to go roundly to work. . . . Valiant working never wanted good fortune.' But the fire-eaters tendered their advice in vain: Elizabeth clung to her policy of mediation, insisting that the Pacification of Ghent was the only sound basis of peace in the Netherlands. If Orange wanted allies, let him form a 'league of association' secretly with the protestant princes of Germany and the protestant cantons of Switzerland, and she would give it her blessing.

Such was the situation in July 1577, when another kaleidoscopic change occurred in the Low Countries, destroying for good the possibility of peace. For five months Don John had chafed at the restrictions imposed upon him by the states-general, and the stubborn recalcitrance of Orange; and at the end of July, on the plea that his life was in danger from plotters —not an improbable story—he seized Namur with a company of Walloons, and swore to avenge his honour by 'bathing in the blood of traitors'. It was the prelude to another attempt to subdue the country by force of arms. Don John had 'reverted to type', as the Dutch stadholder had prophesied.

From blowing cold, Elizabeth now blew hot. Early in August William Davison was sent to the states-general with the urgent message that they should call upon the prince to take command of the situation, coupling with it an offer of armed help from England for the defence of the country. Davison's report, on arrival, was exceedingly pessimistic. 'Both by what I see and find here', he wrote, 'there is no one thing more certainly projected than to shake the state of her Majesty.' To make matters

worse, a rumour was afloat that the duke of Guise was marching
to Don John's assistance with 7,000 or 8,000 foot and 4,000
horse. After some hesitation, the states-general decided to fall
in with the queen's advice. Orange was summoned to Brussels,
and the marquis d'Havré was sent to England to raise a loan
and procure a contingent of troops under the command of
Leicester. On 22 September, 'with such singular demonstrations of joy . . . as if he had been an angel sent down from
Heaven', the prince entered the capital. Havré, meanwhile,
obtained from the queen the promise of a loan of £100,000,
together with the offer of 5,000 foot and 1,000 horse.

At last it seemed as if Elizabeth had determined to join forces
with the rebel cause, and Leicester began to turn over in his
mind the details of the forthcoming expedition. But the decisive
hour had not yet struck. In December the queen decided to
make another appeal for peace, and Sir Thomas Wilks was sent
to Spain with instructions to urge the recall of Don John, whose
conduct had destroyed that 'mutual confidence which is more
than necessary in every well-constituted government'. He was
to point out to the king that, if aid had been promised in men
and money to the states, it was because they might be compelled
by Don John's action to seek the help of some other prince or
lose their liberties. As for herself, the queen was concerned
only to maintain the obedience of the provinces to Spain and
the Pacification of Ghent. Wilks's departure for Madrid was
followed by the dispatch of Sir Thomas Leighton to the Low
Countries with cogent arguments for an armistice, pending the
king's decision. 'When account is made of treasure spent, of the
spoils of cities, of the intercourse of merchandise broken off',
he was to say to the states, 'it will fall out in truth that he that
wins loses, and the conqueror has more cause to lament than
rejoice.' If the states were agreeable to a cessation of arms, he
was to put the case for mildness and conciliation to Don John,
and to point out to him that 'to shed blood, if it can be saved, is
greatly repugnant to the nature of a prince'.

Meanwhile Elizabeth's reluctance to plunge into the fray in
the Netherlands had brought the danger of French intervention
nearer. For two years she had successfully staved it off, partly
by playing on the need for amity between the two governments,
partly by threats that if France intervened England would side
with Spain, and partly by interference in the internal affairs

of that kingdom contrary to the spirit as well as the letter of
the treaty of Blois. Volunteers and munitions had been passed
across to La Rochelle to assist the huguenots, and German
armies under John Casimir, son of the Elector Palatine, had been
subsidized with English gold to invade France as an ally of
Condé and Navarre—in short, the expedients with which Elizabeth had combated the French government in the days before
the treaty were revived again, to the anger of the king and queen
mother. 'My son', said Catherine de Médicis to Sir Amyas
Paulet, Elizabeth's ambassador in Paris, 'may no longer bear
this kind of double dealing: he is ready to be a friend, and if not,
he is able to be an enemy.' 'All your dealings', wrote Paulet to
the queen, 'are registered and laid up against a day, and nothing
is forgotten, no not things done before the last treaty.' The real
danger, however, was not from the French government but
from the factions in France which had mastered the crown and
were trying, in different ways, to force the country into a foreign
policy that boded ill for England. The duke of Guise and the
catholic extremists—the party of the League—were in favour
of supporting the Spaniards in the Netherlands as the best
way of securing the supremacy of catholicism in France. The
huguenot-politique combination under the leadership of Anjou
(Alençon) were for open intervention on the side of Orange—a
reversion, it will be noted, to the policy of Coligny in 1571–2.
Anjou, in fact, was the Coligny of 1578. His plan was to wrest
the Netherlands from Spain, unite them by a marriage alliance
with England, and so create a powerful protestant federation,
under French protection, to balance the designs of the Spaniard
and Guise. The issue at stake was no other than the domination
of northern Europe, for Don John and Guise had entered into
a compact to establish catholic supremacy in the same three
countries. Both schemes were in the highest degree dangerous
to England, although at the moment they existed only in the
brains of their promoters.

Elizabeth was between the devil and the deep sea. Her
appeal to Don John for an armistice met with no success, nor
were the states in favour of it; and months must elapse before
she had a reply from Spain to the Wilks mission. Meanwhile
Don John piled up a great army at Namur, consisting of Germans, Walloons, Burgundians, and Spaniards, and in January
1578, aided by Alexander of Parma from Italy, inflicted a

defeat on the states' army at Gembloux. Further Spanish victories followed quickly in the open country, at Louvain, Tirlemont, Aerschot, and elsewhere; and Orange was compelled to use what troops remained to him for purely garrison work. Davison was in despair at the queen's delays. 'Now seeing that her Majesty cannot abandon them without peril to herself and to them', he wrote to Walsingham in March, '. . . it would in my judgement (under correction) be more profitable and honourable for her Majesty the sooner she gives her determination.' Like the other members of the war party in England, Davison believed with all his heart that the only way to keep the Spaniard 'more terrible in opinion than in effect' was to occupy him in the Netherlands. But Elizabeth was not to be moved. The utmost she would do was to subsidize her faithful John Casimir with £20,000 to hire German horse for a demonstration against Don John; but her name was not to be mentioned in the matter, and the subsidy was to be deducted from the £100,000 she had undertaken to advance to the states. 'By your letter', wrote Davison to Leicester, commenting on the transaction, 'I see that our long doubtfulness and irresolution has at length brought forth an unworthy conclusion.' True enough; but the queen had other things to think about: she was expecting Stukeley in Ireland, and Paulet had bad news to communicate about French preparations in Brittany, Bordeaux, and other parts, the object of which was veiled in profound secrecy, but rumour had it that 'something was intended in favour of the queen of Scots'.

The decision to employ John Casimir had scarcely been taken when Bernardino de Mendoza arrived in England with the eagerly awaited communication from Madrid. As the first resident Spanish representative to be sent to Elizabeth's court since the expulsion of De Spes in January 1572 Mendoza was not a happy choice, being essentially a soldier puffed up with Spanish military pride, and contemptuous of the English. He explained that the king's determination to subdue the Netherlands remained unchanged, but that he intended to recall Don John, and would not abrogate but rather extend the ancient privileges of the country. Meanwhile Elizabeth was to understand that no threats of sending assistance to the rebels from England nor any other consideration 'will cause us to relinquish the determination we have adopted to bring our subjects back to obedience, using against them all the force that human or divine right permit us to employ and our royal dignity demands'. As a friendly neighbour, the queen must prohibit, under severe penalties, any help, direct or indirect, being sent to the Netherlands. At first Elizabeth was inclined to accept this statement of the king's intentions as satisfactory; but when the matter was discussed in the council Mendoza became acutely aware that the majority was hostile to Spain. He learnt also that unless there was an immediate suspension of arms it was the intention of the queen to send 'resolute aid' to the rebels. When he attempted, somewhat foolishly, to shake her decision by alluding to the length of his master's arms, he elicited the sharp retort that she would not allow either the French to set foot in the country or the Spaniards to rule it 'while she had a man left in her country'. On his expressing the hope that at least his dispatches would not be tampered with, he was blandly informed that no such assurance could be given, 'as certain people came here with no very good objects in view'; and as for his complaints about succours being sent to the Netherlands, 'there were so many people leaving and arriving in so large an island that she could not prevent them from leaving without permission'. Intimidation was obviously the wrong way to handle Elizabeth.

Matters were now going from bad to worse in the Low Countries. Two of the provinces, Artois and Hainault, were threatening to go over to Anjou, while great cities like Douai, Arras, Lille, Mons, Courtrai, and St. Omer were strongly in favour of making peace with Don John, and it looked as if the agreement of Ghent would be dissolved. The war party in England, fortified by Davison in the Netherlands and Paulet in France, renewed their clamour for open intervention. 'The loss of a day', wrote Paulet, 'is of great moment; the Spaniard knows it, and therefore seeks to gain time by all feasible means; and when his purpose is served, he may turn his flatterings into threatenings, and rip up old matters to ground his new quarrels.' But Elizabeth, despite her bellicose utterances to Mendoza, was not to be stampeded into war. In April she turned again to Don John, refusing to abandon her efforts 'so long as any spark of hope remains'. At the same time she approached Henry III with a proposal for joint mediation. In neither case was there any chance of success. Don John had become so 'insolent' by reason of his victories that 'he makes his account to be master of all these countries within two months'; and Henry III proved so broken a reed that he could only exclaim, 'You see how I am obeyed: I have no means to stay my brother otherwise than by force, which cannot be done without danger of new civil troubles.'

The resources of diplomacy were almost exhausted, but not quite. In a last effort to avert the peril of French intervention, which was now the only hope of the rebels, Elizabeth decided to send a powerful embassy, headed by Walsingham and Lord Cobham, to moderate the demands of both sides. If they failed to achieve this, they were to come to an understanding with Anjou's representatives that his entrance into the Netherlands would not mean their subjugation by France. 'Do your endeavour for a peace', wrote Wilson to Walsingham, as a final injunction, 'and you shall have thanks on your return. If you tell us of the necessity of war, I tell you plainly that we cannot abide to hear of it.' But the embassy which landed at Antwerp on June 28 soon found the task confronting it impossible of fulfilment. The states, backed by Orange, insisted on the withdrawal of Don John and the Spaniards, together with the acceptance by Spain of the Pacification of Ghent; while Don John refused to consider such terms 'even if he were a prisoner in the Brodhuis of Brussels'. Walsingham's private opinion was that 'it is greatly to be doubted that "Monsieur" will have a further foot in the country than ever the king of Spain will be able to remove, or we to hinder; of which I fear we shall too soon see an experience'. It did not lessen his fear to know that Monsieur had given his assurance that he would do nothing without the consent of the queen. Less than a fortnight later, on 13 August, a treaty was signed between Anjou and the states, giving the duke the title of 'Défenseur de la liberté belgique' and a joint share in the military command.

Elizabeth had undoubtedly contributed to this result by her dilatory policy, by holding back the bonds for the £100,000 she had promised, and by demanding the surrender of Sluys and Flushing as security for the moneys she had already spent in loans. Her apparently irrational conduct evoked from Walsingham the despairing cry, 'The only remedy left to us is prayer.' 'Where the advice of counsellors cannot prevail with a prince of her judgement', he wrote, 'it is a sign that God hath closed up her heart from seeing and executing what may be for her safety.'

The position in the Netherlands was one of increasing chaos. No fewer than four armies were encamped on her soil, for, in addition to the principals in the struggle, John Casimir with his horsemen and Anjou with his motley horde of Frenchmen had arrived on the scene. But since Elizabeth had drawn her purse-strings tight, and Spain could not finance Don John, all four armies were reduced to disorder through lack of pay and supplies. Discipline broke down, and bands of famished soldiers devastated the land as if a cloud of locusts had settled on it. Plague, famine, and pestilence completed the ruin. In September Walsingham and Cobham were ordered to return to England, their presence being no longer useful in the stricken country. A fortnight later (1 Oct.) Don John died a victim to the plague, broken in spirit, as well as in health, by the total collapse of his hopes. Finally John Casimir and Anjou, after taking part in a religious struggle between the people of Ghent and the Walloons, departed the country—the former to be honoured at Elizabeth's court with the Order of the Garter and banqueted by the city of London, the latter to return to France.

While Walsingham and Cobham were in the Netherlands, Elizabeth executed one of her swift and unexpected changes of front in foreign policy—she revived the Alençon (Anjou) marriage project. Her object in doing so was to draw off the restless and ambitious duke from his dangerous designs on the Spanish king's inheritance, or, at all events, to make sure that if he persisted in his enterprise he would be amenable to her advice and control. The bait of the most acceptable marriage in Europe might even turn the would-be liberator of the Nether-lands into a champion of England's interests. It is impossible to say whether Elizabeth intended to go farther than this when she reopened the question. If the love palaver failed to produce the political results she hoped for, there were plenty of loop-holes for escape from undesirable consequences, or diplomacy would provide them. Nevertheless there was a seriousness about this courtship of Anjou in 1578 entirely absent from all previous manœuvres of a similar nature. Apart altogether from the French danger in the Netherlands, a marriage with Anjou seemed to offer a means of egress from many of the fears that haunted the queen's mind, both domestic and foreign. It might, for example, get rid of the otherwise insoluble succession problem by providing an heir to continue the line of Henry VIII —a matter that troubled Elizabeth more acutely as she grew older. It would greatly strengthen the Anglo-French alliance, which was now honeycombed with distrust and sadly in need of repair. It would go far to dispose of the danger from Mary Stuart. It might even compel Philip II to come to terms with his rebels. In fact, when Burghley considered the benefits likely to accrue from it, and set them against the perils that would have to be provided for if the queen did not marry, he was convinced that the advantages were overwhelming. Sussex was also in favour of the marriage. The only obstacles in the way were the queen's age and Anjou's attachment to the catholic religion; but neither was insuperable. If Elizabeth was forty-five—a risky age at which to contemplate matrimony —the risk, after all, was not abnormal; and as for the duke, he was assuredly no zealot, as his championship of the huguenots showed. It remained to be seen, however, how public opinion in England would regard these matters.

The negotiations began in the summer of 1578; and in November Anjou sent over his agent, Jean de Simier, the most accomplished philanderer of the period, to discuss terms and to prepare the way for his master's 'frenzied wooing'. In a remarkably short time Simier won the queen's affections—so much so, indeed, that she added him to the menagerie of her pet 'beasts', nicknaming him her 'monkey' and the duke her 'frog'. The vicarious wooing went on all winter, while Burghley addressed himself to the business side of the transaction. The articles of 1571 were reviewed, and at the end of March 1579 Simier delivered twelve articles in French, three of which were new. So confident was he of success that on 12 April he wrote: 'I have very good hope, but will wait to say more till the curtain is drawn, the candle out, and Monsieur in bed.' Three weeks later, when the French terms came up for discussion in the council, he was somewhat shaken in his optimism, for some members were for rejecting the three new articles outright, and the majority held that, while the second must be disallowed, the first and third could only be dealt with by parliament, and should be kept in suspense until the duke had been inspected by the queen. The substance of the articles in question was as follows: Monsieur was (1) to be crowned king immediately after the marriage, (2) to share jointly with the queen the authority to grant all benefices, offices, and lands, and (3) to have an annual income of £60,000 during marriage, and the minority of any child of the marriage, being heir to the crown. Simier was dissatisfied with the decision of the council; but after consultation with his master intimated that the duke was prepared to leave all to the queen's determination. Thereafter the whole of the articles were gone over, some being accorded and others reserved for 'colloquy'. On 8 July Simier was handed an 'act of council' sanctioning the duke's visit, together with a safe-conduct, and about the middle of August the Frenchman set out on his adventure, travelling incognito to Greenwich. For the next twelve days—17 to 29 August—his courtship was the talk of London. Then he departed, leaving Simier to capitalize the results of his mission and bring the business to a conclusion.

But England was now alarmed. Hardly had Monsieur returned to France when John Stubbs published his *Discovery of a Gaping Gulf Wherein England is like to be Swallowed by Another French Marriage*, an intemperate puritanical pamphlet, grossly libelling the royal family of France, and picturing, in exaggerated language, the disastrous consequences of an alliance with it. Elizabeth was mortally offended at the insult to a neighbouring power whose friendship she desired to cultivate; and the writer, printer, and publisher were arrested, all three being sentenced to have their right hands chopped off and to be imprisoned during the queen's pleasure. Subsequently the printer was pardoned; but when the barbaric punishment was inflicted on the other two it was clear that they had the sympathy of the multitude, for the spectacle was witnessed in dead silence, broken only by the intrepid Stubbs, who, raising his hat with his left hand, cried with a loud voice, 'God save the Queen.'

When the council met to discuss the question, in October, whether the queen should be advised to go forward with the marriage or not, there was a marked difference of opinion. Five voted for it and seven against it. Burghley, who was its main supporter, concluded that unless Elizabeth herself was definitely in its favour he could not advise her to marry the duke. It was agreed, therefore, to ask the queen to 'open her mind' in the matter. When the news of the council's decision was communicated to her (7 Oct.) Elizabeth was greatly distressed: 'not without tears' she complained 'that her councillors should think it doubtful whether there could be any more surety for her and her realm than to have her marry and have a child to inherit and continue the line of Henry VIII'. On the following day the council revised its decision, and offered, after long consultation, to do all it could to further the marriage 'if it so pleased her'. To the surprise of every one the queen, who had now recovered her equanimity, received the offer coldly, trouncing those who had opposed the marriage, and reiterating that she had expected a unanimous request urging her to marry. As a matter of fact she realized that the game was lost. Nevertheless to save her face she played it out to the bitter end. On 24 November the articles were signed by Simier and the royal commissioners, two months being allowed for Elizabeth to dispose her subjects in favour of the marriage. A vain hope! What the council had refused to recommend of its own free will there was little likelihood of parliament agreeing to. So Simier returned to France at the end of November, to while away the time of waiting by inditing love-epistles to the queen, adorned with pink silk ribbon. 'Be assured on the faith of a monkey', he wrote, 'that your frog lives in hope.'

Meanwhile the situation in the Netherlands grew worse instead of better. Under the governorship of Parma, the ablest and most astute of all Philip's soldier administrators, and a brilliant diplomatist to boot, the catholic Walloon provinces made their peace with Spain (May 1579), leaving the calvinist north, leagued in the Union of Utrecht, to continue the grim struggle with the Spaniards. The Pacification of Ghent was now a thing of the past: so, too, was the idea of a united Netherlands, to which Orange had dedicated his life; and there was less hope than ever of a peace by compromise, for the star of the Spaniard was in the ascendant. But the resolute will of the Dutch stadholder remained unimpaired. Beset by perils on all sides, and troubled by Elizabeth's insistence upon repayment of the successive borrowings for which she had stood caution, William now decided to reopen negotiations with Anjou and to offer him the sovereignty. No doubt he was influenced in his decision by optimistic reports of the recent marriage discussions in England: the potential husband of the English queen was an asset of unquestionable value. If, as seemed not at all un-likely, the duke succeeded in his suitorship and accepted the sovereignty of the Low Countries, England would probably rally to his support, and the desperate problem of military defence against the Spaniard would be virtually solved. This, however, was not Elizabeth's way of looking at the matter—she was as opposed as ever, in principle, to any radical change in the status of the Netherlands; she could not contemplate marriage with Anjou in the existing state of English public opinion; least of all did she want a war with Spain. On the other hand, she could not afford to see France step into the breach and profit by the duke's crusade in the Netherlands—assuming he accepted the sovereignty. The readiest way out of the difficulty was to keep the hope of marriage dangling before the eyes of Anjou, thereby securing some sort of control over his actions, so that whatever happened the result would not be unduly harmful to England's interests. If the worst came to pass, and he proved to be un-manageable, she might, as a last resort, defy public opinion at home and marry him.

It was, indeed, an anxious moment. The relative security in which England stood in 1575 was rapidly passing away. Pope Gregory XIII had republished his predecessor's bull of excommunication and declared war on the queen in Ireland: the Jesuit mission was about to leave Rome—to prepare the way, it was believed, for the *empresa*: Munster was in revolt under Fitzmaurice, subsidized by money and munitions from Spain; and Scotland was beginning to drift from its moorings in the English alliance, after the fall of Morton's regency and the assumption of personal power by King James. If the outlook was gloomy nearer home, it was all the gloomier because of the prevalent unsettlement abroad, consequent upon the emergence of the Portuguese succession question. Since the death of King Sebastian at the battle of Alcazar (Aug. 1578) statesmen in all countries had been perturbed about the future of Portugal, for no less than six candidates were preparing to contest this rich prize as soon as the aged cardinal-king Henry, Sebastian's uncle and successor and the last representative of the dynasty, breathed his last. Of the six claimants—Don Antonio, a bastard nephew of the late king, the dukes of Braganza and Savoy, Alexander of Parma, Catherine de Médicis, and Philip II of Spain—the only one who was likely to succeed was the king of Spain, whose right was as indisputable as his sword was sharp. It required no imagination to foresee an immediate *bouleversement* of the existing balance of power in Europe if Portugal with its dependencies, its navy and material resources, became an appanage of the Spanish crown.

The spectre of Spanish aggrandizement was particularly alarming to Elizabeth: the more so, perhaps, because France, the ancient opponent of the house of Habsburg, was not in a position to resume her traditional role. On the contrary, that country was in imminent danger of being submerged beneath a tide of fanatical catholicism, generated by the League and the pro-Spanish Guise faction. Thus when the crisis in Portugal came, with startling suddenness, at the end of January 1580, the Spaniards were ready to overrun the country without fear of serious opposition. By September they had mastered it, and Don Antonio, after a futile struggle, was driven into exile. Only at Terceira, in the Azores, was he acclaimed king of Portugal. Clearly, if Spain was to be checked, France must be pacified forthwith, and the whole force of the kingdom directed to the support of Anjou in his projected enterprise in the Netherlands, and of Antonio in Portugal. It was a strange reversal of previous English policy, that what the queen had feared most in the past should now be accepted as imperative for her security. In 1571–2 she had frowned on the Coligny-Nassau plan: in 1580–1 she gave its Anjou-Orange counterpart her ardent blessing. Circumstances had altered her attitude. A European war in which France and Spain would figure as principals, with its main theatre in the Low Countries and a subsidiary one at the Azores, was the only way of dealing with the threat to her own dominions; and to the attainment of this object Elizabeth now devoted all her diplomatic skill and ingenuity. When her efforts to mediate a peace in France failed, she revived the moribund marriage negotiations as a lure both to the duke and his brother, the French king. In the end she partially succeeded. Anjou, who made the treaty of Plessis-les-Tours with the States, accepting the sovereignty (Sept. 1580), set out on his second adventure into the Netherlands in the following year, a willing servant of the queen. A year later (1582) the French government launched a great offensive against the Spaniards at the Azores. In neither case did England take any material share in the enterprise; but she gained the respite necessary for dealing with her more pressing problems at home, with Ireland, Scotland, and the papal emissaries in England.

The key to the complicated diplomatic history of the period is to be found in the marriage negotiations, which assumed an importance out of all proportion to their intrinsic worth. First came an illustrious embassy from France, in April 1581, headed by the prince dauphin, the duke de Montpensier, the count of Soissons, and Pinart, secretary of state, which was received with a round of banquets, bear-baitings, and 'triumphs', but achieved no positive result. It transpired that Elizabeth could not saddle herself with a war on Monsieur's behalf; for war would be the inevitable result of a marriage, Anjou being committed to the support of the Netherlands. War meant 'charges', and England would not 'digest' charges. It would be better to have a league. The embassy had, of course, no authority to conclude a league: so it returned to France in June, having accomplished nothing. Then came Walsingham's great mission to the French court (July–Sept.) to clear up the matter, or, perhaps, to wrap it in deeper obscurity—probably the most exasperating mission ever committed to this sorely tried statesman. His instructions, as he understood them, were to work for an alliance without marriage, but not to break off the hope of marriage altogether. He was given a free hand, subject to compliance with two conditions: that marriage with a 'present war' could not be entertained, and that the duke of Anjou must not abandon his enterprise for the marriage. If the king was adamant that the marriage must precede the league, Walsingham was to say that 'we shall be content to marry . . . so as the French king and his brother will devise how we shall not be brought into a war therewith, although Monsieur shall continue his actions in the Low Countries'. But the French court was not blind to the purport of this complicated and superfine diplomacy. As Anjou remarked to Walsingham, his brother was afraid that if a league were made without the marriage to consolidate it the queen might 'slip the collar' and leave him 'in the briars'. Walsingham, too, was perplexed by the continually changing instructions he received from England. In August he wrote to Burghley:

'I would to God her Highness would resolve one way or the other touching the matter of her marriage. . . . When her Majesty is pressed to marry, thus she seemeth to affect a league, and when a league is proposed, then she liketh better of a marriage. And when thereupon she is moved to assent to marriage, then hath she recourse to a league; when the motion for a league or any request is made for money, then her Majesty returneth to marriage.'

If Walsingham was mystified by the queen's variability, he had no doubt about the attitude of Henry III and Catherine de Médicis: they insisted on the marriage—no marriage, no league. Anjou, however, was more pliable. About the middle of August, having completed his preparations, he set his army in motion and crossed the frontier. Cambrai fell into his hands almost at once, Parma withdrawing his forces, all 'his swagger gone and vanished into smoke'. But the campaign did not thrive: it stopped from lack of funds, although subsidies of £10,000 and £4,000 were sent to bolster it up from England and France respectively; and, at the end of October, Anjou placed his men in winter quarters at Le Catelet. On 2 November he paid a visit to Elizabeth at Richmond, which turned into a second courtship of three months' duration, more remarkable than the first. 'The queen', wrote Mendoza, 'doth not attend to other matters but only to be together with the duke in the chamber from morning to noon and afterwards till two or three hours after sunset. I cannot tell what a devil they do.' On the 22nd of November Elizabeth staged a superb piece of acting. Exchanging rings with Anjou, she kissed him in the presence of the French ambassador, Leicester, and Walsingham, adding the words (to the Frenchman): 'You may write this to the king, that the duke of Anjou shall be my husband.' The love palaver continuing, the duke forgot all about the Low Countries, and it became difficult to get him out of England; but eventually at the price of a solatium of another £10,000 he was prevailed upon to take his departure. Elizabeth accompanied him to Canterbury, leaving Leicester and a bevy of nobles to see him safely across from Dover to Flushing, where he arrived on 10 February 1582.

At Antwerp he was awaited 'not otherwise than the Hebrews awaited the coming of the Messiah'; but expectations were cruelly disappointed. He found it impossible to collect an army able to cope with Parma, who had captured Tournai in his absence. The States could provide neither men nor equipment nor money, and to make matters worse Elizabeth began to pester them for the repayment of her loans, and the moneys they had borrowed under cover of her credit. Even in France he could not raise recruits, because Henry III would not recognize the enterprise officially. As the year advanced, his position became more and more unbearable. He found that the sovereignty given to him was hedged about with restrictions, that he and his mercenary levies—a motley rabble—were not well liked by the population or by the older huguenots serving with Orange, and that it was difficult even to procure a church for the exercise of his religion. In July Parma captured Oudenarde, and inflicted an overwhelming defeat on the States at Ghent. This was followed by a crop of fresh surrenders to the Spaniard, and a corresponding diminution of Anjou's prestige, who, strange to say, had taken to tennis, riding, running at the ring, and hunting the otter with spaniels, while Parma was capturing cities.

Meanwhile the French government, after prolonged hesitation, had dispatched a naval force under Philip Strozzi to assist the Portuguese pretender, Antonio, at Terceira. It was a foolish expedition, badly organized and practically unsupported by Elizabeth, who, in virtue of Philip's warning (August 1581) that countenance of Antonio would be taken by Spain as a declaration of war, limited her contribution to a few transports. Its destruction by a Spanish fleet under the command of Santa Cruz (July 1582) seemed to vindicate the verdict registered at Lepanto, twelve years before, that Spain was the greatest sea-power of the age. Strozzi was killed in the engagement, and the survivors were hanged as pirates at Villafranca: a testimony to the fact that the Spaniard could be as ruthless as Lord Grey at Smerwick.[1] Coming, as it did, about the same time as Parma's triumph at Oudenarde, the victory of Terceira completed the discredit of French arms, and caused no little consternation in the Low Countries.

Elizabeth was annoyed at the miserable position of Anjou and the lack of support given to him by the States. In August she wrote a sharp letter of remonstrance to Orange. 'If they act in this way', she said, 'and behave so to him as to compel him to withdraw once more, in a manner of speaking in his shirt sleeves, to his dishonour . . . they may be assured . . . we shall be the first to advise him flatly to leave the whole thing

[1] See p. 398.

betimes and to take no more account of them than their ingratitude deserves.' Henry III had also his qualms about his brother's discomfiture. In November he allowed the duke de Biron to conduct reinforcements into the Netherlands. But it was impossible to stop Parma's progress. By the end of the year the Spaniards had recovered all Flanders south of the Lys and the Scheldt with the exception of Alost and Dendermonde: only four cities—Bruges, Ghent, Brussels, and Mechlin—now stood between him and his chief objective, Antwerp. Early in January 1583 Anjou had recourse to desperate measures to re-establish his credit. On a preconcerted signal his troops made an attack on several cities, including Antwerp, Bruges, Ostende, Dixmude, Dunkirk, Nieuport, with a view to getting rid of the limitations placed upon him by the States. The main attack at Antwerp was defeated by the burghers, who afterwards alluded to it as the 'French Fury' and made it impossible for a Frenchman to move about the city without the protection of the burgomaster and magistrates. 'Since France was France', commented Roger Williams, in a letter to Walsingham, 'France never received so great a disgrace.' It practically marked the end of the duke's career as 'liberator of the Netherlands'. Shortly after, he left the country for France, all efforts to patch up an agreement by Elizabeth and Orange proving unsuccessful. He died in May of the following year.

All this time Elizabeth had been keeping a careful eye on Scotland, where the ascendancy she had enjoyed for many years under Morton's iron-handed rule had given place to a feeling of insecurity and foreboding. Trouble began in March 1578, when a combination of the regent's enemies, led by the earls of Atholl and Argyll, succeeded in driving him from office and obtaining control of the young king. Immediately the effect of this revolution was felt abroad. In April the duke of Guise and Vargas, the Spanish ambassador in Paris, were in communication with a view to armed intervention on behalf of Mary Stuart, and Archbishop Beaton, the queen of Scots' ambassador in France, plied both with urgent requests for help. Even the hesitant Philip II began to interest himself in the possibility of using Scotland as a means of checking Elizabeth's interference in the Low Countries. The danger, however, passed. In June Morton recovered his position in the council and there was a

respite for a few months. Then in the autumn of 1579 a person-
age arrived in Edinburgh who was destined to excite the liveliest alarm among Elizabeth's advisers, namely Esmé Stuart,
seigneur d'Aubigny, nephew of regent Lennox and cousin to
the king's father. This new entrant upon the political stage
is one of the most romantic and intriguing figures of the time,
albeit surprisingly little is known about him. His polished
French manners, which screened a subtle if not profound mind,
enabled him to glide with remarkable ease through the inter-
stices of Scottish politics, until he became almost a dictator, out-
rivalling the dour and repellent Morton. The king took to him
at once as a convenient refuge from the imperious mentors who
surrounded him on every side, and loaded him with honours
and offices. He was granted the valuable abbacy of Arbroath,
the earldom of Lennox, the governorship of Dumbarton castle,
and the high chamberlainship of the kingdom. Very soon it was
said that all Scotland was 'running the French course'. Worse
still, there were rumours afloat that the new favourite was an
emissary of the duke of Guise and the pope. Under his leadership a confederacy was formed for the overthrow of Morton;
and in December 1580 the great man was arrested and thrown
a prisoner into Dumbarton castle, despite Elizabeth's frantic
efforts to save him. Six months later (June 1581) he was condemned to death and executed for complicity in the Darnley
murder thirteen years before—again not without vehement protests from the English queen. For a short time, indeed, it
almost seemed as if Morton's fate would plunge the two kingdoms into war, for Hunsdon at Berwick had instructions to
prepare a punitive expedition against Edinburgh. But once
again the tension slackened. There was no party in Scotland
in favour of drastic action, and France was on the watch for a
breach of treaty obligations. Thus Lennox and his coadjutor,
Captain James Stewart, who had been personally responsible
for the prosecution of Morton, became virtually masters of the
kingdom. Stewart was rewarded with the earldom of Arran.

Elizabeth's position was one of acute anxiety, much worse than
it had been at the time of Mary's triumph over the protestant
lords in the autumn of 1566, for the forces of the counter-
reformation were now gaining ground every day, and fully one-
third of the Scottish nobility was catholic. So formidable were
the adherents of the old religion in the northern counties of

Caithness, Sutherland, Inverness, Moray, and Aberdeen, and along the border from Nithsdale to Teviotdale, that it would have required only a small expeditionary army, resolutely handled, from France or Spain to turn the scales definitely against the protestants. Hardly less disturbing from Elizabeth's point of view was the doubtful position of the king. James was now in his sixteenth year, still too young to take a firm grasp of policy, and far from mature in mind and character; but already he was beginning to show traits that set shrewd observers thinking.

According to Fontenay, brother to Claude Nau, Mary Stuart's secretary, who studied him closely in the year 1584, James was a *vieux jeune homme*, a curious blend of youthful audacity and the wisdom of eld, learned in languages, science, and affairs of state, of a quick and penetrating intelligence, clever in debate, whether the subject was theology or any other, and gifted with a long and accurate memory. Supremely confident in his own powers, he hated to be surpassed by others and longed to be accounted a brave fellow, but was really a weakling and a coward. His passion for hunting, in the pursuit of which he spent long hours in the saddle, made him neglectful of affairs and overmuch dependent upon favourites; but he excused himself by boasting that he could discharge more business in an hour than others in a whole day, and that nothing of importance could happen in his realm without his knowing about it, because he had spies at his nobles' doors day and night, who reported to him constantly. His worst fault was his vainglory: he was ignorant of his own insignificance, but esteemed himself superior to all other princes. Not a very pleasing description of a young man in his teens, it may be said, but indicative of latent intellectual force with which the experienced Elizabeth would soon have to try conclusions. More than once his swift and disconcerting manœuvres would outwit her eagle eye and test her patience to the uttermost: so that on one occasion she was heard to remark: 'That villain of Scotland! What must I look for from such a double-tongued scoundrel as this?' No doubt she had a strong hold over him in his hopes of the English crown, but she could not trust his good-will; and although he knew full well the necessity of keeping on good terms with her, he was not prepared, any more than his mother before him, to become her 'thrall'. If she would not accommodate him in the matter of the succession, he was only too ready to play upon her fears by fraternizing with her enemies in the best machiavellian manner of the age. Even his protestantism appeared to sit lightly on him; and many believed, or affected to believe, that if he were withdrawn from the influence of calvinist clerics and tactfully handled by priests, he would willingly proclaim himself a catholic, especially if his conversion materially improved his chances of the English crown.

It is hardly surprising, therefore, that the interest of the whole catholic world turned again to Scotland as the chief hope of better things in Britain. True, England had an unfaltering ally in the kirk, whose sleepless vigilance and powerful influence in politics were worth a host in themselves. But Elizabeth would have been singularly regardless of her interests had she not striven with every device at her command, short of open war, to combat the machinations of her enemies on Scottish soil, to regain the ascendancy she had lost by Morton's fall, and to retain Scotland, as the phrase then ran, 'at England's devotion'. The record of her intrigues, plots, and counter-plots makes this one of the most baffling periods in the reign.[1]

Meanwhile the establishment of Lennox in the seat of power opened the way for the realization of the scheme with which his name is always associated—the conversion of the king to catholicism, the catholicization of Scotland, and the restoration of Mary to the sovereignty of which she had been deprived in 1567. This was to be merely the first stage in the main enterprise against England, on which the duke of Guise, the queen of Scots, Mendoza, the jesuits, the pope, and the king of Spain had set their hearts. It was the jesuits, however, who took the initiative in laying the foundations for this wider and more pretentious plan: Philip and the pope remained in the background. First of all Fathers Holt, Crichton, and Hay went to Scotland during the year 1581–2 to establish relations with Lennox and the Scottish catholic nobles, Huntly, Caithness, Seton, Eglinton, and Ferniehirst, to whom they communicated exaggerated and glowing reports of the readiness of the Spanish king and the pope to lend assistance. Then, in May 1582, a conference was held in France, under jesuit auspices, to which Mendoza was invited, but for obvious reasons could not go, and

---

[1] For details see C. Read, *Mr. Secretary Walsingham*, vol. ii, ch. ix, or P. F. Tytler, *History of Scotland*, vol. iii, chaps. vi, vii.

which Guise alone of the responsible political leaders attended. Here the details of the invasion plan were fully discussed. Lennox with the support of Spain and the pope, to the extent of eight thousand troops, was to secure the king's conversion and the subjection of the kingdom to the catholic faith. After this was accomplished, the combined Spanish and Scottish force would cross the frontier into England under the command of Lennox or the king: the English catholics, of whom, according to Holt, there were 'incredible numbers' ready to rise as soon as the signal was given, would join hands with it; and the mastery of England as well as Scotland would be assured. Guise gave his approval to the plan, and even offered to create a diversion by landing a contingent of French troops on the Sussex coast. But it will be obvious that a more flimsy, fantastic, or immature plot for the subversion of a well-established kingdom could hardly have been devised by men unfamiliar with the technique of military enterprises. Mendoza and Mary were astounded at the political ineptitude of the jesuits, who seemed to believe that armies could be conjured out of the ground by the mere stamp of a foot. Moreover, Philip had said nothing so far of a Spanish expeditionary force; nor, indeed, could such a force be available until the affairs of the Netherlands were settled. The time had clearly come for Mendoza, who ought to have been the co-ordinating brain of the whole conspiracy, to reassert political control over the schemers and bring them back to realities. But before he could straighten out the tangle he was beaten by English intrigues in Scotland.

In August 1582 the anglophil faction, led by the earls of Angus, Gowrie, and Mar, captured the king by a stratagem while he was hunting near Ruthven castle, Gowrie's Perthshire residence. Then followed the arrest of Arran, the break-up of the Lennox confederacy, and finally the flight of Lennox himself to Dumbarton, whence he made his way, by Elizabeth's permission, through England to France. This unexpected blow completely upset the calculations of the conspirators, who now realized that Scotland was a harder nut to crack than they had imagined. 'It would be better', said Guise, 'to effect the enterprise by way of England'—a sentiment shared by Tassis, the Spanish ambassador in Paris, who wrote: 'It appears to me that the true road to success is by England rather than by

Scotland.' Mendoza also agreed, for, as he said, the English catholics were unlikely to support James unless they were sure of his conversion and subjection to Spain. Further support for the reconstructed scheme was forthcoming from Mary's agents —Charles Paget, Thomas Morgan, and Charles Arundel; and the jesuits were apparently 'so confident . . . that it is impossible for any one hearing them to help being convinced'. Plans were therefore made for the abandonment of Scotland and a direct invasion of England. Spain was to provide an original striking force of 4,000 or 5,000 men from Parma's army in the Nether-lands, which would be joined as soon as it landed in England by 20,000 English catholics. Guise was to be commander-in-chief of the expedition. The English catholic exiles on the Continent would accompany him; so too would William Allen, head of the seminary at Rheims, to whom was to be given the title of bishop of Durham and the task of publishing a papal bull justifying the enterprise. Charles Paget was to assure his co-religionists in England that the invading army would be withdrawn as soon as its object was achieved, and Mendoza, after remaining at his post to the last possible moment, was to cross to Dunkirk and join Guise.

Meanwhile Elizabeth, ignorant of the danger threatening from abroad, was anxiously watching Scotland, where another revolution had taken place in July 1583, leading to the over-throw of the Ruthven clique and the return of Arran to power. In September, hearing of a renewal of French machinations in the northern kingdom, she sent Walsingham with a splendid embassy to remonstrate with James, and to see what he could do to restore English influence. The veteran diplomatist was no lover of the Scots, whom he regarded as a mercenary nation to be won only by hard cash—a method of persuasion he knew Elizabeth would never sanction. From the first, therefore, he fought a losing battle. He had nothing to offer—no gifts of money, no promises with regard to the succession: the anti-English party were in a strong position; and the king, though friendly, was in no mood to listen to expostulation on his 'crooked ways'. It is hardly surprising that Walsingham's report was a bitter one. He pronounced James 'an ingrate and such a one as if his power may agree to his will, will be found ready to make as unthankful a requital as ever any did that was so greatly beholding unto a prince, as he hath been unto your

Majesty'. Nay, he concluded that the only way to deal with him was to have him 'bridled and forced, whether he will or no'.

Since Elizabeth was just as averse to the use of force as to bribery, there was clearly nothing to be done but to recall her ambassador, and work for a more favourable turn in Scottish affairs in conjunction with the now exiled Ruthven nobles. Walsingham therefore returned to London in October.

Although the mission to Scotland had been a failure, it was to some extent offset by the immediate discovery of the Hispano-Guise plot. The revelation came as the result of an accident. Hitherto the government had been on the wrong tack, thinking that if trouble came it would come from France through Scotland. But now Walsingham learned from one of his spies in the French embassy of the suspicious movements of a certain Francis Throckmorton,[1] who, in fact, had been drawn into the Guise conspiracy by Morgan, and had become one of the principal intermediaries through whom Morgan communicated with Mary, and Mary with Mendoza. In November Throckmorton was arrested and his papers confiscated. On the rack he confessed everything he knew. So complete was the disclosure that the plot collapsed on the stocks before it was ready to be launched. Throckmorton was executed for treason: others were arrested and imprisoned; and the government decided to take drastic measures against Mendoza. On 9 January 1584 he was summoned before the council and ordered to leave the realm within fifteen days. It was the second time that a Spanish ambassador had been expelled from England for alleged treason. But Mendoza was no craven. 'The insolence of these people', he wrote to Idiaquez, the Spanish secretary of state, 'has brought me to a state in which my only desire to live is for the purpose of avenging myself upon them, and I pray that God may let it be soon and will give me grace to be His instrument of vengeance, even though I have to walk bare-footed to the other side of the world to beg for it.' When he applied for a ship to carry him across the Channel to France it was refused him; the queen, it was explained, reserved such privileges for her friends. His last words were almost a declaration of war. 'Don Bernardino de Mendoza', he remarked to the officer who brought the message of refusal, 'was born not to disturb

[1] A nephew of Sir Nicholas, and a catholic.

countries but to conquer them.' If war did not follow between England and Spain, it drew appreciably nearer.

There was no denying the fact that the peaceful days of Elizabeth's reign were rapidly passing away. During the year following Mendoza's expulsion, Parma continued slowly but surely his triumphant career in the Netherlands, 'invading deeper and deeper into the land', and drawing his toils closer and closer round Antwerp. In April he recovered Ypres, in May Bruges, in August Dendermonde, and in September Ghent. By the early autumn of 1584 all west Flanders except Ostend and Sluys was in his hands, and the Spaniards had begun to penetrate into the land of Waes, thus cutting the communications between Antwerp and the great cities of Brabant and Flanders. Meanwhile, at the end of May, Anjou had died at Château-Thierry, and a month later Orange fell a victim at Delft to the pistol of Balthazar Gerard, a fanatical 'Burgonian', who had wormed his way into the prince's household for the express purpose of murdering him and winning the reward set upon his head by King Philip in September 1580. It thus seemed only a question of time when the United Provinces, bereft of their native leader and the only foreign prince who was committed to their defence, would be reduced to the Spanish obedience. In their despair they turned again to France, offering the sovereignty to Henry III. But France could not and would not take up the burden, either alone or in conjunction with England. Henry of Valois, a wastrel and a pietist, and the last of his line, was spending his time in devising new spiritual ecstasies, and squandering his treasure on unworthy favourites. 'If he had a foolish toy in his head,' wrote Sir Edward Stafford, 'or a monk's weed to make, or an *Ave Maria* to say, he would let his state go to wrack.' What with his new Order of Jeronomists, his pilgrimages, and his 'mignons', the king of France was fast becoming a mere *roi fainéant*, while the duke of Guise and Henry of Navarre contended with each other over the succession—Navarre for his legitimate rights, and Guise in order to prevent the accession of a huguenot to the throne. In vain Elizabeth tried to rally the miserable king to a sense of his duty by pointing out the danger from Spain and the house of Guise. 'Jesus!' she wrote, 'Was there ever a prince so smitten by the snares of traitors without the courage or counsel to reply to it? . . . For the love of God rouse yourself from this too long sleep.' In January 1585 Guise made the secret treaty of Joinville with Philip of Spain, and in March, subsidized by Spain, unfurled the banner of the Holy League, professing as his aim the coercion of the king, the enforcement of the Tridentine decrees, and the exclusion of Navarre from the succession. Truly the Tagus had at last poured into the Seine and the Loire. France was a negligible factor in the affairs of Europe, or of value only as a satellite of the Spanish monarchy.

The time had come for Elizabeth to bestir herself. During the previous autumn Burghley was of the opinion that if France could not be relied upon to send assistance to the Netherlands, England must enter into the war without delay, if only for the purpose of keeping the inevitable conflict with Spain confined to the dominions of her enemy, and retaining the still unconquered Holland and Zeeland as her allies. In November 1584 Elizabeth sent over Davison with a tentative offer to the states that if the towns of Flushing, Brill, and Enckhuizen were handed over as caution for the recovery of expenses and a guarantee that no peace would be made with Spain without her approval succour would be given by England. Negotiations, however, moved slowly, partly because of the conflicting reports as to the resisting power of Antwerp, and partly because of the difficulty in reaching a decision as to the nature of the power Elizabeth was to assume in the Netherlands. The states proffered the sovereignty: the queen would accept only a protectorate. It was June 1585 before the commissioners from the Low Countries reached England, and another seven weeks elapsed before a satisfactory agreement was obtained. Even so, when the treaty was eventually signed on 2 August, it was concerned merely with the relief of Antwerp and not the general defence of the provinces. Only after the fall of the great seaport (7 August) did Elizabeth sanction the conclusion of a general treaty (10 August), binding her to provide an army to serve in the Netherlands at her cost until the end of the war. The states, for their part, undertook to hand over Flushing, Brill, and the fort of Rammekens as 'cautionary towns' to be held by England until the expenses incurred were repaid in London. In September the earl of Leicester received his commission as lieutenant-general, and shortly after the middle of December set sail with a splendid retinue for the Netherlands. His army totalled 6,000 foot and 1,000 horse.

The campaign that followed—if campaign it can be called—began under brilliant auspices. The earl was welcomed with the cry 'God save Queen Elizabeth!' uttered as heartily, he noted, 'as if the queen herself had been in Cheapside'. Bonfires, fireworks, pageants, and banquets awaited him in every town he visited. The triumphal pomp of his reception at The Hague was celebrated in engravings. At Amsterdam he was greeted with a display of aquatic monsters—'whales and others of great hugeness'—who seized his ship and towed it to the shore. 'Never was there a people in such jollity as these be,' was his gratified comment. Alas for his hopes! Leicester was neither a great general nor a wise administrator. Ominously enough, he began by flagrantly disobeying his instructions and the letter and spirit of the queen's express declaration that she had no intention of governing the Netherlands either directly or indirectly. He accepted the title of 'His Excellency the Governor and Captain-general'. Elizabeth was furious at his indiscretion. 'It is sufficient', she remarked, 'to make me infamous to all princes, having protested the contrary in a book which is translated into divers languages.' In her anger she would listen to no expostulations, either by the earl himself or by his friends among her ministers: the obnoxious title must be rescinded at once, or she would withdraw him. Only after Burghley had threatened to resign and the states had made an earnest supplication for the title to be allowed for the time being, did she relent and relax. It was a bad beginning; but worse was to follow when the question of finance arose. Elizabeth had undertaken to maintain her expeditionary force for the duration of the war, at a cost, it is calculated, of £126,000 per annum—a very considerable sum when her other commitments, actual and prospective, are taken into account, and her entire ordinary revenue was not more than £300,000. Stringent economy in expenditure, always a fundamental matter with Elizabeth, was now more imperative than ever. Yet in spite of instructions to handle the disbursement of the queen's treasure with care, and warnings from Walsingham to beware of 'charges', Leicester appears to have used part of the money transmitted to him to alleviate the condition of the English troops in the service of the states, of whom there was a considerable number, and to

have taken no steps whatever to check the widespread pecu-
lation that went on among the captains, although this, too,
was part of his instructions. Nay, he even increased the officers'
pay, his own included. Can it be wondered at that Elizabeth
soon grew 'weary of the charges of the war'? Her lieutenant-
general had become, in effect, a 'conduit along which English
treasure passed into the abyss of Dutch bankruptcy'.[1] To
crown all, the queen was supremely anxious for peace: she did
not accept the view that she was irrevocably at war with Spain.
On the contrary, now that she had the 'cautionary towns' in
her possession, she was hopeful that she might use them to
bargain for a general peace, safeguarding the interests both of
the Netherlands and of England. It did not trouble her that
this might be considered a gross betrayal of the people she had
promised to help. 'We are so greedy for a peace', wrote Walsingham to Leicester, 'as in the procuring thereof we neither
weigh honour nor safety.' Nor did Leicester improve matters
by quarrelling with Sir John Norreys, the best of his subordinates, or by his bickerings with the states over the extent of the
power committed to him. The English army rapidly dwindled.
'To say rightly', wrote Sir William Pelham, marshal of the host,
'such are their miseries as I know not how to turn to satisfy
them, for some wanting wherewith to feed them, others almost
naked, many falling sick daily, and all in general barefoot
wanting hose or shoes, do by hundreds flock about me if I stir
abroad amongst them, crying for relief of their troubles.'

In brief, the campaign was a miserable failure. Apart from
a few brilliant exploits like the capture of Axel, the check to
Parma at Zutphen, the taking of Doesburg—all of them due
to individual bravery rather than the scientific organization of
the generalissimo, the war added nothing to the stability of
the states, and brought no glory to the name of Leicester. The
true hero of the campaign was Sir Philip Sidney, who died of a
gunshot wound received at Zutphen.[2]

If Elizabeth had been reluctant to strike a blow against Spain

---

[1] J. E. Neale, *Queen Elizabeth*, p. 289.

[2] Leicester returned to England in November 1586. Seven months later he
was sent back to the Netherlands, with a fresh army of 5,000 and £30,000 to
prepare the way for peace with Spain, and to secure a cessation of arms. But his
temperamental unfitness for leadership, combined with the stubborn resistance of
the Dutch to a policy of peace, rendered his mission again a failure. Soon after
his arrival he fell foul of the states-general, weakened his own forces by cashiering

in the Netherlands, Philip had done his best to incur her hostility on the sea, with dire consequences to himself. On 19 May 1585 an order had been issued to the corregidor of Biscay to seize any shipping lying in Spanish ports belonging to England, Holland, Zeeland, and 'the other states and seignories that are in rebellion against me'. As there were a considerable number of English corn-ships at Bilbao and other ports, the seizures that followed resulted in great losses to the merchant community, together with the imprisonment of hundreds of English sailors. Only a few captains succeeded in getting their ships away, among whom was the captain of the *Primrose* of London, who brought the news to England, along with a copy of the writ authorizing the seizures, and the corregidor of Biscay himself, whom he had captured during the scuffle in the port. The outcry against this act on the part of the Spanish king was so general in England, even among the merchants, who had hitherto been afraid of violent measures, that a retaliatory embargo was at once placed on Spanish goods: letters of reprisal were issued to the merchants; and Drake was ordered to proceed with a fleet for the release of the confiscated vessels. On 14 September the great corsair set sail with a composite flotilla consisting of 30 ships, mainly merchantmen from London and the west country, and 2,300 men—ostensibly for northern Spain; but his real objective was the West Indies and the *flota*. At Vigo, his first landfall, he took plunder to the extent of 6,000 ducats. Then, leaving the Spanish coast, he made for the Canaries and Cape Verde islands, where he burnt Santiago and Porto Praya. Late in November he stood away across the Atlantic, touched at Dominica and St. Christopher, and launched himself with all his force at San Domingo (Hispaniola), the richest and finest city in the Indies, and the 'Queen of Spain's colonial empire'. San Domingo was ransomed for 25,000 ducats, part of it being burnt in order to extort the payment quickly; and on 1 February Cartagena on the mainland was similarly treated, its shipping burnt, and a payment of 110,000 ducats

his old enemy, Sir John Norreys, and irritated the Dutch commander, Maurice of Nassau, with whom he attempted to relieve Sluys, then besieged by Parma. The enterprise failed, and Sluys fell into the hands of the Spaniards in August. In November Leicester was recalled to England, and never again set foot in the Netherlands. His repeated failure did not, apparently, weaken his position at court, for in the following year he was appointed commander of the English forces assembled at Tilbury to meet the armada.

exacted. At the end of March 1585 Drake put to sea again with a fresh cargo of plundered merchandise, guns, and all the church bells and metal he could lay hands on, together with a great quantity of furniture stolen from private houses. Altogether his loot amounted to more than half a million of our money. After hovering about the Yucatan channel and Havana, in the hope of intercepting the *flota*, he bore off to Florida, destroyed fort St. Augustine, and finally visited Raleigh's colony at Roanoak, where he rescued the disappointed colonists and brought them back to England in the midsummer of 1586.

This remarkable exploit filled Europe with amazement, not so much by the quantity of plunder it produced as by the moral damage it did to Spain. Spanish credit was temporarily shattered, and the sources of King Philip's wealth and power seemed to be absolutely at Elizabeth's mercy. The aggressive power of England, hitherto not fully realized, was seen to be so great that the Antwerp merchants, on whom both Parma and his master had relied for loans, tightened their pursestrings. If Leicester had failed to prick the bubble of Spanish might, Drake had shown his countrymen how and where to do it.

Meanwhile the situation in Scotland had righted itself in England's favour. After three years and nine months of continuous intrigue and counter-intrigue, and at least two abortive rebellions, the Arran régime had been at last destroyed by the combined efforts of the exiled Ruthven nobles, the Master of Gray, one of the wiliest intriguers of the age and Arran's deadly enemy, and Sir Edward Wotton, the English ambassador in Edinburgh. In July 1586 James was induced to sign the treaty of Berwick, thereby becoming a pensioner and ally of England. The two monarchs undertook to maintain the religion professed in their respective realms, to assist each other with armed forces in the event of invasion by a third power, and to make no alliances prejudicial to each other with any foreign state. In return for this agreement the king received a pension of £4,000 per annum from the English exchequer; and although no reference was made to the succession, it was understood that the queen would not permit any measure to be brought forward in parliament detrimental to his right. The name of the queen of Scots was not mentioned either in the discussions leading up to the treaty or the treaty itself, which was simply a mutual insurance arrangement between Elizabeth and James.

Note on Scotland

The account given in this chapter of the issues at stake in Scotland, though brief, is more or less the accepted version. It should be noted, however, that a good deal of mystery still hangs about the period. We have no adequate biography of James VI and none at all of Lennox; and although Conyers Read's account of the various English embassies to Edinburgh between 1580 and 1583 is admirable, it lacks the force it would otherwise have if we knew more of the actual state of affairs in Scotland. The first requisite is a reasoned inquiry into the attitude of both the king and Lennox to the catholic intrigues of the time. Until this is forthcoming it will be impossible to arrive at any reliable conclusion as to the merits or demerits of Elizabeth's later Scottish policy.

Note on the change in the Calendar, 1582

In order to rectify the error of ten days which had crept into the Julian calendar, Pope Gregory XIII, by the Bull *Inter gravissimas*, ordained that 4 October 1582 should be followed by 15 October, the intervening days being dropped. The alteration was adopted at the exact time in Italy, Spain, and Portugal. Brabant, Flanders, and most catholic countries followed suit shortly afterwards. Protestant countries, however, with the exception of Holland, Zeeland, and Denmark, held to the 'old style'. The change was not made in England until 1752. We have decided, therefore, to keep the old style of dating when referring to foreign events in the period 1582–1603. The only exception to this general rule will be found on pp. 348, 349, where both styles are given.

X

THE EXECUTION OF MARY STUART: THE SPANISH ARMADA

WE have now arrived within measurable distance of the supreme crisis in Mary Stuart's fortunes. The chain of events leading to it forms one of the most melancholy dramas in history—a drama, on the one side, of egregious folly and unscrupulous, self-destructive plotting, and on the other of illogicality, irresolution, and timorous clemency. For many years Elizabeth had shielded the queen of Scots from the wrath of the English protestants, who made no secret of their opinion that so long as the 'monstrous and huge dragon' lived, not only their own queen's life was in danger, but also the security of the state. Parliament in 1572 would have attainted her of treason in the highest degree, or, failing that, would have disabled her from the succession for her complicity in the Ridolfi plot; and convocation plied the queen with many 'godly arguments' to set aside her scruples. It was only by a supreme exercise of will-power that Elizabeth saved Mary from the scaffold on which Norfolk perished, and vetoed the retaliatory measures against her title to the English crown which both houses had formally approved. There is no doubt that Elizabeth could have put her rival to death with impunity at this time, had she so wished, for even France would have raised no serious protest. 'They will put her to death', exclaimed Charles IX, when informed of the plotting: 'I see it is her own fault and folly.' But it was not to be. For one reason or another, possibly because she was temperamentally averse to shedding a kinswoman's blood, or afraid of the imputation of cruelty, or simply because she thought that Mary alive in an English prison was a better security for the future than Mary dead with a martyr's halo round her head, Elizabeth refused to 'put to death the bird that had flown to her for safety from the hawk'. She may even have hoodwinked herself into the belief that time and diplomacy would solve the problem. There were moments, of course, when she wavered in her attitude. After the massacre of St. Bartholomew, when 'all men cried out' against the Scottish queen and her Guise relations, she would only too willingly have handed her over to Regent Mar to be executed in

Edinburgh for her share in the Darnley murder. But the regent stood upon conditions; and rather than be held in any way responsible for the deed, Elizabeth allowed the matter to drop. She reverted to it in 1574, during Morton's regency, but again was unsuccessful; and Mary remained for the next twelve years the great 'untouchable' in English politics—rebellious, defiant, incorrigible in her hope of ultimate victory. The tragedy of her position lay in the fact that, while there was a chance of elevation to power by way of conspiracy, war, and revolution, she would not honestly accept the *pis-aller* of a settlement by treaty. The hope of the English succession, always the dominant ambition of her life, was far too precious a thing to be bartered away for personal security or even freedom. With amazing effrontery she said she would not leave her prison save as queen of England.

From 1569 to 1585 the earl of Shrewsbury acted as her jailer; and except for short intervals when her apartments were being cleansed, or when she visited Buxton for health's sake, her abode was Sheffield castle. Life was far from intolerable, although the restrictions were irksome. She dined as a queen, under a cloth of state. Her servitors, of whom there was a considerable number, were chosen by herself, and Shrewsbury was a man of honour, who interpreted his trust in a just and reasonable spirit. He looked after his prisoner so carefully that escape was out of the question 'unless she could transform herself into a flea or a mouse', but he could not exercise an equally close supervision over her communications with the outside world, and was sometimes unfairly charged with laxity. The fact is that although all correspondence was supposed to pass through his hands, there were numerous ways by which cipher messages could be smuggled into and out of the castle, and Mary was able to keep herself fairly well informed of events not only in England but also abroad. Moreover, since the cost of her maintenance was borne principally by the English treasury (£52 per week), she could use the proceeds of her dowry, which yielded annually £12,000, to finance an intelligence service, subsidize her partisans, and, generally speaking, weave plots with Elizabeth's enemies.[1] While she occupied herself with needlework, the results of which she sometimes sent to the

[1] But the dowry was systematically plundered by Mary's friends and relatives in France; and Elizabeth cut down her allowance to £30 per week (1575).

English queen as an earnest of her goodwill, or amused herself with pet birds and small dogs, or listened to court scandal purveyed to her by 'Bess of Hardwick', Shrewsbury's terrible spouse, she kept an eye constantly lifting for the propitious moment when, as the cardinal of Lorraine assured her, her 'patient dissimulation' would be rewarded. Her indefatigable ambassadors, the bishop of Ross and the archbishop of Glasgow, advocated her cause in Rome and Paris. She corresponded with the pope, the kings of France and Spain, the emperor, and a number of potentates great and small. Indeed her intrigues ramified through the length and breadth of Europe. Always professing the most amicable intentions, she constantly plotted to bring the whole fabric of society in ruin about Elizabeth's ears. When brought to book—for sometimes her correspondence miscarried into Walsingham's hands—she lied volubly but unconvincingly. She was a credulous and impulsive conspirator, full of exaggerated optimism, unburdened by a conscience, but void of a sense of reality, which is the basis of success in plotting as in all other things. It was a wearing and tearing existence. Disappointment rendered her irritable and frantic: she grew prematurely old: her hair whitened; and her health suffered from the prolonged confinement. The surprising thing is not that she failed, but that Elizabeth persisted so long in the attempt to seek a *modus vivendi* with her.

Of the English queen's desire to reach a settlement without bloodshed there is ample evidence. Even during the dangerous plotting which began in 1580 and culminated in the expulsion of Mendoza she made a serious effort at conciliation, on the plea, put forward by Mary herself, that a joint sovereignty comprehending both mother and son could be established in Scotland (April–May–June 1583). The plan had much to commend it, especially if France as well as Scotland would agree to enter into a mutual insurance treaty with England. But it proved to be mere castle-building in the air. The confusion of Scottish politics, the jealousy of James, who would rather have built a monument to his mother's memory than signed a treaty for her release, and the opposition of the lords who surrounded the court in Edinburgh, both catholic and protestant, to the exile's return under any conditions—all this destroyed the possibility of an agreement. The final blow to the negotiation was struck by Throckmorton's confession on the rack. 'All this shows', said Elizabeth, summing up her impressions in March 1584, 'that her intention was to lull us into security, that we might the less seek to discover practices at home and abroad.' Yet Elizabeth was slow to admit failure in this, as in all her diplomatic ventures. Only after a further effort had been made and her commissioners, Beale and Mildmay, reported that they could not 'by all the cunning that we have, bring her to make any absolute promise' (May 1584), did she give up the struggle. To have proceeded any farther would have been merely fatuous. At the same time as the negotiations were going on, Mary was in active communication with William Allen about her release by force, pointing out how easily a *coup de main* could be effected at Sheffield!

What was to be done with such a woman? Elizabeth was unable to offer any solution of the problem. She had the 'wolf by the ears' with a vengeance! To let go would be fatal: she could only hold on. And doubtless she would have held on indefinitely, for the long duration of the imprisonment created, so to speak, a kind of vested interest in its continuance. Events, however, were at last beginning to move towards a very different conclusion.

The papal sanction of assassination as a lawful weapon against the heretic ruler had already worked havoc with public morality. In 1582 William of Orange had been shot at and severely wounded by a fanatic, only escaping death by a miracle. Soon afterwards the murder campaign was extended to England. In October 1583 a certain John Somerville, a Warwickshire papist, whose head had been turned by the exploit against Orange, swore he would kill Elizabeth with a pistol, 'for she was a serpent and a viper'. Happily this murderous intention was revealed by his own incautious and boastful words, and he was arrested, sent to the Tower, and executed. The loyalty of the nation was roused to a high pitch; and Mauvissière, the French ambassador, noted that when the queen journeyed to Hampton Court in November great throngs of people knelt by the roadside, wishing her a thousand blessings and expressing the hope that evilly disposed persons who meant her harm would be discovered and duly punished. The portent was unmistakable. The time had come to take stern measures to protect the queen. Consequently, less than four months after the second and successsful attempt had struck down the Dutch leader (June 1584), the council drew up the so-called Bond of Association, pledging those who signed it—and thousands of signatures were soon forthcoming from many parts of the country—in the event of an attempt on the queen's life, not only to prevent the succession of any person by whom, or in whose interest, the attempt should be made, but also to pursue the said person to the death. No clearer intimation could have been made to Mary that, if the plotting continued, her life would be in danger, if not actually forfeit. There was a considerable conflict of opinion when the question of legalizing this appeal to Texan justice came before parliament, towards the end of November. The commons were unstrung at the prospect of utter chaos likely to supervene if the events envisaged in the Bond took place. The queen, on the other hand, was averse to having any one put to death 'for the fault of another', but would not give parliament the satisfaction it wanted in regard to the succession. Confusion was increased, and excitement grew, when it transpired, in January 1585, that another plot—the Parry plot—was in train for the murder of the queen. Eventually a bill was passed (27 Eliz., cap. i), incorporating the substance of the Bond so far as 'pursuing to the death' was concerned, but making provision for a proper trial and judgement, followed by disablement from the succession, of the person *with whose privity* the attempt was made, prior to the infliction of the death penalty. The Bond of Association was consequently modified in the same way. In February Parry was sent to the scaffold. Immediately afterwards a motion was made in the commons to revive the proceedings against Mary begun in 1572; for Parry's examination elicited the fact that the Scottish queen's agent in Paris, Morgan, had encouraged the plot. At this point, however, Elizabeth interposed her veto: the matter had been carried far enough for the present.

Meanwhile the watch over the captive was intensified. Early in January 1585 she was removed from Sheffield to Tutbury; and shortly afterwards the puritan Amyas Paulet replaced Shrewsbury as her keeper. The new jailer, a man after Walsingham's own heart, at once instituted a régime of stringent surveillance. The armed guard at the castle was strengthened: no stranger was permitted to enter it on any pretext whatever; and 'laundresses, coachmen, and the like' were accompanied by soldiers when they left its precincts. Gradually but surely

Mary was cut off from her friends. For a short time she was permitted to send and receive her letters through the French ambassador, subject, of course, to inspection by either Walsingham or Paulet. But by September even this luxury was denied her: she was informed that all her correspondence, to or from France, must pass through government channels. The cordon was now complete; but Walsingham was not satisfied. His object was not so much to procure Mary's isolation as to collect information about her intentions that would ensure her trial, condemnation, and death. In order to do this it was necessary, first, to establish and maintain the impression that Mary's correspondence was open to inspection, while, at the same time, creating the illusion that a new and secret post had been invented, whereby letters could be carried into and out of her prison without the knowledge of her jailer. To this task he now addressed himself with characteristic ingenuity. It is not necessary to suppose that he invented the plot that lured the queen of Scots to her doom; but the measures he adopted were such as to make it practically certain that, if Mary was bent upon the queen's destruction, she would become hopelessly entangled in a conspiracy against her life, and therefore liable to the penalties laid down in the statute. In this sense he was the *agent provocateur* in the remarkable sequence of events that culminated in the tragedy of Fotheringay.

About the end of December 1585 the Scottish queen was taken from Tutbury to Chartley, a manor belonging to the young earl of Essex, situated some twelve miles from Tutbury, in south Derbyshire. Here, with the help of Paulet, Walsingham constructed his trap for intercepting, or, as we should say, 'listening in' to the secret plans of Mary and her partisans. By means of a brewer of Burton, who supplied the manor with beer, and Gilbert Gifford, a renegade catholic who consented to play the part of 'confidence man' and intermediary, Mary was led to believe that a perfectly safe line of communication had been found connecting Chartley with the outer world. Since Gifford bore credentials from the redoubtable Morgan, she leapt at the opportunity without the slightest hesitation. The mechanism of the trap was simple in the extreme. Letters coming to Mary were wrapped in a waterproof case (*une petite boîte ou sac de cuir*), inserted through the bung into a cask of ale, and carried into Chartley: out-going letters, enclosed in the same way, were smuggled out in the returned 'empty'. Each packet, before it was placed in the cask for conveyance to Mary, was carefully examined, deciphered, and copied by Thomas Phelippes, Walsingham's secretary; and the same procedure was applied to the outgoing packets before they were passed on to their destination.

Walsingham had not long to wait for results. His first haul was a batch of old letters, hung up at the French embassy since the Throckmorton conspiracy, of value only for the light they shed on Mary's part in the plotting of the previous two years. In May came a letter from Mary to Charles Paget, urging him to hasten on the invasion project, and about the same time a letter from Paget to Mary stating that one Ballard, a priest, had just returned to France with a report that the English catholics were ready to rise as soon as help arrived from abroad. Ballard, it appeared, was sent back to England for more precise information. Then, in June, Walsingham became aware that a new plot was taking shape—the Babington conspiracy, so called from its chief executant agent, Anthony Babington of Dethick, Derbyshire, who had formerly been a page in Mary's household at Sheffield. This ardent, vainglorious, but essentially simple-minded youth of twenty-five had been brought into the deep waters by the persuasive tales of Ballard, and now offered, with five companions, to attempt the murder of the queen and the release of Mary. They were to be supported by an invading army. Meanwhile a broad hint of what was coming was conveyed in a letter from Morgan to Curle, Mary's English secretary. 'There be many means in hand', wrote Morgan, 'to remove the beast that troubles the world.' On 7 July Babington imprudently disclosed the whole plot in a letter to Mary, asking for her approval and advice, and for assurance of rewards for his associates, or for their dependants, if they themselves lost their lives in the attempt. This fateful letter, too important to be entrusted to other hands, was carried to Chartley by Phelippes in person, and delivered to Mary through the secret post. Everything now turned on her response. Would she commit herself? She could hardly avoid it: the temptation was great. On 14 July Phelippes wrote anxiously to Walsingham: 'We attend her very heart at the next.' He was not disappointed. Three days later came Mary's reply—a full and frank approval of the plot, accompanied by an earnest recommendation that her friends should study carefully the number of troops to be raised in the counties, the towns, and ports for the reception of succour from abroad, the place of the general rendezvous, the strength and resources of the invading army, and, lastly, the manner in which the 'six gentlemen' had decided to proceed. She also entered into details as to the time and method of her own release after the accomplishment of the 'design'. Before passing on this damning epistle to its destination, on 29 July, Walsingham caused a forged postscript to be added to it, in the same cipher, asking for the names of the six associates.[1] He knew two of them already—Babington and Ballard, and he probably suspected who the others were; for all six had been swaggering about London, toasting the success of the enterprise at tavern dinners, and had even committed the incredible folly of having their pictures taken, in a group, with Babington in the centre and subscribed with the motto: 'Hi mihi sunt comites, quos ipsa pericula ducunt.' But like the good sleuth he was, Walsingham wanted to complete his *dossier* of the plot before striking: it was his duty to the queen and the state.

But time pressed: the birds might take fright and fly. So, early in August, without waiting for Babington's reply, he struck his blow. Mary's papers were seized at Chartley, and her secretaries arrested. Babington and his confederates were rounded up and lodged in the Tower, where they admitted their guilt. On 20 September, amid general rejoicing, the would-be assassins were dragged across London on hurdles to St. Giles's Fields, to be executed with the utmost barbarity which the treason law permitted. Their deaths were the occa-

---

[1] Mary's *knowledge* of the plot, which earlier catholic apologists denied, on the ground that the vital passages affecting the murder of the queen were probably interpolations by Phelippes (see Chantelauze, *Marie Stuart, son procès et son exécution*, pp. 57–64), is now generally admitted. J. H. Pollen writes: 'I accept the *textus receptus* [i.e. of the letter] throughout, apart from errors of translation. I believe the postscript was the only forgery which Phelippes was allowed' (*Mary Queen of Scots and the Babington Plot*, App. I, p. 32). Pollen also states: 'If the assassination was a crime, Mary was not free from guilt. If it was not a crime, but an inevitable incident in the struggle for liberty, Mary was free from blame' (ibid., Introduction, p. cxliv). Walsingham's biographer sums up the case as follows: 'On these grounds one is justified in concluding that unless further evidence can be produced to the contrary, the official version of Mary's letter to Babington is in all essential respects the correct version, and that Mary stands guilty of being implicated in Babington's conspiracy to murder Queen Elizabeth' (C. Read, *Mr. Secretary Walsingham*, ii. 44).

sion of numerous ballads, showing the widespread interest and emotion the plot had created.

The prisoner at Chartley still remained to be dealt with. It was the old problem back again in an acuter form, and Elizabeth was gravely perplexed. That Mary was guilty of being consentient to a plot against her life she had no doubt: the straight course was to commit her to the tender mercies of the law. The nation demanded it as a measure of self-preservation, and the statute of 1584 would provide sufficient legal cover. Other considerations were really irrelevant. Nevertheless Elizabeth was tormented by scruples. Mary was still a queen regnant in the eyes of the law, and the trial of a prince was without precedent in English history. Moreover, what would France and Scotland say if justice took its course and Mary was condemned? She was a queen dowager of France, and, if for no other reason, France had an interest in keeping her alive as a bar to the pretensions of Philip II. The king of Scotland, too, would certainly raise the point of honour. He could not be expected to consent to his mother's death, however lightly he might view her strait confinement. Altogether Elizabeth would have preferred, now as earlier, to let Mary live.

Meanwhile protestant England became a prey of rumours, circulated, no doubt, by Mary's enemies. It was reported that the queen had been assassinated, that civil war had broken out, that Parma had landed at Newcastle and Guise in Sussex. The musters were called out for the defence of the coast. The fleet was ordered to sea. The hunt for priests increased in vigour. Paulet became nervous and refused to be responsible for his charge if she remained at Chartley. The council was in favour of sending Mary to the Tower. Elizabeth would not hear of the Tower, nor would she hear of any other place, although many were suggested. Finally, however, she decided on Fotheringay, a royal castle in Northamptonshire. Early in October, after prolonged deliberation, she appointed a commission consisting of thirty-six judges, peers, and privy councillors to meet there for the trial. It assembled on the 11th of the month, with instructions, laid down in accordance with the statute, to try and judge the queen of Scots. Mary, of course, refused to recognize the jurisdiction of this court, making the most of her unique position as a sovereign anointed prince, and descanting at great length on the injustice of her

long imprisonment. Her remonstrances, however, had all been
discounted, and the trial went forward notwithstanding.

In accordance with invariable practice in treason trials, Mary
was denied the help of counsel, but she conducted her own case
with superb confidence and not a little ability. She denied all
cognizance of the plot, denied the authenticity of her letter to
Babington, denied receiving letters from Babington. 'I would
never make shipwreck of my soul', she said, 'by compassing the
death of my dearest sister.' The only admission she would make
was that she had sought help where she hoped to find it. She
even rallied the lawyers, these 'sore fellows' who often make a
good cause seem a bad one. But it was all in vain. The evidence
against her—the confessions of Babington and of her own secretaries, not to mention the fatal letter she wrote to Babington—
was overwhelming, irrefutable.

The court was about to proceed to judgement when Elizabeth's scruples again got the better of her: she prorogued it for
ten days, summoning its members to London. Here, in the
star chamber, the evidence was gone over again, and the
commissioners unanimously found Mary guilty, not only as
accessory and privy to the plot, but also as an imaginer and
compasser of her majesty's destruction. Under the statute she
thus became *ipso facto* incapable of succeeding to the crown.
The queen and parliament were left to decide what should be
done further; but the commissioners took occasion before they
dispersed to relieve the anxieties of the king of Scotland by
declaring that 'the sentence derogated nothing from him in
title or honour'.

The verdict of parliament was a foregone conclusion. It met
on 29 October, settled down to business on 5 November, and
after several consultations both houses decided, as they did
fourteen years before, to make a joint petition to the queen
asking that 'a just sentence might be followed by as just an
execution'. This was done on 12 November at Richmond.
Elizabeth's reply was that of an embarrassed woman. She did
not know what to do, and virtually said so. Constitutional
timidity in the face of an irrevocable act struggled with the
desire to perform a plain duty which her people demanded.
After three days' consideration, she sent a message asking
whether 'some other way' might not be found. The only other
way was to settle the crown on James and keep Mary in per-

manent solitary confinement. Parliament, however, confirmed its verdict, and a deputation returned to Richmond to uphold it with compelling arguments. Still Elizabeth was unconvinced. She delivered another speech, exposing the grounds of her scruples—the fear to spill the blood of a kinswoman—and ended by saying that they must be content with 'an answer answerless'. On 2 December parliament was prorogued to February, on the assumption that in the meantime the queen would publish the sentence against Mary. This she did without delay, to the great joy of London, which was illuminated to mark the occasion.

Mary had been shown the sentence before its publication, and warned that parliament had ratified it. But she showed no alarm at her desperate position. She would confess nothing, she said, because she had nothing to confess. 'Utterly void of all fear of harm,' according to Paulet, she defied the government to do its worst. Perhaps she believed that Elizabeth would not dare to execute the sentence, or imagined that her hesitation, which was daily becoming more and more apparent, bespoke a cowardice on which she might safely rely. More probably, if we may judge by her actions and utterances, she had already made up her mind that death was now inevitable; and if die she must, she would do so in the role of a martyr. When Paulet removed her cloth of state, as a sign that she was now dead in the eye of the law, she hung a crucifix in its place. Becoming transfigured in the light of the coming 'martyrdom', she wrote to the pope committing into his hands the spiritual welfare of her son, and to the duke of Guise declaring that she was about to die for 'the maintenance and restoration of the catholic church in this unfortunate island'. The only thing she feared was that fate, or the malice of her enemies, would deprive her of this glorious and spectacular end.

While Mary thus reached her mount of transfiguration, Elizabeth wrestled alone with her unhappy lot, a victim to her own morbid imaginings. Ambassadors from France and Scotland presented themselves at court with vigorous pleas and petitions for clemency, which could not be ignored, however irritating they might be. It was only too apparent that the affairs of Mary Stuart, in this last phase as in all previous phases, were inextricably bound up with the general European situation. To endanger the completed alliance with Scotland or the goodwill of France, at a time when the great duel with Spain was looming over the horizon, was a heavy price to pay for ridding England of the enemy within the gate. To some extent Elizabeth's anxiety, in regard to Scotland, was relieved by the king's obvious and clearly expressed desire to retain her favour in view of his interest in the succession, and the master of Gray whispered in her ear the significant words: 'Mortui non mordent.' In January 1587 the intercession of France was largely nullified by the discovery (or was it the invention?) of a fresh plot against the queen's life, emanating, it was said, from the French embassy in London. A 'great fear' swept over the country. It was reported that the Spaniards were at Milford, that Mary had escaped, that the northern counties were in revolt, that the capital was on fire. Elizabeth was heard to murmur the words: 'Aut fer, aut feri; ne feriare, feri.'[1] On 1 February she signed the death-warrant.

Even so, she could not bring herself to part with the document. She tried, in a moment of weakness, to get Paulet to play the executioner on his own responsibility, as a signatory to the Bond of Association; but the 'nice fellow' refused to soil his conscience with murder. At last Burghley took the intolerable situation in hand, and solved Elizabeth's dilemma for her. At a meeting held in his house on 3 February the privy council decided to brave the royal displeasure and to send off the warrant on the next day. Thus, on the night of the 7th, all unknown to the queen, Mary was told to prepare for death at eight o'clock the following morning. She received the news calmly, supped cheerfully, busied herself with correspondence most of the night, giving instructions for the disposal of her personal effects and jewellery. At two in the morning she slept for a little; and then, at the appointed hour, she was summoned to the scaffold in the great hall of the castle. Here she died, in the presence of her ladies-in-waiting, her surgeon, her apothecary, and Andrew Melvil, the master of her household. In London, when the news arrived, on the following afternoon, the church bells were rung, bonfires lit, and people held banquets in the streets. The festivities lasted for more than a week, in token of the great deliverance.

Elizabeth took no part in the rejoicing. She grieved: she wept: she stormed at the ministers who had acted without her authority. Whether her emotions were real or simulated it is

---

[1] i.e. Suffer or strike; in order not to be struck, strike.

impossible to say. The most charitable explanation is that they were partly true and partly fictitious. Certainly the victims of her wrath had no illusions about the vigour with which it was expressed. Hatton, who as chancellor had affixed the great seal to the warrant, was petrified with fright. Burghley was excluded from the royal presence, and sorely troubled. So, too, was Leicester. And Walsingham only escaped the tempest by temporarily taking refuge in his house at Barn Elms, under pretext of sickness. Beale, who had carried the warrant to Fotheringay, was relegated to a subordinate post at York; while Davison, the unfortunate secretary of state, who had allowed it to pass out of his keeping, was heavily fined and committed to the Tower, where he remained for more than a year and a half.[1] These punitive measures were necessary if Elizabeth desired to give substance to the legend that the execution of Mary was a 'deplorable accident', for which she at least was not responsible. It was an unworthy subterfuge, no less specious than Catherine de Médicis's after the massacre of St. Bartholomew. The play-acting was continued almost to the point of absurdity when the distracted queen went into mourning, and authorized a public burial for Mary's body, with royal honours, in Peterborough cathedral.

Europe was not deceived. In Scotland there was a violent explosion of anger against Elizabeth, and a cry arose for reprisals. Expecting immediate war, the borderers raided northern England, and for a time communications were interrupted between the two countries. Sir Robert Carey, who was dispatched with an exculpatory letter from Elizabeth, was stopped at the frontier by the king's command, because of the 'fury of the people'. James, however, was astute enough to realize that however much the occasion might demand a display of filial wrath, war with England would only jeopardize his own interests and ambitions. After a brief interval he professed himself satisfied with the official English version of what had happened. He was probably strengthened in his decision to avoid war by a pointed and significant letter from Walsingham to the Scottish chancellor, Maitland of Thirlstane, in which the veteran diplomatist showed that neutrality alone would safeguard the Stuart title to the English succession.

[1] His fine was subsequently remitted and he retained his fees as secretary till his death.

The uproar in Scotland was small compared with the whirl-wind of passion that shook France. In Paris the English am-bassadors, Stafford and Waad, were afraid to be seen in the streets, and were denied access to the court. Jacques de Guesle, the *procureur-général du roi*, drew up a juridical indictment of Mary's trial, which was forwarded by Henry III to England. Funeral orations, books, pamphlets, and poems flowed from the press, commemorating the 'martyrdom of the queen of Scotland', and attacking venomously the character and career of the 'Jezebel' who had put her to death—'this singular bastard and shameless harlot'. The chivalrous youth of the nation were urged to avenge the 'outrage':

> Ravagez-moi leur terre et faites abîmer,
> Sous le faix de vos pieds cette île dans la mer.

All this was to be expected from a country dominated by the catholic League and the duke of Guise. But Henry III had no desire, any more than James of Scotland, to be involved in a war with Elizabeth, especially when his own kingdom was torn by internal convulsions, and the Spanish menace, which threatened him even more than it did the English queen, was daily becom-ing graver. Still the fear of France operated strongly on Eliza-beth until after the defeat of the Spanish fleet at Gravelines.

The Gordian knot had been cut. The death of Mary Stuart had a simplifying effect on the problems with which England was confronted, for the catholic cause lost most of its driving force. Except for unforeseen circumstances, the succession was now assured to James, and the king of Scotland was certainly not a catholic. Consequently the hopes of the jesuits turned to Philip II and the infanta. This, in turn, had a repercussion on the attitude of the English catholics, who, while they might conceivably fight for Mary, had no intention of helping to seat a Spaniard on the throne. Their patriotism, which was never really in doubt, received a brilliant vindication a year later, when the supreme question of the hour was the preservation of national independence.

The decision to launch a great armada against England probably took tangible form in the mind of Philip II during the summer of 1585. In June of that year Olivarez, the Spanish ambassador at the court of Pope Sixtus V, offered his master's forces for the enterprise, provided the pontiff was willing to contribute generously to the equipment of the expedition— 'much more than any one has hitherto imagined'. The gist of the Spanish plan was to represent the undertaking as essentially altruistic and unworldly. Much stress was therefore laid on the good of religion, the conversion of England, the avenging of the blood of the martyrs, the restoration of Mary Stuart to her rights, and the other advantages that might be expected to carry weight with the head of the church; and a discreet veil was drawn over the political motives actuating the king. But the pope was difficult to move. He was not only sceptical and distrustful of Spanish promises: he suspected that under the cover of 'serving the Lord' Philip really intended to reap some personal gain for himself, or, worse still, use the threat of invasion in order to extort a favourable peace from the English queen. As a matter of fact, if Sixtus had been able to read the dispatches that passed between Madrid and the Spanish embassy in Rome, he would have discovered that the king's main concern was to secure a controlling hand in the disposal of the English succession, in the event of Elizabeth's overthrow. For more than a year Olivarez played his game of mystification at Rome, 'weaving his web' and 'placing his snares' about the pontiff, until at length, in December 1586, the contest ended with the conclusion of a contract pledging the Holy See to contribute a million crowns to the Spanish war chest. The cautious pope, however, stipulated that the payment of the money must depend upon results: the first instalment of 500,000 crowns being payable only after the armada landed in England, and the remainder in smaller sums of 100,000 every four months thereafter.

This was hardly what Philip wanted. He would have preferred to have the money in advance, and an absolutely free hand to determine the time of the expedition. But since the pope was inflexible on the point that the invasion must precede payment, he pocketed his disappointment and proceeded with the next stage of his machiavellian plan. In February 1587 Olivarez was instructed to obtain a secret Brief declaring that, failing the queen of Scots, the right to the English succession fell to the king of Spain, in virtue of his descent from the house of Lancaster, and a will[1] which, it was alleged, had been drawn

---

[1] In a letter to Mendoza (20 May 1586) Mary stated that she intended to make a will ceding her right to the English succession to King Philip, 'en cas que mon

up by Mary in his favour. If the pope showed any scruple in the matter, it was to be pointed out to him that his majesty could not undertake a war with England merely for the purpose of placing 'a young heretic like the king of Scotland' on the throne; but that there was no intention of adding England to the Spanish dominions, since the right was to be exercised not in the interest of the king himself, but in that of his daughter, the infanta Isabella. As for Mary Stuart, who had figured prominently in Philip's previous utterances on England's future, he was cynical enough to suggest that the heretics in England would almost certainly put her to death as soon as the invasion project materialized. Perhaps the wish was father to the thought; for with the mother out of the way the exclusion of the son on account of his heresy became a more practical proposition, and the king could go forward with his design in comparative confidence.

But the exclusion of the king of Scotland was a question on which there was bound to be a sharp division of opinion in the catholic world. France could not stand by neutral while Spain interfered with the natural order of the English succession to her own advantage. Nor was the duke of Guise, humble suitor to Spain though he was, likely to acquiesce in the deprivation of his Scottish kinsman without at least making a further attempt to secure his conversion. Europe also would be alarmed at the threat of a Spanish domination. Philip therefore addressed himself to meet these contingencies with accustomed subtlety and foresight. At Rome, Olivarez and the pro-Spanish cardinal, Caraffa, plied the pope with arguments to prove that the conversion of James was neither practicable nor wise, and that reliance on France was dangerous because of the understanding between that country and England. William Allen, who was shortly to be raised to the cardinalate at the request of Spain, was also briefed to plead the Spanish case in the name of the English catholics. In Paris, Mendoza spread the view that the exclusion of the Scot was just as important for the catholic cause in Britain as the exclusion of Henry of Navarre was for the triumph of catholicism in France. At the same time he urged Guise, with substantial offers of assistance, to focus his attention on the overthrow of the huguenots. In

dit fils ne se réduise avant ma mort à la religion catholique'. This will, however, was never made (see *Scot. Hist. Rev.* xi. 338–44).

short, every possible means was used to bend public opinion in those centres where opposition showed itself, or was feared, into acceptance of the Spanish policy. Eventually the superior weight of Spanish diplomacy bore down most of the obstacles; but no decision was arrived at with regard to the English succession, the matter being deferred, at the express wish of the pope, until after the conclusion of the war. It cannot be said, therefore, that the outcome of the negotiations was altogether satisfactory to Spain. If the expedition failed, the burden must fall entirely on the king; and if it succeeded, the fruits of victory would have to be shared with the pope.

While these diplomatic exchanges were going on, Philip was busy with his other preparations relative to the invasion, poring over memoirs on previous invasions and studying charts of the chief ports and landing-places. He was not without expert advice, for the question of how best to set about the enterprise had been settled by Santa Cruz in a detailed memorandum which this famous admiral had drawn up after his victory over Strozzi at Terceira in 1582. Santa Cruz favoured a combined naval and military attack, delivered directly from Spain.[1] The advantage of such a plan was that it ensured unity of command during the most critical part of the operations, and involved no difficult system of co-ordination between the naval and military factors. Its disadvantage lay in the fact that it was exceedingly costly, and made no use of the splendid army of Parma in the Netherlands. Three and three-quarter million ducats—the sum demanded by Santa Cruz—was more than Philip could afford to spend on the reduction of England. So this promising scheme was laid aside as impracticable. Even shorter shrift was given to another which emanated from the pen of Mendoza in December 1586. Mendoza was a soldier of great experience in continental wars, but he was little versed in major operations of the kind now contemplated, nor could he draft a scheme with the meticulous minuteness required by the king. Nevertheless the plan he submitted had merits of its own, and was well worth attention because of its salutary comments on the difficulties of sea enterprises. With commendable insight the ambassador pointed out that it was to the enemy's advantage to be attacked by a force 'which needs great sea fleets for its transport and maintenance'; that fleets

[1] 556 ships, carrying 85,000 landing troops, and 200 flat-bottomed boats.

are more subject to delays and disasters than land armies; and that in the event of a disaster the loss in ships, men, and guns cannot be replaced except with great difficulty. He therefore advised the king that the naval forces of Spain, already too exiguous for the empire's needs, should be kept for defensive purposes alone, and suggested that a comparatively small military force should be sent to assist the Scottish catholics in the catholicization of their country. If this were successful, and if, later, the English catholics rose, further measures might be taken to increase the strength of the Spaniards with a view to subjecting the whole island. 'It will be no small advantage to your Majesty', he commented, 'that the game should be played out on the English table, just as she [i.e. Elizabeth] has tried to make Flanders and France the arena.' But Philip was evidently not impressed with Mendoza's arguments. He had already made up his mind that the blow must be struck in England and not through Scotland, and that Parma's army was to provide the striking force. A military expedition from the Low Countries supported by a powerful naval armament from Spain —this was the plan which the king especially favoured, and to which he stood committed.

In making his decision to stake everything on a single throw of the dice, Philip relied to some extent on the help he believed he would receive from the catholics in England. His information on this very important point was derived from mentors like William Allen and Robert Persons, who, in the absence of more direct informants, constituted themselves the spokesmen of their fellow religionists within the enemy country. But neither of these men, however much they were cognizant of the opinion prevailing among the exiles on the Continent, really knew the temper and bearing of the catholics at home during the critical years immediately before the invasion. Thus it was the biased views of those whom self-inflicted banishment had soured and embittered that reached the king's ears rather than the authentic voice of the English catholics. Both Allen and Persons held the view that to be a good catholic was synonymous with being pro-Spanish in politics. Of the intense patriotism which now surged through England they knew and cared nothing. Their own love of their native country was eclipsed by a deeper attachment to the cause for which they would willingly have given their lives. As early as 1582 Persons had asserted in all seriousness that 'almost all catholics without a single exception regard the invasion with approval. Nay, they even burn with longing for the undertaking.' Consequently by 1585 it was generally believed among Spanish diplomatists that if an invasion of England took place it would be supported by a general rising within the country; and a year later a list was compiled for the king of Spain stating the numbers likely to take part in it. The 'Catholic army' was placed by the anonymous intelligencer at the remarkable figure of 25,000! The grotesque character of the report is apparent from its confident assertion that 'almost the whole' of Yorkshire was ready to take up arms. If the true opinion of the English catholics is to be sought for anywhere, it is to be found, not in the biased reports of Spanish intelligencers, but in the glowing words of the Campion ballad.[1]

So badly did Allen misinterpret the feelings of his countrymen in England that after his promotion to cardinal rank in August 1587 he composed a Manifesto justifying the invasion, and addressed an Admonition to the nobility and people of England, urging them to assist the invader on peril of their souls. He was repeating, though probably unaware of it, the tactics of the French Leaguers and of Condé and Coligny, when they bartered the integrity of their country for the help of the foreigner. Nevertheless it was on such false hopes that the king of Spain partially depended for the success of his expedition.

The first reports of Spanish preparations reached England in December 1585; but they were so uncertain that the council did not begin to take precautionary measures for another six months. In July 1586 orders were issued for the mustering of foot-bands by the lords lieutenant of the shires, the training of men in the use of the arquebus, the setting of watches in the towns by night, the preparation of beacons, the rounding up of seminarists and all who spread seditious rumours. In August alarm was created by the report of a French landing in Sussex, and emergency mobilization orders were issued. Another scare followed in September, when rumours were circulated of a Spanish fleet off Conquet in Brittany. But by November the tension had subsided, and England passed into the new year still nominally at peace. In the meantime, however, Philip had sent instructions to the duke of Parma to carry out his preparations

[1] See p. 150.

on the coast of Flanders, and to be ready by the summer
of 1587; and for nearly a year thousands of workmen were
employed in deepening and widening the canals between the
Scheldt above Antwerp and the towns of Ghent, Bruges, Nieuport, and Dunkirk, improvising dockyards for the construction
of flat-bottomed boats, and making the other arrangements for
the transport of the expeditionary force to England. At the
same time the mobilization of the fleet in Cadiz harbour was
proceeding according to plan, and Philip was making frantic
efforts to learn from Mendoza the state of the English preparations. This last point was of capital importance, because the
assembling ships were absolutely at the mercy of a raid from
England. But in spite of all Mendoza's devices he could obtain
very little news from the English ports, so great was the secrecy
preserved, and so strictly were strangers watched. On the other
hand, Walsingham's excellent spy service kept England closely
in touch with every stage of the Spanish preparations. The
result was that when the concentration at Cadiz was well
advanced in the spring of 1587 Drake was released from
Plymouth on 12 April, with a roving commission to 'distress'
it in every way possible—to impeach the assembling of the king
of Spain's ships, to cut off their supplies of victuals, to follow
them up if they had already set sail for England and fall upon
them, to intercept galleons from the East and West Indies, and,
above all, to do what damage he could 'within the havens
themselves'. Mendoza only learnt of the intended raid on the
very eve of Drake's departure; and his warning letter was not
received by the king until the English had entered Cadiz
harbour, or possibly a little later! Thus the execution carried
out among the defenceless Spanish ships was considerable.
Thousands of tons of shipping and a vast quantity of stores
were destroyed: Santa Cruz's own galleon was captured, gutted,
and fired; and six vessels laden with provisions were taken as
prizes. But the destruction was probably of less importance than
the chaos created in the Spanish naval administration, for
while Drake was on the coast, from April to June, not a ship
could move without danger of capture. From his improvised
base at Sagres Bay, the English commander was master of the
situation. Finally, after challenging Santa Cruz at Lisbon to
come out and fight, which the Spaniard refused to do, Drake
stood away for the Azores, and when sixteen days out from

Cape St. Vincent captured the Portuguese galleon, the *San Felipe*, worth nearly a million of our money. On 26 June he returned to Plymouth, conscious that if he had not ruined the king of Spain's plans he had at least delayed the sailing of the armada for another year.

As soon as the terror and confusion inspired by the raid was dispelled, work was resumed on the armada, losses were made good, and the concentration continued with accelerated pace. It would appear that the fear of renewed raiding in the spring decided Philip to dispatch the fleet against England in January (1588); but he was dissuaded from this folly by Santa Cruz, who asserted that March was the earliest practicable date. The plan of the campaign was communicated to Parma in September; and so badly was the secret kept that between this date and the end of the year all Europe was talking of the details of the Spanish strategy. 'The affair is so public', wrote Parma late in January, 'that I can assure your Majesty there is not a soldier but has something to say about it, and the details of it.' Less than three months later the despondency of Parma reached a climax. In a letter of 10 March he counselled the king to come to terms with the English, and so 'escape the danger of some disaster, causing you to fail to conquer England, while losing your hold here'. This was due not merely to the fact that the English had been forewarned, but also to the mortality among the troops in the Netherlands, and the shrinkage of their numbers from 30,000 to 17,000. But Philip was not to be turned from his purpose. Nor was he much disconcerted by the death of Santa Cruz, which had taken place early in February, although it was a patent fact that no other commander had the same experience in handling a great fleet in action. A new admiral was found in the duke of Medina Sidonia, a landsman without experience in sea-faring or war. 'I possess neither aptitude, ability, health, nor fortune for the expedition,' was his comment on learning of the king's decision. Nevertheless the luckless duke was ordered to take command at Lisbon on 8 February, and to be ready to start by 19 February. Finally, after much delay, the armada was got ready for sea in May, and put out from the Tagus in the last days of the month, wafted by the prayers and hopes of the whole Spanish nation.

There is little doubt that the king viewed its departure with

some anxiety and misgiving. The year 1588 was regarded by
Spanish soothsayers as pregnant with misfortune for their country; and the nervousness of the government is reflected in the
Easter proclamation that masses should be celebrated throughout the country for the safety of the fleet. The king himself
became as ascetic as a monk, spending many hours daily before
the blessed sacrament, commending the enterprise to God.
Before the stupendous nature of the issue, the religiosity of the
Spanish temperament attained an intensity unparalleled since
the days of the crusades. The banner borne by the fleet and
blessed by the church was inscribed with the legend 'Essurge,
Domine, et judica causam tuam!' Very different was the
attitude in England, where the appeal to the God of Battles
was mingled with a secular optimism based upon the consciousness of superior naval power and skill. Unlike Spain, England
pinned her faith to the patriotism of her people. The defence
of the gospel against antichrist, the cry of the puritan section
of the community, was certainly not that of the queen or the
government. The difference in the mentality of the two peoples,
each at the crisis of its fate, is depicted in the very names of
their ships. The Spanish fleet was a calendar of saints—the *San
Mateo*, *San Cristobal*, &c.: the English fleet boasted the purely
secular names of the *Lion*, *Tiger*, *Dreadnought*, *Revenge*, &c. In
fact the other-worldly piety of the middle ages was in conflict
with a new order of things which shifted the centre of gravity
from heaven to earth. When the decisive moment arrived,
King Philip entered the sanctuary of his chapel; but Queen
Elizabeth rode among her troops at Tilbury.

It is important to bear in mind that Spanish strategy in 1588
arose out of the fact that the king contemplated not a naval but
a military conquest of England. The fleet was to be merely the
instrument whereby the forces of the duke of Parma could be
brought to English soil. This was in keeping with the Spanish
king's confidence in his army, and also with his knowledge that
England was militarily weak. Consequently the instructions
issued to Medina Sidonia were so devised as to render it impossible for him to take the initiative in an independent naval
engagement with the English fleet. He was to proceed to the
English Channel and 'ascend' it as far as the 'Cape of Margate',
where the junction with Parma might most easily be effected.
No diversions, feints, or descents on the coast of Portugal, or

of Spain, or at the Azores by Drake were to be allowed to deflect him from this prime objective. If, however (and here Philip showed his strategical grasp), the enemy followed, or came within striking distance, he was to be attacked, particularly if the encounter took place at the mouth of the Channel, because this would mean that the English forces were divided and could be separately beaten, or at least prevented from uniting. Finally, if he met with no obstruction until he encountered the united squadrons of Drake and Howard in the narrow seas, he was to give battle, 'trying to gain the weather gage and every other advantage'. The king adds that the enemy will seek to fight at long range, and that the Spanish tactics must consist in drawing him to close quarters. Supplementary instructions authorized Medina Sidonia, in the event of Parma being unable to cross, or of his army being unable to join him, to consider, in communication with Parma, whether the Isle of Wight might be seized and converted into a base for future operations. But on no account was this to be done during the passage up the Channel: it was merely a precautionary measure to be put into operation if the main design failed.

Many grave flaws have been pointed out in these famous instructions, such as the absence of accurate information where the actual junction with Parma was to take place, the belief that a battle could be joined in the Channel whenever the Spaniard desired it, and the lack of a sound system of co-ordinating the movements of the fleet and the army. But as a general strategical scheme it represents the Spanish school of thought to perfection. If the chief consideration in strategy be to overwhelm the enemy at a decisive point by superior numbers, and if the decisive point for the Spaniard was the passage between the Belgian and Kentish coasts, Philip was not far wrong when he insisted upon the attainment of this point as a principal objective. As an essential preliminary to further movements, his secondary schemes of battling in the Channel and of seizing the Isle of Wight were also perfectly feasible in case of failure in the major action. The real weakness of the plan, taken as a whole, was that it left out of account, and did not provide for, the use of a suitable fortified port on the coast of the Netherlands, where the entire expedition could assemble in safety. Possibly the Spanish king forgot that heavy warships could not

THE SPANISH ARMADA

approach the shallow Flemish ports and so give Parma the protection he needed, or rest at anchor until he was ready to come out. That Philip was informed of this all-important point is clear from a warning he received from Cabrera of Cordova: in whose opinion the junction of the armada and Parma's flotilla could not be achieved unless such a port was available, and if it was not available the enterprise ought to be abandoned. Of course, the only port capable of fulfilling the requirements was Flushing—and Flushing was in the hands of the enemy.

How did England propose to meet the threat of invasion? Charles, Lord Howard of Effingham, was appointed commander-in-chief on 21 December 1587; and in January three separate fleets were placed in commission for the defence of the realm: a grand fleet, under the lord admiral, consisting of sixteen powerful ships, with its base at Queenborough in the Thames, a light squadron under Sir Henry Palmer to patrol the narrow seas, with its head-quarters at Dover, and Drake's smaller scouting force at Plymouth to watch the entrance to the Channel from the west. When reports began to filter in from the Spanish coast giving clearer information as to the size and probable date of departure of the armada, Howard moved the grand fleet from Queenborough to the Downs, and finally to Plymouth, where he joined Drake (May 1588). This rearrangement of England's forces held good until the two fleets met in the Channel in July. So far nothing was wrong with the English naval dispositions: Plymouth was the obvious base for a fleet operating against Spain, and Dover the proper place to keep an eye on Parma. But these were only the preliminary moves, on the completion of which the real strategical battle would begin. What use was to be made of the massed squadrons at Plymouth? Or, putting it otherwise, what counter-strategy did the English government adopt to check or defeat the Spanish scheme of attack? Obviously the only effective reply was to deny the Spaniards the advantage of time and place which they sought to secure; that is, to force the conflict at a point as far away from the critical area as possible—in short, in Spanish waters. This was the uniform advice of the naval commanders. In particular it was urgently put forward by Drake in a celebrated letter to the council. Nevertheless, in the face of expert counsel, the government remained wedded

to the landsman's fear that if the naval forces of the realm
were on the high seas or the Spanish coast England would be
exposed to the enemy. The seas were broad, the possibility of
accidents was great; and the Spaniards might slip past Howard
by sailing westward to latitude 50° and then 'shoot over to this
realm'. Consequently the utmost liberty the council would
allow Howard was that he might make reconnaissances in
force as far as the entrance to the Channel. The system of
provisioning the fleet from month to month, against which
Howard complained, also helped to reduce his radius of action,
and some stress should perhaps be laid on the prevalence of
foul weather during the early summer.

The principal reason for the queen's objection to the salutary
measures proposed by her sea captains must be regarded as
conjectural. It was probably dictated by military as well as
political considerations. Elizabeth was firmly convinced that
Anglo-Spanish differences could be liquidated without war,
and to the very last minute she pinned her faith to the conferences taking place between Parma and her representatives
at Brussels. Every one else dismissed these as merely a Spanish
stratagem: better an open trial by battle, they said, than
clandestine and uncertain peace palavers. 'Therefore in my
mind', wrote Hawkins, 'our profit and best assurance is to seek
our peace by a determined and resolute war.' And again: 'In
open and lawful wars, God will help us, for we defend the
chief cause, our religion, God's own cause.' It was not until
17 June, when the negotiations with Parma broke down irre-
mediably, owing to the publication of Cardinal Allen's manifesto calling upon the catholics in England to co-operate with
the invading forces from Spain, that Howard at last received
the liberty he had so long sought. On the 22nd he acknowledged the dispatch: on the following day his victuallers arrived
at Plymouth; and shortly afterwards he was at sea on a grand
reconnaissance with his squadrons strung out between Ushant
and Scilly. While in this position, on 5 July, a Rochellese bark
brought him news that the Spanish fleet had left the Tagus in
May, had been dispersed by a gale, and was now taking refuge
for repairs among the harbours of the Biscay coast of Spain.
The wind being in the north, Howard at once held a council
of war, and at 3 p.m. on 7 July gave orders to proceed to the
coast of Spain. If the wind had held for two days and nights,

and if his supplies had been sufficient, he would have fallen upon the helpless Spaniards with more disastrous effect than Drake did during the previous year. Unfortunately when he was eighty leagues south-west from Ushant the wind veered to the south, and fearing to be caught with insufficient provisions on the enemy coast, he turned back and made for Plymouth.

But if the wind balked Howard of his prey, it hastened the emergence of the armada from Corunna (12 July[1]) and on the whole favoured it during the remainder of its journey to the English Channel. At 4 p.m. on 19 July[2] Medina Sidonia sighted the Lizard, and at 7 p.m. gave orders to shorten sail three leagues from the shore. At dawn on the following day he had a glimpse of the beacon fires lit by the English; but it was late in the afternoon before he descried Howard's fleet through the mist and rain, slowly warping its way out of Plymouth. Contact was established between the two fleets on the 21st, and from then until 30 July,[3] when the armada escaped in a battered condition into the North Sea, a more or less continuous battle took place. It is not possible here to do more than chronicle the outstanding features of that fateful ten days' struggle; nor perhaps is it necessary, for the story has been told fully and impartially by both Spanish and English naval historians.

Owing to the dispersion of the English forces required by the defensive strategy of the council, Howard could not afford at first to provoke a decisive fight with the Spaniards, but was compelled to hang on their heels, 'plucking their feathers', as he expressed it, when occasion offered. The consequence was that the Spanish fleet, keeping a close and compact array, continued its advance up the Channel as far as the roadstead of Calais without incurring serious material loss. Up to this point (27 July)[4] the vitality of the armada was the dominating fact of the situation. 'This is the greatest and strongest combination, to my understanding,' wrote the English admiral, 'that ever was in christendom,' a remark that seems to reveal his chagrin at not being able to break it up sooner than he did. But although Medina Sidonia was now well on the way to his objective, with his fleet practically intact, his difficulties and troubles were only beginning. Already his morale had been badly shaken by the elusive tactics of the English, who steadily

---

[1] 22 July according to the reformed Spanish Calendar (N.S.).
[2] 29 July (N.S.).   [3] 9 August (N.S.).   [4] 6 August (N.S.).

refused to come to the 'handstroke'. He was worried also by the fear of being caught by foul weather without a safe port to fly to; and he was far from confident about effecting his junction with Parma, on which the success of the expedition depended. From the moment he entered the Channel this matter had caused him no little anxiety; and the nearer he approached the decisive theatre of action the stronger became his conviction that the safety of his ships was of even greater importance than that of Parma's army. On 28 July[1] he wrote to the duke stating that further advance of the fleet was impossible unless he came out to its rescue, 'so that we may go together and take some port where this armada may enter in safety'. Clearly Medina Sidonia was forgetting his instructions; for nothing was more certain than that Parma could not venture into the open with the English and Dutch squadrons blockading his ports, and the mastery of the sea still undecided, or in the hands of the enemy. In fact, the Spanish plan of campaign had come to a complete deadlock: the expeditionary force being bottled up in Flanders, and the necessary naval escort immobilized in the Calais roadstead some hundred miles away.

It was this unfortunate position of the Spaniards that gave Howard the opportunity of taking the initiative and of planning his master-stroke of the fire-ships. On the night of 28 July,[1] after being reinforced by Seymour and Winter, he sent in eight ships blazing with pitch and other combustibles among the anchored Spanish galleons, creating such confusion that a veritable *sauve qui peut* ensued. On the following day a general engagement took place—the decisive battle of Gravelines. The armada streamed off from Calais along the coast in the direction of Dunkirk, with the wind at south-west; and a running 'off-fight' was maintained from daybreak to dusk, of the kind that suited the genius of the English captains. So effective was the English gunfire that every great ship in the armada was 'very much spoiled'; and, to make matters worse, the wind shifted to the west-north-west, threatening to drive the entire fleet on the treacherous Ruytingen shoals. On the 30th[2] Medina Sidonia was informed by his pilots that if the wind held, only the mercy of God could save him from destruction. Five of his principal ships were now out of action, 'all their people being either slain or wounded', and the *San Felipe* and the *San Mateo*

---

[1] 7 August (N.S.).   [2] 9 August (N.S.).

were wrecks on the Nieuport and Blankenbergh banks. Fortunately, when things were at their worst, and the sounding-lead of the flagship registered only six fathoms, the wind went back again to the south-west, and the desperate commander managed to pull the remainder of his fleet out of the shallows into the comparative safety of the North Sea. But his plight had now become such that no recovery was possible. Medina Sidonia therefore resigned himself to a northward course, hoping to reach Spain by circumnavigating the British Isles. Howard followed him as far as the Firth of Forth, and then broke off the pursuit.

Such, in brief, is the story of the first great sea-fight in which sailing-ships only were used; and the victory had gone to those who had mastered the science of the new type of fighting. The essential difference between the two fleets did not lie in the tonnage, which was about equal, but in the superior speed and seaworthiness of the English ships, and the heavier ordnance with which they were armed.[1] It must have been a bitter thought to the Spanish commander that long before he could lure the elusive enemy within range of his less effective armament the decks of his ships had become a shambles by artillery fire. But the English practice of salvo-firing without any attempt to grapple and board—the *gran furia de artilleria* of the 'off-fight'—was a new thing on the sea, a revolutionary departure from the accepted usage of centuries. The amazement of the Spaniards when they encountered it only serves to show how far behind they were in the technique of naval warfare.

As soon as the Spanish fleet vanished into the mists of the North Sea the wildest speculation prevailed not only as to its whereabouts but also as to what had happened in the fighting between Plymouth and the Belgian coast. The only people

---

[1] For a comparison of the two fleets see J. K. Laughton, *The Defeat of the Spanish Armada*, i, Introduction, pp. xxxix–lii; also J. S. Corbett, *Drake and the Tudor Navy*, ii. 179–90. The strength of the Spanish fleet, when it left the Tagus, was 130 ships and 30,000 men, of whom 19,000 were soldiers. It is doubtful, however, whether more than 120 ships entered the Channel, and the number of men probably did not exceed 24,000. The English fleet, reckoning all ships registered as in the queen's service during the season, consisted of 197 ships and 17,000 to 18,000 men, practically all of whom were seamen. Of the 197 English ships by far the greater number neither had, nor were expected to have, any part in the fighting. But the gun power on the English side was overwhelmingly preponderant, not only relatively to the tonnage of the ships, but also absolutely in the actual weight of the projectiles they could throw.

who knew the truth, so far as the truth could be known, were the English and the Dutch; but their knowledge was slow to gain credence in catholic countries, where the might of Spain was regarded as precluding any possibility of disaster. At first, rumour had it that a great victory had been won by the Spaniards, and there were rejoicings in Madrid and among the English, Scottish, and Irish catholics at Rome. Mendoza was probably responsible for this lying report, albeit he based his surmises on what purported to be information from authentic sources. Further heartening news to the Spaniards came from Dunkirk on 20 August, to the effect that a battle had taken place between the armada and the English fleet off Newcastle, in which the English lost forty ships and Sir John Hawkins went to the bottom with all his crew. On the following day a message from Calais stated that the armada was 'at a very fertile Norwegian island, where they will find abundance of victuals without resistance'. Still more fantastic was the belief entertained in Prague, on 29 August, that 8,000 Spaniards and 30 guns had been landed at Plymouth. By this time, however, responsible persons in the Spanish and papal diplomatic services had become alarmed at the continued absence of authentic news from Medina Sidonia, and anxiety began to replace the earlier optimism. In Madrid, at the beginning of September, the first disturbing tidings that a disaster had taken place caused the nuncio to write: 'God grant that the evil tidings are false and the good ones true!' But the king maintained a brave countenance, and gave no outward indication of what was passing through his mind. Meanwhile Medina Sidonia was ploughing his way southward along the western seaboard of the British Isles, leaving ship after ship behind him on the rocky Irish coast. It was late in September before he reached Santander with the first battered remnant of his fleet, full of sick and wounded—a vivid and heart-breaking testimony to the toll of the sea and of the English guns. When the king received the report from his grief-stricken commander of the disaster to his ships, he accepted it without rancour, gave orders for the hapless survivors to be properly looked after, and consoled himself with the melancholy reflection that the catastrophe had not been worse than it was.

Greater perhaps than the actual material loss,[1] which, after

[1] The loss sustained by the Spaniards cannot be accurately stated; but it has

all, was not irremediable, was the shock to Spanish pride and the general stupefaction at the mysterious ways of Providence. In that superstitious age, when calamities were interpreted as judgements of God, it was difficult to avoid the conclusion that the hand of God seemed to be against Spain. Philip withdrew himself more than ever from the world, and tried to puzzle out the meaning of the calamity that had befallen his kingdom in the seclusion of the confessional. But his father confessor could supply him with no useful answer, nor was any sympathy vouchsafed to him either at Rome or elsewhere. On the contrary, the pope, who had not shown any degree of warmth for the crusade against England, expressed his admiration for the courage and bravery of the English queen, who, he said, would be his favourite daughter if only she were a catholic. Mendoza was ridiculed in the streets of Paris by workmen and children, who followed him with cries of 'Victoria, Victoria!' In the anti-Spanish world, again, jubilation at Spain's discomfiture was unrestrained. The Dutch struck a medal to commemorate England's victory, representing the terrestrial globe slipping from the Spanish grasp; and caricatures of the armada were printed with the legend: 'She came, she saw, and fled.'

been estimated that 63 ships, or approximately half the armada, suffered destruction, as follows:

Abandoned to the enemy	2
Lost in France	3
Lost in Holland	2
Sunk in the battle	2
Wrecked in Scotland and Ireland	19
Fate unknown	35

The English loss in ships was nil. 'They did not, in all their sailing round about England, so much as sink or take one ship, bark, pinnace, or cockboat of ours, or even burn so much as one sheepcote in this land,' says an anonymous but reliable contemporary writer.

XI

THE LAST YEARS OF THE REIGN

SELDOM has the strategic situation in a great war been so manifestly favourable to England as on the morrow after the battle of Gravelines. The enemy had struck his blow, and the blow had totally miscarried. The command of the sea rested with the English fleet; and a resolute, well-sustained counter-offensive, followed by occupation of the Azores or some part of the Spanish coast, might have laid King Philip's empire prostrate at Elizabeth's feet. So might a modern critic argue, and, having argued, proceed to judge the Elizabethan government for failing to deliver the *coup de grâce* when every circumstance was in its favour. But it should be remembered that combined military and naval operations, carried out at a distance of a thousand miles from home, were not understood in the sixteenth century, when standing armies were unknown and national resources were very limited. A rising in Portugal would undoubtedly have rendered the problem easier of solution; but in spite of Don Antonio's hopes and assurances in this direction there was little likelihood of such a rising taking place: the yoke of the Spaniard was too firmly planted on the neck of that country. Moreover, it may be doubted whether the queen really desired to fight her adversary to a finish. A war of annihilation, like that which the Venetians waged against the Genoese in the fourteenth century, was alien to the mind of a princess who believed in the virtue of the old maxim—'Sufficient unto the day is the evil thereof'. Never in her career did Elizabeth risk her all on a single throw of the dice. Thus England's war policy continued in the old groove of fighting the Spaniard mainly on land in alliance with the Dutch, and the struggle on the sea went on spasmodically without any approach to finality. The result was that Spain was given time to recover from her humiliation and to become far stronger as a sea power in the later years of the century than at any previous period in her history.

This inconclusive and unsatisfactory state of affairs is reflected in the atmosphere of uncertainty that prevailed in England during the post-armada period and lasted unabated to the end of the reign. The confidence inspired by the triumph of 1588 was soon qualified by the presentiment that the

A a

Spaniard would come again—as come he did. The general uneasiness was probably increased by the continuance of jesuit propaganda, and the publication of Persons' tract (1591),[1] comparing England to the barren fig-tree of the gospels, which the husbandman spared for another year to see if it brought forth fruit. Only far-seeing statesmen like Francis Bacon could afford to allude to Spain contemptuously as a 'barren seed plot of soldiers . . . not, in brief, an enemy to be feared by a nation seated, manned, furnished, and pollicied as is England'. Lesser minds turned to the perusal of military treatises, with a feeling that the land defence of the country required attention, and that the winds and waves might not be so friendly to England in the next encounter. Sir Henry Knyvett's pamphlet on *The Defence of the Realm* (1596), which advocated the compulsory military training of all men aged eighteen to fifty, is indicative of the widespread anxiety regarding the nation's second line of defence. To the same cause may be attributed the prevalence of invasion scares which kept the government continually on the *qui vive* throughout the nineties. In 1597, for example, a situation developed similar in some respects to that which existed two centuries later, when Napoleon's expeditionary force lay encamped at Ambleteuse and Boulogne. The offices of earl marshal and master of the ordnance, temporarily in abeyance, were re-created and invested with a new significance; and a committee of national defence sat in London, under the presidency of the earl of Essex, to co-ordinate and systematize the mobilization of the county levies. Efforts were made to organize the entire country south of the border defences into military districts under superintendents, so that a mobile army could be placed at the government's disposal for immediate service at the point of danger. The Thames defences were likewise overhauled, and a new system of signals prepared, by which the approach of the enemy could be communicated rapidly to Chatham, the strong point and rendezvous of the London area.

In addition, moreover, to the danger of a direct invasion from Spain, there was also the further possibility, after 1590, that the encroachment of the Spaniard on the Breton and Picardy coast might lead to the establishment of a Spanish naval

---

[1] *Elizabethae Angliae Reginae hæresim Calvinianam propugnantis sævissimum in Catholicos sui regni edictum* . . . (the author used the pseudonym Andreas Philopater).

base within easy striking distance of England—a contingency that roused a lively fear in the queen's mind, and led her to embark on military expeditions to France and to advance considerable sums of money to Henry IV. The situation in Ireland also had to be carefully watched for much the same reason; for a widespread rebellion there would provide Philip II with the opportunity he had long sought after, viz. of converting Ireland into a 'catholic Holland' on England's flank. The danger from this source reached a climax when the earl of Tyrone, in 1596, entered into military relations with Spain. Nor was Scotland wholly dependable, despite the king's expressions of goodwill, as the abortive conspiracy of the 'Spanish blanks'[1] clearly demonstrated. Altogether it is difficult to believe that the international problems confronting Elizabeth in the last years of her reign were appreciably less grave than at any previous period.

The government's anxiety also showed itself in other directions less directly connected with the military question. Catholics were watched with a vigilance and suspicion that made their lot more miserable; and the pursuit of priests was intensified. In 1591 a royal proclamation caused the establishment of commissions in every shire, with sub-commissions in the parishes, for the weekly examination of householders in respect of their beliefs and church attendance. Two years later (1593) parliament passed its last and severest statute against those who professed the Roman faith. By this act catholics were placed under a system of restriction not unlike that which a modern state adopts in time of epidemic, being prohibited from approaching within five miles of any corporate town; and catholic parents were in certain cases deprived of the right to educate their children. At the same time precautionary measures were taken to guard the queen from the shadowy figure of the assassin, and it became difficult for petitioners and even ambassadors to gain access to her presence. The execution of Dr. Lopez, the royal physician, for an alleged attempt to murder the queen by poison (1594) shows the strength of the public apprehension on this score, even if the evidence that sent him to the block was not conclusive.

From an economic point of view, too, the period was far from happy. The country seemed to emerge from one crisis only

---

[1] 1592–3: *vide* T. G. Law, *Collected Essays*, pp. 244–76.

to fall into another. The eleven years of booming trade and general prosperity (1576–87) were followed by a series of lean years, in which industrial and commercial undertakings of all kinds languished, and losses were incurred by companies trafficking with overseas markets. This was partly due to the disturbing effects of the war, partly to the depredations of the Dunkirk pirates, who became more and more active, and partly to the competition of the Dutch as an exporting community. Harvests were frequently bad during the years 1594–8, and again in 1600, causing 'dearths', when the price of corn and provisions rose to unprecedented heights; while recurrent visitations of the plague in 1592, 1602, and 1603 swept off great numbers of the population, producing temporary cessations of trade. There was distress in the clothing districts of Gloucestershire and Wiltshire; sheep farmers complained that they could not find a market for their wool at 'any reasonable price'; merchants were falling into bankruptcy from 'loss of traffic'; clothiers could not find a 'vent' for their cloth; and there were occasional bread riots. The chaos was increased by the heavy taxation necessitated by the growing war expenditure in the Netherlands, France, and Ireland, which was greater in the last fifteen years of the reign than in the preceding thirty years. The pressure of taxation is shown by the fact that between 1589 and 1603 the annual burden imposed by parliamentary subsidies was treble that of the earlier part of the reign. Parliament was fractious under the government's constant demand for money and more money. Yet, in spite of the greater sums raised by taxation, they only covered half of the war expenditure, and the queen had to sell crown lands in order to keep solvent.

Let us take first of all the naval war. For the reason already stated it was neither given pride of place in the defensive plans of the queen and her councillors, nor was it waged with sustained and concentrated vigour. There was no falling off in valour or seamanship among those who conducted the enterprises —Essex at Cadiz in 1596 and Cumberland at Porto Rico in 1598 worthily maintained the highest traditions of Elizabethan heroism on sea and land. But a certain lack of simplicity, coherence, and co-ordination in the more pretentious undertakings, due to misconceptions and faults in the drawing up of plans, dogged them with failure from the very start. The question of victualling and military hygiene was little studied and quite unscientifically handled, men and treasure being squandered with a lavish incompetence that almost baffles description. Nor were the troops employed in attacks on the Spanish coast of the right sort or sufficiently well trained and equipped for their task. They were for the most part raw levies, 'pressed' into the service for the occasion, and stiffened with a leavening of veterans from the Low Countries. And in addition to the lack of organization and technique on the English side must be reckoned the fact that the growing efficiency of the Spanish coastal defences in Europe and America made it difficult to repeat the *coups* of pre-armada days, however cleverly planned. In fact the only branch of naval activity that offered much chance of success was the raiding of commerce and the capture of carracks and treasure-ships; although even here the gains were small in proportion to the outlay, owing to the adoption of the convoy system by Spain, and the efficiency of her intelligence service. All things considered, commerce-raiding was singularly unprofitable, only two carracks being captured in ten years, in 1592 and 1602 respectively; while not one treasure-consignment found its way into the hands of the raiders.

In 1589, when circumstances were especially favourable for a swift and decisive blow at the remnant of the armada lying helpless in the northern part of Spain, an expedition was prepared by Drake and Sir John Norreys, consisting of 150 ships and 18,000 men, and set sail for Lisbon in April, with the following elaborate instructions: to distress the ships of the king of Spain in their harbours, to attempt, if they judged the circumstances favourable, the reinstatement of Don Antonio in Portugal, and to seize an island of the Azores and hold it until the end of the war. In point of size it was the most formidable armament that assembled at Plymouth in the reign of Elizabeth; but in actual fact it was a joint-stock piratical enterprise rather than a properly equipped government undertaking. It was financed partly by the queen, partly by London merchants, and partly by the promoters themselves. Only seven ships of the royal navy accompanied it, the rest being merchantmen and transports; and Don Antonio was present in person to give the intended operations against Lisbon a quasi-legitimate appearance and to rally the expected Portuguese malcontents as soon as the army set foot on his native soil. The first objective selected by the leaders was Corunna, which was plundered and burnt; then the troops were landed at Peniche to be conducted overland to the gates of Lisbon by Norreys, while Drake undertook to co-operate with the fleet in the Tagus. But the exhausting march across the burning sands, the outbreak of disease, the terrorized passivity of the Portuguese, and the failure of supplies, together with the lack of siege-artillery and the refusal of Drake to risk his ships under the guns of the Lisbon fortifications, prevented the attack on the city from being more than an impressive demonstration. The quixotic valour of the young earl of Essex,[1] who, having stolen away from court, joined the expedition off Corunna and challenged the Spanish commander of the Lisbon garrison to single-handed combat, did nothing to mitigate the bitterness of the retreat. After burning Vigo on the way homeward, the disappointed commanders reached England in the autumn, only to fall victims to the displeasure of the queen.

While Drake and Norreys were on their way home, Frobisher and Cumberland set out from Plymouth with privately organized squadrons for the Spanish coast and the Azores, where they cruised about until the late autumn in search of the West Indian *flota* or other windfalls from the East Indies, Brazil, or the Rio de la Plata. But their main quarry escaped them, and although they made many captures of less value, some of these were lost in the return journey to England. The following summer saw Hawkins and Frobisher with fresh squadrons in Spanish waters; and Philip became so anxious for the safety of that year's *flota* or treasure-fleet that he took the unprecedented step of ordering it to remain in the West Indies until the spring of 1591. In the interval he equipped a powerful fleet under the command of Alonzo de Bazan to act as its convoy through the danger zone. This fleet sailed from Ferrol in August (1591), and making its way to the Azores fell suddenly

---

[1] Essex (b. 1567; d. 1601), son of Walter Devereux, first earl of Essex, and Lettice Knollys, became Leicester's step-son when, in 1578, the latter married the countess of Essex after her first husband's death. He served with his step-father in the Netherlands as a cavalry officer, distinguished himself at Zutphen, and was created a knight banneret. At the age of twenty he became such a favourite with the queen that it was said 'he cometh not to his own lodging till the birds sing in the morning'. In 1587 he was appointed master of the horse, and became a bitter opponent of Raleigh and Charles Blount, with the latter of whom he fought a duel. During the armada campaign he was detained, by the queen's orders, with the army at Tilbury.

on Sir Thomas Howard's squadron at Flores, where it was lying in wait for the *flota*. Howard managed to extricate the bulk of his ships without difficulty; but Sir Richard Grenville who commanded the *Revenge*, either because he did not see the signal for retreat in time or simply out of bravado, allowed himself to be trapped, and for fifteen hours fought a single-handed action against the massed galleons of Spain. With forty men killed and almost all the survivors wounded, with six feet of water in the hold and all powder spent, the crew eventually surrendered against Grenville's orders, who wished to blow up the ship, but being mortally wounded could not enforce his will. This remarkable display of berserker valour, which cast an immortal glamour about the name of Grenville, has to some extent obscured the real meaning of the episode, which ought to be regarded as the first triumph of Spain's new naval policy. The adoption of the convoy system was the solution of the *flota* problem.

There was little likelihood, however, of the plan being extended to all Spain's commercial routes; and the English continued their hunt for carracks unabated. In August 1592 a squadron captained by Sir John Burrows and financed by the queen, Raleigh, and London merchants intercepted and captured the *Madre de Dios*, a Portuguese galleon from the East Indies of 1,600 tons burden, off the island of Flores. In the indiscriminate plunder that ensued the English crews got out of hand, the ship's bills of lading disappeared, and a considerable part of the rich cargo, whose total value was estimated at £800,000, passed into the possession of ignorant seamen, who sold their takings to land-sharks at ridiculously cheap prices. For days Plymouth was like a Bartholomew Fair, the atmosphere reeking with the perfume of cinnamon, cloves, musk, and other oriental spices—so much so that Sir Robert Cecil, who went down to the west country to look after the queen's interests, wrote to his father that he 'could well smell' the ships before he reached Exeter. 'There never was such spoil,' he commented. In the final distribution of the prize-money Elizabeth netted between £60,000 and £90,000, her original contribution being £3,000.

In the meantime the centre of interest had veered from the high seas to France, where the protestant Navarre, who had

succeeded to the throne in July 1589 on the assassination of
Henry III, was struggling to assert his rights against the catholic League. So desperate were the prospects of Henry of Bourbon that he could maintain himself in the field only by the
friendly assistance of Elizabeth in men, money, and munitions.
A loan of £20,000 in September, followed by an expeditionary
force of 4,000 men under Lord Willoughby, did something to
restore the situation in the king's favour, and prepared the way
for his victory over the League at the battle of Ivry (March
1590). But the triumph was short-lived. England's intervention provoked an appeal by the League to Philip II, who now
came forward as the protector of the Crown of France (Jan.
1590), and offered his help on condition that certain towns
were placed in his hands, and all harbours within the control
of the League opened exclusively to Spanish warships. His
intention was not merely to manœuvre himself into a commanding position with respect to the disposal of the French Crown,
but also to secure a suitable base for the invasion of England.
After the conclusion of the agreement considerable Spanish
forces began to move into France in support of the League. In
August Parma swept down on Paris, where Henry IV had
concentrated his army after the battle of Ivry, broke up the
siege, and revictualled the city. A few months later (Oct.)
Don Juan d'Aguila landed with 3,000 Spaniards at Blavet in
Upper Brittany, and proceeded in conjunction with the duke
de Mercœur, the League commander in the province, to capture the town of Hennebon. These places were some distance
to the south of Brest, whose governor, Sourdeac, was a loyal
supporter of the king; but they were sufficiently near the channel coast to cause anxiety in England. Consequently, when the
Breton estates applied to Elizabeth for help against the Spaniards, she dispatched Sir John Norreys with an army of 3,000
to Paimpol (May 1591) to co-operate with the royalist leader,
the prince de Dombes. A detachment of this force, consisting of 600, was sent to Dieppe under Sir Roger Williams to
keep open the main port of entry into Normandy for further
English succours. Later, in August, the earl of Essex joined
Williams with 3,600 well-equipped troops; and the entire body
awaited the king's arrival from Noyon, where he had gone to
meet his German mercenaries, before proceeding to the siege
of Rouen.

Elizabeth was no philanthropist. Every penny she spent on France, either by way of loans to the king or for the equipment of expeditions, was carefully debited to the recipient's account. Her sole interest was to prevent the Spaniard from mastering Brittany, and the League from gaining control over the Norman ports. Norreys's task was to guard the coast from Brest to St. Malo, and Essex was instructed, much against his sense of chivalry, to retain his troops at Dieppe until the king was ready to invest Rouen. The result was that both armies quickly deteriorated—the one through excessive marching and counter-marching in 'Base' (or lower) Brittany, the other through prolonged encampment in one place, desertions, and sickness. When at length Henry IV was ready to begin the attack on the city (Oct. 1591) he found it impossible, owing to lack of artillery and man-power, to make any impression on its fortifications; and in January 1592 Essex was recalled to England. Three months later Parma, making another whirl-wind raid from the Netherlands, broke up the siege, and strengthened the garrison of Paris with 1,200 Walloons. It was the last exploit of this great soldier, who died shortly after his return to Brussels. Meanwhile the Spaniards in Brittany inflicted a defeat on Norreys at Craon (May 1592).

It was now fairly evident to the French king that the English alliance alone could not save him from disaster; and the persistent demand of Elizabeth for the concession of a seaport in Brittany, which might be fortified and used as a base of operations, or as a place of retirement in case of need, roused his suspicions of her good faith. Neither would he sanction the dismemberment of his country, nor would he hand over an important town to be converted into another Brill or Flushing, even if the refusal involved him in difficulties with his imperious ally. Thus, while he continued to receive succour from England both in Normandy and Brittany during the year 1592–3, he decided that the only sure way to break the power of the League and rid himself of the Spanish menace was to proclaim himself publicly a catholic (July 1593). The political character of the conversion, summed up appropriately in the celebrated phrase 'Paris is worth a mass',[1] was apparent from the first; but it caused great excitement in protestant countries, especially in England, where it was held to presage an early settlement

[1] 'Paris vaut une messe.'

between France and Spain and the downfall of the 'common cause'. Actually it made no immediate change in the prospects of the king, nor did it affect the international relations of France for another four or five years.

The rebirth of patriotism and the expected rally of the catholics to the king's banner proved to be a very slow process; and before the year ended the alliance with England was renewed. It was almost immediately justified by events in Brittany. During the winter and the early spring of 1593–4 Don Juan, the Spanish commander at Blavet, moved some of his troops northwards to Crozon, where they constructed a fort with a view to cutting off Brest from the sea. In May they began to bring up their siege-guns in the direction of the fortress, and Sir John Norreys, writing to the queen, commented as follows: 'I think there never happened a more dangerous enterprise for the state of your Majesty's country than this of the Spaniard to possess Brittany, which under humble correction I dare presume to say will prove as prejudicial for England as if they had possessed Ireland.' Needless to say the danger was fully understood by the queen, who was never a sluggard when dangerous enterprises were afoot; and in July preparations were made for the dispatch of a relief expedition under the joint command of Frobisher and Norreys, consisting of 8 warships and 4,000 men. On 10 August the contract was drawn up pledging the French king to bear all expenses of the undertaking: by mid-September the investment of Crozon by land and sea began; and on 7 November, supported by Frobisher's guns, Norreys carried the fort at the second assault. The 350 Spaniards who had defended it for some seven weeks were either put to death by the sword or perished in the sea, and the fort itself was razed to the ground. The death of Frobisher, who was mortally wounded in the attack, was the only serious loss the English force sustained.

Some idea of the strain which these expeditions to France, from the autumn of 1589 to the fall of Crozon, had imposed on England's resources may be gathered from the fact that 20,000 men had been dispatched overseas, and only about half the number returned. The effort is all the more remarkable when we remember that the army in Ireland had also been kept up to strength, and 4,000 troops maintained in the Netherlands, either in the permanent garrisons at Brill and Flushing or in the service of the Dutch. It was fortunate that affairs in the

Netherlands were prospering under the leadership of Maurice of Nassau during this period, for again and again the expeditions to France had to be stiffened with contingents of veterans who had learnt their craft in the contest with Parma.

Owing to the queen's preoccupation with the war in France there had been a lull in major operations against Spain since the return of the Portugal expedition in 1589. But the year after the fall of Crozon saw the offensive resumed with the departure of a powerful squadron under the joint command of Drake and Hawkins for the West Indies. This expedition, which sailed in August 1595 from Plymouth, consisted of 6 royal ships, 21 other ships, and 2,500 men, the latter being under the command of Sir Thomas Baskerville. Experience had shown that if the treasure-fleet of Spain was to be taken it would have to be attacked, not at the Azores or in Spanish waters, where adequate protection was provided, but at its point of departure in the Caribbean sea. Moreover, information had come to hand that the principal galleon of that year's *flota*, said to be worth £600,000, had been compelled to put in to the harbour of San Juan in Porto Rico, owing to injuries at sea. To capture this ship was part of the task allotted to Drake and Hawkins, after which they were to attack Panama and the Spanish settlements in the Pacific. For this latter task the army was provided.

For some reason or another, however, Drake and Hawkins made bad yoke-fellows; and calamity after calamity dogged their footsteps. First of all they attacked Las Palmas in the Canaries, but the attack failed; and when they crossed the Atlantic they found that news of their intentions had been circulated before their arrival among the Spanish settlements. Hawkins sickened and died as the ships reached Porto Rico, and was buried at sea. Drake, who then assumed sole command, completely failed to master the fortifications of San Juan, and made for the mainland, plundering and burning Rio de la Hacha, Santa Marta, and Nombre de Dios, none of which yielded much in the way of spoil. At the last-named place he landed Baskerville and his troops for the march across the isthmus to Panama. But again the English were balked of success; for they found the way blocked by the Spaniards in considerable numbers, and had to retreat to their ships. This final disappointment made a deep impression on Drake, who soon after succumbed to dysentery at the unhealthy island of Escudo and died at Porto Bello. It was a tragic end to the most brilliant career in Elizabethan history, and the tragedy was unrelieved by the consciousness of purpose achieved. Thus the second offensive against the dominions of King Philip proved to be as great a failure as the first; for with the death of the leaders, the enterprise collapsed, and Baskerville brought the ships back to England in the spring of 1596.

While the fate of the expedition was still unknown in England, the earl of Essex, Lord Charles Howard, and Sir Francis Vere, commander of the English army in the Netherlands, were hatching an ambitious plan for the invasion of Spain. In March the first moves were made to collect the necessary ships and to transport a strong force of veterans from the Netherlands; but the actual destination of the expedition was kept a profound secret, it being given out that its purpose was to cover the return of Drake and Hawkins from the West Indies. Suddenly, however, the most alarming news arrived from France, and for a brief space it appeared as if every ship and man would be required for a renewal of the war in that country. On 29 March the cardinal archduke Albert of Austria, Parma's successor in the governorship of the Netherlands, who had captured Doullens and Cambrai from Henry IV in the previous year, took advantage of Vere's withdrawal to make a swift descent on Calais, and seized the Risbank fort dominating the entrance to the harbour from the sea. By 7 April the town was in his hands and the attack on the citadel began. The speed and ease with which the success had been accomplished caused surprise and consternation in London, where the liveliest anxiety was expressed for the safety of the narrow seas and the security of the realm. Essex was immediately ordered to suspend his preparations for the projected expedition to Spain, and to proceed to Dover to take charge of a Calais relief force. Yet although the urgency was great, Elizabeth hesitated to sanction its departure until she was assured by the king that her troops would be received into the garrison of Calais. It was her last effort to recover control over the seaport which her sister had lost, and she fondly believed that in his extremity Henry IV would consent to her desire. Down at Dover Essex was in a fever of impatience to be off. He could hear the Spanish guns 'playing with great fury' on the citadel; but week after week passed in futile palaver and still the relief force remained confined to port. At length on 29 April the garrison capitulated, and the archduke entered into possession of the most coveted fortress on the northern coast of France, within easy striking distance of England.

The most effective reply to this new threat was to release Essex with all speed for his enterprise against Spain. News of Drake's failure at Porto Rico also emphasized the need for an immediate offensive. Consequently on 3 June the best equipped of all the expeditions that left England for Spanish waters in the reign of Elizabeth set sail for Cadiz. It comprised 17 ships of the royal navy, 47 other warships, and a Dutch squadron of 18, together with 6,000 or 8,000 well-trained troops under the command of Sir Francis Vere. Apart from the three leaders, Essex, Howard, and Vere, who formed a remarkable trio of initiative, prudence, and science, the subordinate commanders —Sir Walter Raleigh, Sir Thomas Howard, Baskerville, Amyas Preston, Sir Conyers Clifford, Sir George Carew, Sir John Wingfield, and Sir Charles Blount—were all men of mark in the naval and military world of the time. So well was the secret of the expedition kept that when it reached Cadiz, on 20 June, the Spaniards were completely taken by surprise. Not only was the city captured by assault, but the Spaniards had to burn the *flota*, which was sheltering in the harbour, in order to prevent it from falling into the hands of the English. Twelve million ducats were thus lost to King Philip. Essex originally intended to remain in occupation of Cadiz for an unlimited period, and doubtless would have done so, but his supplies failed him and there was a general desire to return to England on the completion of the operations: consequently he contented himself with the sack of Cadiz, which he left in flames on 5 July. On the homeward journey two other towns, Faro and Loulé, were subjected to the same fate. It was a brilliant display of the might of England on sea and land, a revival of the Drake 'touch' at its best.

Essex was now at the height of his fame. Twenty-nine years of age, he was the first courtier in the realm and the greatest 'swordsman' of the age, on whom seemed to have fallen the mantles of both Drake and Leicester. His rise had been meteoric but not unmerited, for he had many admirable qualities. Men of action adored him for his courage and magnanimity: the queen found the court insipid without his cultured and fascinating personality; and sycophants courted him for his liberality. He was the chartered libertine of the period. Nothing he did was on a small or mean scale. Even his gestures were grandiose. But already his dazzling success bore within it the seeds of decay. He was a child of impulse—overbearing, intemperate, quick to take offence, prodigal, unruly. When things went well with him at court he irritated the queen by his domineering attitude: when they went wrong he sulked in his tent like Achilles. He must be *aut Caesar aut nullus*.

For some time he had been engaged in laying the foundations of a 'domestical greatness' that would enable him to seize the reigns of power as soon as they slipped from the nerveless grasp of the now aged Burghley. Death had removed from his path all possible competitors among the elder generation of statesmen, who had borne the burden and heat of the day; and the new 'jousters for power' who were pressing forward into the magic circle were all young men, in many respects less happily circumstanced than himself. The brilliant but unstable Raleigh,[1] whose rise to place and power had been almost as meteoric as the earl's, and whose versatility was probably greater, ostracized himself from the queen's good graces for five years by his secret marriage to Elizabeth Throckmorton, one of her maids of honour. The two Bacons, Anthony and Francis, were handicapped by the lack of a patron, since their uncle, Burghley, on whom they mainly relied, was devoted to the advancement of his own son Robert. In fact Robert Cecil was Essex's only dangerous opponent; but he, too, suffered from a defect that might, in other circumstances, have proved fatal to his chances —he was a hunchback.

The one factor in the situation which Essex seems to have ignored in his quest for power was the queen herself. Elizabeth

---

[1] Raleigh was not a really serious competitor. Born some fifteen years before Essex (*c.* 1552), he had served in the huguenot wars, in the Netherlands, and in Ireland; became a great favourite at court after 1582; and was appointed captain of the guard in 1587. But he never reached the rank of privy councillor, or any place of first command. He was the great 'improviser' of the age—poet, philosopher, historian, soldier, privateersman, shipbuilder, member of parliament, vice-admiral, explorer, colonizer, and empire-builder. Every activity he took up seemed to 'sparkle at his touch' and then, unaccountably, to lose its fire as soon as his hand was withdrawn. It was not until death removed Essex from his path (1601) that Raleigh received his chance, and then it was too late: the glory of the reign had departed, the queen was an old woman, and Robert Cecil was in power.

was undoubtedly ageing; but declining years did not bring with them any decay of her intellectual faculty. She was as little likely as ever to succumb to the control of a favourite, or to entrust the destiny of the state to a man whose genius shone on the battle-field and not in the counsels of peace. She had trusted Burghley for his faithfulness and strong common sense, and she would trust his son, when the time came for changing the pilot, for much the same reason. Essex probably realized the difficulty with which he was confronted; for, in 1592, with the aid of Anthony Bacon, he had begun to build up a vast intelligence system rivalling that of the Cecils, which would show the queen that he, too, had his finger on the pulse of Europe, and could play the statesman as well as any one. In the following year he was admitted to the privy council, and soon 'all matters of intelligence' were 'wholly in his hands, wherein the queen receiveth great liking'. His feet seemed to be firmly planted on the path to 'domestical greatness'.

But the earl could not change his nature. He could not school his imperious mind to the give and take of politics, or share responsibility at the council-table. The main advantage that accrued from his promotion was that he was now able to keep an eye on 'Robert the devil', who had become a councillor in 1591. When he attempted, however, to foist his partisans into high offices in the patronage of the Crown, he found the influence of the Cecils too strong for him. The attorney-general-ship went to Sir Edward Coke, a Cecilian candidate, and not to Essex's protégé, Francis Bacon; and the same story was repeated later, when Bacon was again passed over for the solicitor-general-ship, although in this case it was the queen herself who blocked the way, not the Cecils, who were friendly. Doubtless Bacon had himself to blame for his patron's failure, because his speech against the subsidy in the parliament of 1593[1] profoundly annoyed Elizabeth. But these and other pinpricks of disappointment were not without their effect on the sensitive Essex.

The Cadiz expedition was, or at least seemed to be, a welcome turn of fortune, for it won him great popularity with the people; and Anthony Bacon was quick—too quick, perhaps—to record his impressions of its effect on the political situation. 'Our earl', he wrote, 'hath with the bright beams of his valour and virtue scattered the clouds and cleared the mists that malicious envy

[1] See p. 190.

stirred up against his matchless merit; which hath made the old fox [i.e. Burghley] to crouch and whine.' But Francis, wiser in his diagnosis, foresaw danger ahead in the very eminence of his patron. In a letter packed with candour and prescience he advised him to shun the appearance of martial greatness and concentrate on winning the queen's approval by studied humility and dissimulation. Rather seek the safe office of privy seal, he said, than that of earl marshal or master of the ordnance. But the oracle spoke in vain. The effort to establish 'domestical greatness' had not been a conspicuous success; and in 1597 Essex was back again on the quest for military renown. He accepted both offices, feeling, no doubt, that another *annus mirabilis* was at hand.

Although Spain was now on the verge of a fresh bankruptcy, the sack of Cadiz spurred on the king to revenge. In the summer of 1596, after the English left the coast, he began to collect ships and men for a second armada, and late in October the *adelantado* (or governor of Castile) set sail from Ferrol with a fleet of over a hundred ships and some 16,000 men. Rumour had it that part of the armament was destined for Calais, where the archduke was preparing to co-operate with it, and part for Ireland to support the earl of Tyrone. But the winds and waves effectually destroyed the hopes of the Spaniards. Before the *adelantado*'s fleet had cleared Cape Finisterre it was dispersed and partially wrecked by a gale; and the lateness of the season prevented any possibility of renewing the attempt that year. Meanwhile elaborate preparations had been made in England to cope with the danger; and by July of the following year (1597), in anticipation of a further effort by the *adelantado*, Essex, Howard, Vere, and Raleigh were once more on the sea with 98 sail and 5,000 men *en route* for Ferrol. Their instructions were to burn the armada in its own harbours or meet and destroy it on the high seas, and thereafter proceed to the Azores and intercept the *flota*. Bad weather, however, drove them back to port; and it was not until 17 August that the expedition really got under way. By this time news had come to hand that the *adelantado* would not sail that year, and it was decided to abandon the attack on Ferrol and make direct for the Azores. September saw the English squadrons cruising about the islands in search of the *flota*; but, as bad luck would have it, the Spanish ships slipped past them and ran under the guns of Angra, where they were perfectly safe from attack. Thereupon the English commanders, considering the season too late for any longer stay at the Azores, resolved to return.

On the very day on which they sailed, the *adelantado*, contrary to expectation, put out from Ferrol with his reconstituted fleet of 136 ships and troops to the number of 9,000. His intention was to land at Falmouth and establish a Spanish naval base on English soil—a plan very similar in idea to that of Essex in the previous year. The two fleets thus ploughed their way towards the English channel on converging courses but quite oblivious of each other's existence; while a feeling of profound security prevailed in England. Then suddenly, on 23 October, London was thrown into a state of consternation by a report from Sir Ferdinand Gorges at Plymouth that the Spanish fleet was on the high seas. As a matter of fact it had been seen off Blavet on 12 October; but immediately afterwards it ran into a storm, which scattered it and drove the main body in full retreat to Spain. Only a few ships which failed to see the signal to retire reached the English coast, and they made off again when they realized that they were isolated. The safe return of Essex when the panic was at its height helped to restore public confidence; and by the middle of November all anxiety was at an end for the year.

In May of the following year France abandoned the common cause and made peace with Spain at Vervins, leaving England and Holland to continue the struggle alone, or settle their accounts as best they could with Philip. There was great debate in the council as to the future policy of the country. Essex was unhesitating in his advocacy of war: Burghley was for peace. It was the last appearance of the lord treasurer. When argument failed him he drew out a prayer book and pointed to the verse: 'The bloody and deceitful men shall not live out half their days.'[1] But he was unable to carry the council with him, and in August he died without seeing the fulfilment of his hopes. Little more than a month later he was followed to the grave by the king of Spain. It was as if fate had suddenly swept the pieces from the chess-board and reset them for a new game. But there was to be no new game. The Dutch were adamant in their determination to fight their adversary to a

---

[1] Psalm lv, verse 25.

conclusion, and England was too deeply committed to withdraw without loss of honour. So the war went on. There were brilliant successes like the battle of Nieuport (1600) and the defence of Ostend (1601), in which the English played a noble part; but everything else was overshadowed by the Irish question and the last act in the amazing drama of Essex's career.

Ireland was now uppermost in Elizabeth's thoughts.[1] She wanted to send over Sir William Knollys or, failing him, Sir Charles Blount to crush the arch-rebel Tyrone. Essex objected to both; but in criticizing their qualifications for the office he seemed to indicate that he himself had the only real claim to it, although, it would appear, he had no desire to seek renown in so forbidding a field. He played his cards, in fact, like a novice; and early in 1599 he found himself invested with the Irish command. To make matters worse, he had succeeded in so irritating the queen by his petulant behaviour during the discussions that she boxed his ears—a humiliation he never forgot. He would not have taken such an affront, he said, at the hands of Henry VIII himself. More than ever she was resolved to 'break him of his will and pull down his great heart'. For the moment, however, the earl was to have his chance in the most difficult and dangerous task ever offered to him.

He fully realized that Ireland would try him to the uttermost. Criticism would be prolific; his every action would be brought under the microscope, his mistakes trumpeted abroad, his efforts belittled; while his absence from court would remove any hope of personally correcting the misrepresentations of his enemies. 'I provided for this service', he commented, in a letter to the council, 'a breastplate and not a cuirass; that is, I am armed on the breast but not on the back.' Others, however, took a different view of the matter. 'Observe the man who commandeth and yet is commanded himself,' wrote Robert Markham to John Harington, who accompanied the expedition: 'he goeth not forth to save the queen's realm, but to humour his own revenge.' In such circumstances the only wise thing for Essex to do was to bring back 'rebellion broached on the point of his sword'. He was verily at the crisis of his career. If he failed, even his military reputation would collapse like 'an unsubstantial pageant faded'.

[1] See chapter xii for a consecutive account of the Irish question.

But Ireland baffled him. After a few months' campaigning Essex stole back to London in September, contrary to the queen's explicit orders, and with little to show for the large expenditure incurred but an inconclusive and suspicious-looking treaty with Tyrone. Elizabeth was pardonably annoyed. 'By God's son, I am no queen,' she exclaimed. 'That man is above me. Who gave him command to come here so soon? I did send him on other business.' In June 1600 Essex was tried before a special commission for his misdemeanours in Ireland—his disobedience of orders, excessive creations of knighthoods, and parleyings with Tyrone—and sentenced to be deprived of his various offices. His misery was increased by the fact that Francis Bacon took part in the prosecution! But he had not yet reached the end of his misfortunes: at Michaelmas he lost the monopoly of sweet wines, on which he largely depended for his income. 'An ungovernable beast', said Elizabeth, 'must be stopped of his provender.' There were moments, of course, when the queen relaxed; but the breaking and pulling down were progressing.

It was impossible for such a situation to continue. The humiliated earl must either bow or break under the strain; and bow he would not. Despair overcame him; he gave way to uncontrollable rage; he was like one, says Harington, 'devoid of good reason as of right mind'. His intemperate language concerning the queen disturbed his intimate friends. If his enemy Raleigh is to be believed, he declared her mind to be as 'crooked as her body'. But the end was not yet. He would crush his enemies like Samson by bringing the whole fabric of the state about their ears. He would force his will upon the queen by raising the city against her, capture the court, and dictate his terms. He tried to get Mountjoy, who had taken over the Irish command, to bring an army to Wales. He approached James VI with certain promises concerning the succession if he would support him. He threw open his house to discontented persons of all kinds, adventurers, soldiers out of employ, puritan preachers. In February 1601 plans for the *coup d'état* were concocted at Drury House, the residence of his principal supporter, the earl of Southampton. It was believed that he could calculate on a following of 120, composed of two earls, Southampton and Rutland, several barons, and a large number of gentlemen. By this time, however, the court had learnt of the strange goings-on at Drury House, and on 7 February the earl was summoned to appear before the council. It had probably come to Elizabeth's ears that the play of *Richard II* had been staged, with the sanction of his supporters, at the Globe theatre—an ominous event in itself. But Essex refused to obey the summons, pleading sickness. Then, realizing that the game was up, he decided to stake everything on a swift and sudden irruption into the city at the head of his men. On Sunday 8 February he summoned his party to Essex House, explained that Raleigh and Lord Cobham were plotting to take his life, and imprisoned the lord keeper, the earl of Worcester, Sir William Knollys, controller, and lord chief justice Popham, who arrived from the court while the meeting was in progress, to learn the cause of the assembly.

Leaving the queen's representatives locked up in a room, he then dashed into the city, accompanied by 200 men-at-arms, crying as he rode: 'For the Queen! For the Queen! The crown of England is sold to the Spaniard! A plot is laid for my life!'[1] But the people were more amazed than alarmed, and refused to move. Meanwhile the government proclaimed him a traitor, barricades were thrown up across the principal thoroughfares, and troops under Sir John Leveson occupied Ludgate. The attempt had failed; and Essex, finding his way blocked by land, was compelled to return to his house by water. Here he found his hostages flown: they had been released by Sir Ferdinand Gorges, one of his own supporters, who went back with them to the palace. There was nothing to be done but surrender to the queen's troops, who now surrounded the house, and stand his trial for sedition.

By the unanimous verdict of a court of peers he was condemned to death for treason, and executed on 25 February. Of his principal associates, five were either hanged or beheaded, Southampton was pardoned, and the others were either fined or imprisoned. Altogether the episode is one of the most pathetic in the reign; and the inevitability of the end in no way detracts from its pathos. If Essex was to a large extent the architect of his own fate, there is, perhaps, some truth in the

[1] When Essex was questioned, at his trial, as to the meaning of his wild cry that the crown had been sold to the Spaniard, he had no better excuse to offer than that he had heard that Mr. Secretary (Robert Cecil) had said to one of his fellow councillors, 'The infanta's title comparatively is as good in succession as any other.' Cecil easily cleared himself of the slander. The alleged plot against his life had no better foundation in fact.

indulgent words written by his secretary, Sir Henry Wotton, many years after the event: 'He had to wrestle with a queen's declining, or rather with her very setting age, which, besides other respects, is commonly even of itself the more umbratious and apprehensive, as for the most part all horizons are charged with certain vapours towards their evening.' When he died he was only thirty-three and Elizabeth sixty-seven—a contrast in years that suggests, it has been said, the perennial conflict between youth and crabbèd age.

The death of Essex removed not only the leader of the war party in England and the 'protector' of the puritans, but also one whom even the catholics appear to have trusted because of his generous outlook and professed attachment to toleration. For some time a movement had been afoot in the catholic camp to procure from the state some relaxation of the severe penal laws to which they were exposed, in return for a public avowal of their loyalty and disavowal of Spain as the national enemy. About 1590 a secular priest, Wryght, framed the question thus: 'Is it lawful for catholics in England to take up arms and in other ways defend the queen and the realm against the Spaniards?' It was a courageous thing to formulate such a question, but Wryght did more: he answered it in the affirmative, basing his argument on the demonstrable truth that the king of Spain was no true crusader for the faith but a time-server, who used religion merely as a cloak for political designs. Rome had therefore erred in making him the executor of the bull, and it was not incumbent upon catholics to follow the papal instructions. Rather, said the priest, let them defend their country, yield to the time, and make the best of their lot. In putting forward these views Wryght simply voiced an opinion that had been gradually leavening the whole lump of catholic discontent in England. Experience had shown that Elizabeth's throne was too securely established to be overturned, and the effort to destroy her had only made the situation harder to bear. Why, said the sufferers, in effect, must we be denied the comforts of religion, which we might enjoy in secret, and that freedom in temporal affairs, which would be ours as soon as the plotting ceased—for the sake of some hypothetical catholic England of the future? Could not a catholic church be created in England from the priests and the laity now in the country, with full recognition of papal supremacy in spiritual matters, but pledged to non-political action in the temporal sphere? Between this point of view and that of the jesuits there was a wide gulf fixed; but the opinion of those who stood for present alleviation, albeit they still were in a minority, was at least based upon tangible realities, while that of their opponents, the jesuits, was the outcome of a wild romanticism.

From the government standpoint, this division in the catholic ranks was eminently satisfactory; for a deadlock was reached during the nineties in the handling of the catholic question. Despite the increased severity of the penal laws, the revival of catholicism in England, which began with the foundation of the seminarist and jesuit missions, had continued to make headway,[1] and the flow of priests from the Continent had not been checked. In nearly all parts of the country men were to be found who acknowledged the pope's supremacy and had their spiritual needs attended to by a regularly ordained itinerant catholic clergy. Secret roads connecting the coast with the inland districts, along which priests could travel in comparative safety, and a secret postal service for the dissemination of news and the conveyance of money contributions abroad existed in defiance of all efforts to track them down. In some places justices of the peace and other local officials, themselves secret catholics, winked at breaches of the law, while in the north and west prisoners were rescued from pursuivants in broad daylight, and houses barricaded against the sheriff and his posse. Even in the streets of London, in the spring of 1602, the arrest of a catholic gentleman, Henry Ponod, led to a fracas in which a man was mortally wounded. A point had been reached when the government, with the imperfect machinery at its disposal, could not carry coercion any farther without the risk of provoking rebellion. Yet it had been demonstrated, not once but many times, that both the catholic laity and the priests whom they supported were overwhelmingly loyal to the queen.

[1] The number of catholics in England towards the close of the reign cannot be even approximately determined. We know that the recusants in 1603 totalled 8,630 for the whole country; but this figure takes into account only those who were indicted or presented. Secret catholics who evaded the law must have been many times more numerous. R. G. Usher's guess that 'there were very likely between 750,000 and 1,000,000' is probably an exaggeration. A. O. Meyer seems to accept the figure of 120,000 as the nearest approximation. (See Usher's *Reconstruction of the English Church*, i. 157–9); also Meyer's *England and the Catholic Church under Queen Elizabeth*, pp. 62–3.)

Obviously the time had come to consider the question of extending some measure of toleration, with suitable safeguards, to the
large body of non-political catholics in England.

Other influences were shaping men's minds towards the same
general thesis. After the collapse of the Spanish attack, in 1588,
the papal interest in the *empresa* faded. Neither Sixtus V nor
Clement VIII thought of renewing the extremist policy of
Gregory XIII; on the contrary, both put their trust in spiritual
weapons, and gave only a perfunctory attention to Spanish
designs for the conversion of England by force. Many of the
English catholic exiles also found themselves assailed by a longing
to return to their native land, and were prepared to gratify it,
if only they were given the least assurance that they would be
permitted to practise their religion in peace. Even the veteran
Cardinal Allen seems to have recovered something of his lost
patriotism, for he protected English protestant travellers in
Rome from the inquisition, making himself responsible for their
safety provided they gave no outward cause for scandal. The
splendid victory which Elizabeth had won over the Spaniard,
and the consequently enhanced prestige of her country, may
partly account for this remarkable change of attitude; but it was
also due to the fact that England was obviously not withering
under the papal curse. The government had taken good care
that Europe should know the real facts of the situation. Burghley, in his *Execution of Justice* (1583), laid stress on the material
well-being of the realm, and Bacon adopted a similar vein in his
Observations on a Libel (1591). In those days, when every one was
scanning heaven and earth for a sign, such an appeal to the
evidence of the senses was not without its weight. Elizabeth
herself took up the pen of the propagandist. Writing to the duke
of Florence, in April 1591, to confute the misrepresentations of
the Florentine historian Girolamo Pollini, she pointed out the
moral of England's prosperity. 'It is clear as daylight', she said,
'that God's blessing rests upon us, upon our people and realm,
with all the plainest signs of prosperity, peace, obedience, riches,
power, and increase of our subjects.'

The main interest, however, lies in the quarrel between the
seculars and the jesuits, which became more and more pronounced as the last decade of the reign advanced. The field of
controversy was wide: it embraced the seminaries, the question
of the English succession, and the organization of the catholic

priests in England. In each case jesuit internationalism and centralization collided with the nationalist and separatist policy of the seculars. The jesuits wanted to make the seminaries instruments for the mastering of England by Spanish forces, to place the infanta Isabella on the English throne, and to subject the priests working in England to a control in which the chief say would rest with the Society of Jesus. Spiritual pride played a part, too, in the struggle; for the jesuits claimed a superior sanctity, which the seculars would not brook. 'So holy, so godly, so religious would they seem to be', said one, 'as nothing is holy that they have not sanctified.' But the vital issue was the constitution of the catholic community in England.

Trouble began in earnest in 1594, when Persons published his *Conference about the Next Succession*, defending the infanta's claim in preference to that of James VI and Arabella Stuart, Elizabeth's blood relations. The book was hailed in anti-Spanish circles among the English catholics as 'the most pestilential thing ever written', and provoked a fury of excitement in the English seminary at Rome, where anti-jesuit feeling already ran high. It was the third outburst against the jesuits within fourteen years. Simultaneously the so-called 'stirs' began at Wisbech castle, in the diocese of Ely, occasioned by the attempt of Father Weston, a jesuit, to set up a moral censorship over the little community of priests who were imprisoned there. Thirty seculars refused to acknowledge this assertion of jesuit supremacy (Feb. 1595). The point at issue was perhaps trivial enough; but out of it came the archpriest controversy, and, as we shall see, the eventual submission of a number of the seculars to the queen in the spring of 1603.

Meanwhile the death of Cardinal Allen (1594) had raised the question of appointing some one to supervise the catholic community in England. The seculars would have liked a bishop, chosen by themselves and invested with the usual diocesan authority, under whom they might form themselves into a semi-independent national organization. But jesuit influence prevailed at Rome, and Cardinal Caetani, with the pope's sanction, appointed an archpriest, George Blackwell, with far-reaching disciplinary powers and instructions to consult the jesuits in all important matters (1598). Blackwell was not himself a jesuit; but his arbitrary character and dependent position roused the ire of the seculars who were placed under his authority.

Into the merits of the dispute it is unnecessary to enter. The significant fact is that the centralization for which Blackwell stood precluded any hope of the seculars obtaining a reasonable settlement of their position *vis-à-vis* the government. They therefore sent a deputation to Rome, in the autumn of 1598, to plead their case before the pope, coupling with their request for a bishop a plea that jesuit influence might be excluded from the Roman seminary, and a strict control exercised over the publication of defamatory books against the queen and the English government. To bring forward such a scheme during the heat of the counter-reformation was to court disappointment: the appellants were dismissed from Rome without being allowed to interview the pope; and Blackwell was confirmed in office (April 1599). The archpriest then demanded an acknowledgement of schism from the defeated party, and when they received the support of the university of Paris, and remained contumacious, threatened them with an interdict. This only added fresh fuel to the fire of anti-jesuit feeling among the seculars, who now resolved to appeal for help to the government.

Here at last was the heaven-sent opportunity for which the government had been patiently waiting—the opportunity, namely, of destroying for good the unity of the catholic party, and of discomfiting the jesuits by drawing the seculars to its side. In March 1601 Bishop Bancroft, with the knowledge of Cecil and the queen, summoned Father Bluet, the secular leader, to a series of conferences at his palace of Fulham; and after some five months' discussion an agreement was reached, which, if it could have been carried out in every particular, would have gone a long way towards solving the catholic problem. A petition was to be drawn up in the name of the catholics promising complete fidelity of both the clergy and laity in all temporal matters, and requesting the queen to grant liberty of conscience and to suppress jesuit books. It was a case for heroic action on the part of the government. But when the petition was presented to Elizabeth she bristled: 'If I grant this liberty to the catholics,' she said, 'by this very fact I lay at their feet my honour, my crown, and my life.' She was too old to adapt her mind to the changing circumstances of the time. Nevertheless at Bancroft's instance she consented that Bluet and a few companions should be permitted to present a second appeal to Rome (Nov. 1602). Meanwhile Bancroft secretly relaxed the censorship so as to allow a number of books written by seculars and printed in London to be circulated, in order to foment the case against Blackwell and the jesuits. It was part of the campaign.

What the government wanted, now that toleration was out of the question, was that the appellant priests should procure a papal bull, or brief, prohibiting or discountenancing catholic plots and insurrections against the state. If they secured this, it was to be understood that certain favours would be extended to them. Bancroft and Cecil, who were behind the movement, were not interested in the religious issue at all: they were anxious to secure the support of the seculars, and through them of the catholic laity, for the succession of James VI, to whom they were already deeply committed. Such support they believed might well prove to be the decisive factor in the crisis following the queen's demise. But the reed broke in their hands. In the autumn of 1602 the pope gave his answer to Bluet's petition. He trounced the archpriest for exceeding his powers, but would not hear of a submission, which he described as dangerous to the catholic cause, and peremptorily forbade all intercourse with heretics to the detriment of catholics. Clearly the petition was a failure from the standpoint of both the government and the seculars. Subsequently, in November 1602, the queen issued a proclamation ordering all jesuits and secular priests to leave the realm, but granting six weeks' grace to the latter to acknowledge their allegiance to the government. This was Elizabeth's last pronouncement on the catholic question, and it shows that nothing was farther from her thoughts than toleration. She would admit the catholics to her grace only on her own terms—absolute submission.

It remained to be seen what the seculars would do. Could they find a formula expressing their attitude to the government that would pass muster, without at the same time sacrificing their catholicity? They were fully prepared to profess their loyalty, to reveal conspiracies, and even to disregard any papal excommunication of the queen, but they could not, and would not, disavow the apostolic authority: they would rather lose their lives, they said, than infringe the lawful authority of Christ's catholic church. Bancroft was not satisfied with this important reservation. But after Chief Justice Popham had interrogated the priests more particularly, and learnt that they would give their written assurance not to obey the pope if he excommunicated Elizabeth, Cecil decided that this was sufficient. Thus on the last day of grace, 31 January 1603, thirteen secular priests signed a conjoint declaration of loyalty, pledging themselves to obey the queen in all things temporal, to defend her and the state against all enemies, irrespective of any papal command to the contrary, but reserving their spiritual obedience to the catholic church. They were only a fraction of the number then working in England, and their submission was of small consequence; but it was at least an indication that a new age was at hand.

For the problem of puritanism,[1] on the other hand, no solution could be found. As a practical movement for reforming the polity of the church, by introducing 'presbytery in episcopacy', it had been tracked down and suppressed by Bancroft and the high commission. The separatists had also been accounted for. But the intellectual basis of presbyterial puritanism still remained unshaken. Neither Whitgift nor Bancroft had found it possible to give an adequate reply to Cartwright's contention that, if the Scriptures were the authentic Word of God, they must necessarily be a better guide to the right ordering of the church, on its disciplinary as well as its doctrinal side, than the opinion of men. Moreover, so long as the *Book of Discipline* was believed to rest upon the impregnable rock of Holy Writ, and the anglican argument rested only on expediency, it was futile to expect the puritan to see the error of his

---

[1] The number of puritans in England towards the close of the reign is a very disputable point. It cannot be too strongly emphasized that there was no such thing as a puritan party, in the sense of a unified organization of men pledged to the same aims. There were only groups whose sole element in common was opposition to the form of church government established by law. The strength of the most important, and indeed the only significant of these groups, the presbyterians or disciplinarians, perhaps amounted to between 281 and 350 ministers, and from 50,000 to 100,000 laymen. This estimate is highly conjectural, though not unreasonable; but if it be anywhere near the truth it shows conclusively that the puritans as a body did not number more than an infinitesimal fraction of the total population. Furthermore, it has been calculated that of the 281 puritan ministers whose names have been preserved, only 105 had university degrees, and of these 'degree men' only 31 had higher degrees than master of arts. Hence there is some justification for concluding that the 'party' was not in any way outstanding for its learning. Its real leaders were the thirteen divines who drew up the Millenary Petition at the beginning of James I's reign. (For a full discussion of the subject see Usher, op. cit., vol. i, ch. xi.)

ways. It was to remedy this defect that Richard Hooker embarked upon his immortal treatise *Of the Laws of Ecclesiastical Polity*.

As a controversialist he had no equal in his time, and his learning was encyclopaedic. During the period 1584–91, when he occupied the office of master of the Temple, he had exercised his skill in a prolonged gladiatorial contest with Walter Travers. In the latter year he retired to the country parsonage of Boscombe, in Wiltshire, to begin the composition of his treatise. The first four books appeared in 1594, the fifth in 1597; but the remaining three, though completed before his death (1601), did not see the light until well on in the next century.

The importance of Hooker's intervention in the politico-ecclesiastical arena lies in the fact that he set the whole controversy, for the first time, in a wide framework, and brought it into vital touch with the learning of the renaissance. 'Let the vulgar sort amongst you know', he wrote, 'that there is not the least branch of the cause wherein they are so resolute, but to the trial of it a good deal more appertaineth than their conceit doth reach unto.' After discoursing at length on the nature and function of law in general, he descends to the particular issue at stake between the church and the puritans, pointing out the fallacy of supposing that the only law which God 'appointeth unto man' is the sacred Scripture. The Scriptures, he affirms, are sufficient for the purpose for which they were intended, viz. to reveal the supernatural way of salvation, undiscoverable by the light of reason alone; but they have a limited sufficiency. They neither abrogate the use of human reason nor discountenance it; on the contrary, they presuppose it, being themselves interpretable by reason. Where they are silent, the church, which is invested, like any other society, with the power to make laws for its own well-being and government, may establish what form of polity she pleases, provided she does not contradict or infringe upon the laws and commands laid down in Holy Writ. Furthermore, Hooker demonstrates fully and conclusively that no one form of church government is prescribed as an immutable pattern by the divine Word. Whence it follows that 'in these things whereof the Scriptures appointeth no certainty, the use of the people of God or the ordinances of our fathers must serve for a law'.

Hooker did not write merely to confute the puritan platform.

'My meaning', he said, 'is not to provoke any, but to satisfy all tender consciences.' The *Ecclesiastical Polity* was intended to be an *irenicon*, a book of reconciliation, which all would accept as the basis of agreement. As has been finely said, 'he contrived to unite and hold in a real equilibrium a deep sympathy with the three great spiritual currents of his time. He was sincerely and deeply an evangelical, a catholic, and a rationalist.' The peace of the church was the be-all and end-all of his endeavour.

It may well be doubted whether this noble aim could have been achieved in the acidulated atmosphere of the sixteenth century. Bancroft, the practical administrator, who knew human nature and the puritan mind better than the scholarly recluse, avowed himself convinced that the gulf was too wide to be bridged by any philosophic discussion, and that nothing would induce the presbyterians to abandon their belief in the divine sanction behind their Genevan system. Even Hooker himself seems to have realized this before he died. 'All is vain', he wrote, 'which we do to abate the errors of men, except their unruly affections be bridled, self-love, vainglory, impatience, pride, pertinancy, these be the bane of our peace. And these are not to be conquered or cast out but by prayer. . . . There is no way left but this one; pray for the peace of Jerusalem.'

Thus Hooker's treatise passed into the realm of literature, and Bancroft's estimate of the situation, that the puritans were fundamentally irreconcilable and deserved to be driven out of the establishment, became the basis of church policy in the future. For the moment, however, comparative peace descended on the ecclesiastical field. Both parties turned their eyes speculatively towards the king of Scotland, whose sympathies were believed to be with the reformers—the anglicans with some trepidation, the puritans with hope.

XII

THE IRISH PROBLEM

THE pacification of Ireland was the last great achievement of Elizabeth's reign, and the final act in Irish history
during the Tudor period. For the greater part of a century
English statesmen had wrestled with the problem, only to retire
baffled by the difficulties encountered and the inadequacy of
the resources at their disposal. Those who crossed St. George's
channel as viceroys or lord deputies sometimes cried to be
delivered from 'this hell' of a country, that seemed to devour
reputations with the rapacity of one of its own bogs, or came
home broken in health and spirit by the severities of the service;
and few could be prevailed upon to return for a second spell
of office. There was no glamour, no uplift, no satisfaction in
their work to compensate them for the hardships endured; on
the contrary, much of it was sordid and demoralizing. Humane
men like Anthony St. Leger or Sir Henry Sidney, men of iron
like Lord Leonard Grey, had to admit themselves beaten by
this strange island, whose people, while acknowledging England's sovereign rights, refused either to govern themselves or
to be governed. Can it be wondered at that, as Spenser tells
us, many wished 'all that land were a sea-pool'?

Yet the blame for the confusion cannot be ascribed solely to
the Irish. English statesmanship, even when judged by the
standards of that age, was short-sighted, shifty, and incoherent.
Tudor Ireland was like some bankrupt estate, which successive
owners despairingly thrust upon the shoulders of their heirs,
content if only they could, by temporary measures, avert the
final day of reckoning. Efforts to eradicate the evils that corrupted and disordered the country were neither comprehensive
enough nor sufficiently sustained. Plans were made to be abandoned for, or supplemented by, others of a totally different
nature. Policy was woven and unwoven in accordance with
the needs of the hour, the convenience of statesmen, and,
above all, the limited financial resources of the government.
Doubtless there were times when gleams of a higher statesmanship penetrated the gloom, as, for example, when Henry VIII,
in the later years of his reign, fathered a scheme for the conversion of the Irish land system into English tenures, with the

collaboration of the chiefs. But the spirit of co-operation did not survive him. It was followed by the Edwardian reversion to conquest and extermination, which in its turn gave place to a policy of expropriation, under Mary, and the settlement of English military colonies in Leix and Offaly. Meanwhile the introduction of the religious question still further complicated the issue, and led on to that 'Babylonish confusion' which, in the opinion of Fynes Moryson, 'made the Irish like wild colts having unskilful riders to have all their jadish tricks'.

Of Henry VIII's work in Ireland little remained, when Elizabeth took up the sceptre, save the three earldoms he had conferred on MacWilliam Burke (Clanricard), Murrough O'Brien (Thomond), and Con O'Neill (Tyrone), together with the knighthoods distributed among the lesser chiefs. Mary's reign had witnessed the carving out of King's County and Queen's County from the O'Connor and O'More tribal lands in Leix and Offaly. A number of passes had also been cut for military purposes to the west of this area, in the O'Molloy country and near Athlone on the Shannon, giving access to Connaught. But these were relatively small gains, unlikely to breast the vigorous tide of celticism unless they were consolidated by further advances under Elizabeth.

The Ireland she inherited in 1558 was, as it had been since the beginning of the century, a land without political cohesion or racial homogeneity. A great part of it was not yet reduced to shire ground, and two whole provinces, Ulster and Connaught, were more or less formless, their names being simply geographical expressions for the land north of the Pale and west of the Shannon respectively. Twenty counties are mentioned as being in existence, but only half that number were effective units for administrative purposes, viz. the five counties of the Pale (Dublin, Louth, Meath, West Meath, and Kildare); Kilkenny and Carlow in Leinster; Tipperary, Waterford, and Cork in Munster. The Irish revenue, amounting to less than £5,000, was barely sufficient to meet the maintenance costs of the four or five hundred soldiers who formed the garrison at Dublin, let alone the civil administration. Even in time of peace it had to be heavily augmented from England. As for the Irish parliament, it had only begun to take up its permanent abode in Dublin: as late as Mary's reign it had met wherever the lord deputy found it convenient to hold it—at Limerick and

Drogheda as well as Dublin. Nominally it represented all the interests in the island, but actually only the great landed proprietors of the Anglo-Irish race, the dignitaries of the church, and the counties, cities, and boroughs that lay within the 'obedient' districts. Of the native chiefs only those who had accepted earldoms attended it, and their attendance was not regular. Broadly speaking, therefore, the Irish parliament was an organ of English rather than of Irish opinion; its sphere of influence was small, and its freedom of action was curtailed by the operation of Poynings' laws and the dominant position of the lord deputy, who took his instructions from the privy council or the queen.

The racial map of Ireland presented a still more remarkable picture of arrested development. The population consisted of three different types or groups, bracketed together not by any community of interest but by the mere fact that they lived in the same island. First came the English of the Pale and the chartered towns of the southern and western seaboard—Waterford, Youghal, Cork, Limerick, and Galway. This group preserved at least a tincture of English usages in regard to speech, dress, and behaviour; but their position was that of an insufficiently reinforced outpost in a savage land, constantly exposed to the insidious pressure of the surrounding barbarism, which threatened to engulf them. Except in the Pale, where order was maintained by the presence of an English garrison, their loyalty was always precarious and suspect, and in moments of crisis they could not be depended upon. Fynes Moryson noted that the citizens of Cork refused to allow their wives to speak English with the English soldiers in the streets, but insisted on their using Erse as a badge of their independence. The English of the chartered towns, in short, were a source of weakness rather than of strength to the ruling authority in Dublin, and every English visitor to Ireland regarded them with dislike and contempt. The second group, the Anglo-Irish, were a hybrid race, who made no pretence of upholding English traditions. Domiciled in the most fertile lands of Leinster and Munster and southern Connaught, they lived 'Irishly', having become so intermixed with the aboriginal celtic population that they had lost practically every trace of their Norman origin. In the phrase of the time, they were *ipsis hiberniores hibernis*, as 'very patch-cocks as the wild Irish themselves'. To this group belonged the Butlers of Kilkenny, Tipperary, and Wexford, the Northern Geraldines of Kildare, and the southern Geraldines of West Munster, who, in the persons of their respective heads, held the great earldoms of Ormonde, Kildare, and Desmond, and exercised a palatine jurisdiction covering the whole south-west of Ireland from the border of the Pale to the extreme limits of Kerry. Southern Connaught was similarly within the jurisdiction of the earldom of Clanricard, more recently created (by Henry VIII) in favour of the Burkes. Altogether the Anglo-Irish were a powerful group, on whose co-operation the Crown perforce depended for the parts beyond the Pale; but their help was fitful, sometimes disingenuous, and seldom hearty. The third element in the population, the pure Irish, were to be found in all parts of the island where the natural features favoured their existence—in the Wicklow highlands, in the boggy lands of the central districts, in Thomond, in northern Connaught, and in the inaccessible wilds of Ulster. Ulster indeed was the stronghold of the clans, where no fewer than a dozen rival groups headed by the O'Neills of Tyrone, the O'Donnells of Tyrconnel, and the Scoto-Irish septs of Antrim parcelled out the territorial lordships between them. It was a land of untempered anarchy, given over to the pursuit of plunder, war, and barbaric glory, and so little accustomed to civilized ways that its rude inhabitants were as surprised to see an English official in their midst as the shades in Virgil were to see Aeneas alive in Hell. Truly Spanish America with all its savage wonders could not offer a greater contrast to England than the province of Ulster in the sixteenth century. Elizabeth had a Mexico at her very doorstep. It should be noted, however, that two chiefs of the Irish race, Con O'Neill of Tyrone and Murrough O'Brien of Thomond, promoted by Henry VIII to the rank of earl, held jurisdictional rights over their respective territories similar to those exercised by Kildare, Ormonde, Desmond, and Clanricard. This extension of privileges and responsibilities to the heads of native Irish clans was in the nature of an experiment, the value of which had still to be proved.

From this brief account of its condition it will be seen that if the queen had left Ireland to take care of itself, in all probability the celticization of the whole island would have taken place within a relatively short time, with disastrous results to

the maintenance of English sovereignty. Such a contingency
was sufficient to warrant a more than usually drastic intervention in Irish affairs; for public opinion, already irritated at the
loss of Calais, would not have brooked a further withdrawal
from Dublin. But although the desire to save the Pale from
destruction may be said to have forced the Irish question into
prominence, it did not acquire its subsequent importance in
English politics for that reason alone. The moment Elizabeth
reformed the church in England, a new complexion came over
Anglo-Irish relations. The establishment of a protestant England over against the catholic monarchies of the Continent was
in many ways a dangerous and doubtful experiment; but its
continuance side by side with a catholic Ireland was virtually
inconceivable. Indeed, before anything like assurance could be
felt in government circles, the whole of the British isles must
present a solid phalanx of opposition to possible catholic machinations. Fortunately the danger of Scotland's becoming a *point
d'appui* of England's enemies was quickly averted, or rendered
highly improbable, by the success of the reformation in that
country. Ireland, on the other hand, was a positive menace so
long as her people remained spiritually dependent upon Rome
and loosely attached to the English Crown. In fact, the protes-
tantization of Ireland must have seemed a necessary measure for
England's security. It was perhaps more than this. The anglican church inculcated obedience to the powers that be, and re-
probated rebellion; and it was at least arguable that if the wild
tribes of Ireland could be brought under its influence the civil
magistrate would find his arm lengthened and his authority
mightily increased. As one of the lord deputies put it, 'Where
there is no knowledge of God, there can be no obedience to
the magistrate.' Nevertheless the attempt to make Ireland protestant by parliamentary enactment—which was the method
adopted—only serves to show how defective were the ideas
animating Tudor statesmen in regard to religious policy.

It is impossible to describe with any degree of accuracy the
condition of the catholic religion in Ireland at this time, but
there is abundant evidence to show that it had reached a very
low ebb. The demoralization of the clergy, caused by the confused revolutions of the preceding reigns, had resulted in the
cessation of preaching and teaching; and the people, growing
up in ignorance of the sacraments and even of the creed

itself, abandoned themselves to all manner of gross excesses contrary to the spirit as well as the letter of the christian religion. The ecclesiastical artillery which, in the last resort, might have kept the anarchy in check played harmlessly over the tribal chieftains, who plundered church property with as little compunction as they plundered other property in their rivals' domains. Nor did it matter what dogma was held in England —the chaos continued uninterruptedly, the catholic prelates of Mary's reign being as helpless against it as the protestant bishops of Edwardian times. Ireland was the despair of the saints.[1] So bad was the spiritual condition of the country that almost any change would have been a change for the better. Circumstances were soon to show, however, that if Ireland was to be saved it would not be by the imposition of protestantism.

The first Irish parliament of the reign met at Christ Church, Dublin, on 11 January 1560, and was dissolved on 1 February 'by reason', apparently, 'of its aversion to the protestant religion'. There is no record of what took place; but there is also no doubt that the Acts of Supremacy and Uniformity were subsequently placed upon the statute book. It was a poor augury for the success of the measures. The true state of Irish opinion soon revealed itself, when only five of the reigning catholic bishops, whose sees lay within the Pale or in close proximity to it—Dublin, Ferns, Leighlin—and in Waterford and Cork, consented to accept the settlement. The rest of the episcopate, who were beyond the reach of the government and consequently escaped deprivation, contented themselves with taking the oath of fealty and allegiance to the queen, and continued to celebrate mass as before. Thus a kind of ecclesiastical dyarchy came into being—a protestant state church functioning in the obedient areas, and a papal church in all the rest of Ireland; but neither exercised much influence over the people. Nor were the prospects of the former sensibly improved

[1] When the holy woman Brigitta 'inquired of her good angel, of what christian land was most souls damned? the angel showed her a land in the west part of the world where there is most continual war, root of hate and envy, and of vices contrary to charity, and without charity the souls cannot be saved. And the angel did show till her the lapse of the souls of christian folk of that land, how they fell down into hell as thick as any hail showers. . . . For there is no land in the world of so continual war within himself, ne of so great shedding of Christian blood, ne of so great robbing, spoiling, preying, and burning, ne of so great wrongful extortion continually, as Ireland.' (Quoted by J. S. Brewer, *Carew Papers, 1575–88*, Introduction, pp. xv–xvi.)

by the concession that Latin might be used in the liturgy instead of English, or the subsequent publication of a catechism (1571) and certain articles of religion in Erse (1566). If Ireland set little store by the ancient faith, it had less use for protestantism in any shape or form. So slowly did the reformation progress that in 1565 Loftus, dean of St. Patrick's, Dublin, confessed that the gentry of the Pale were going to mass and giving bed and board to 'massing priests' as if they had never heard a word about reform. A year later Sir Henry Sidney, who had just entered upon his first spell of office as lord deputy, stated that only three sees were occupied by zealous protestants, viz. Dublin, Armagh, and Meath; while in the more remote parts 'order cannot yet be taken . . . until the countries be brought into more civil and dutiful obedience'. Ten years later, when Sidney was again in Ireland on his third spell of office, he found to his surprise that the prospect was as gloomy as ever. In his report (1576) he commented that even in Meath, Bishop Brady's diocese and the best ordered in all the country, no fewer than 105 parish churches were without a resident clergyman, and the pluralist curates who served them were in some cases 'Irish rogues having little Latin, less civility[1] and learning'. Wasted patrimonies, ruined churches, and insufficient supply of trained pastors confronted him on all sides. 'Upon the face of the earth', he concluded, 'there is not a church in so miserable a condition.' But if the position in the Pale was bad, it was much worse in the purely Irish districts. 'So profane and heathenish are some parts', remarked the lord deputy, 'that it hath been preached publicly before me that the sacrament of baptism is not used among them.' No doubt the spiritual needs of the native Irish were to some extent provided for by the ministrations of the jesuit David Wolf, who laboured in Ireland as papal commissary from 1561 until his imprisonment in Dublin castle in 1569, and of wandering friars who lurked about the countryside. But the papal bishops who occupied the remoter sees were by all accounts feeble and ineffective in the discharge of their spiritual functions. It was not, in fact, until after 1580, when the seminaries on the Continent began to pour their scholars and missionaries into the island, that there was any serious attempt on the part of the catholic church to meet the needs of the native Irish. By that

---

[1] i.e. culture, civilization, refinement.

time, however, the situation had undergone a great change, and the cause of catholicism had become interlocked with a political revolt against English rule. We shall return to this catholic revival after we have considered other aspects of Elizabeth's Irish policy. For the present we may note simply the failure of protestantism to secure its position even in the parts subject to inspection and control by the Dublin authorities.

Whatever importance the queen may have attached to the introduction of a protestant state church into Ireland, it was not the whole or even the main object of her intervention. The really fundamental question was not how the Irish were to be proselytized, but how they were to be civilized. However much the new bishops might be regarded as the pioneers of political and social progress, they were evidently unsuited for missionary work in a barbarous society that spurned their advances and plundered their sees with impunity. Civilization must precede the church rather than the church civilization. This was the considered opinion of all competent observers who studied the problem at first hand, and it was fully borne out by the disturbed condition of Ulster and Munster.

In 1559, on the death of Con O'Neill, the province of Ulster was plunged into chaos by a contested succession to the earldom of Tyrone. According to the settlement of Henry VIII (1542) the legal heir was Brian, second baron of Dungannon, whose father, Matthew 'Kelly', despite his illegitimacy, had been promoted by the king to the barony and the position of heir designate. But the heir by celtic law was Shane, the dead earl's eldest legitimate son, who was elected head, on his father's death, of the clan O'Neill. A year before, this able and ambitious chieftain had removed Matthew 'Kelly' from his path by murder; but the support given by the English government to Brian still blocked the way to his ultimate objective, the earldom. After several futile efforts had been made by the lord deputy, the earl of Sussex, to procure Shane's overthrow, both by force and by fraud, the queen decided to compromise, and summoned the rebel chief to London under promise of pardon. In 1562 a settlement was reached by which Shane made his submission, took the oath of allegiance to the Crown, and received in compensation the captaincy of Tyrone. This agreement was precipitated by the news from Ulster that Brian had

been murdered during the chief's absence by his tanist, Turlogh
Luineach; and a clause was introduced into the agreement with
Shane expressly reserving the rights of Matthew's only remaining son, Hugh, who was now removed to England for protection
and education. In later years he was to become Elizabeth's
most resolute opponent in Ireland; but for the time being the
substance of power in Tyrone passed into the hands of Shane.

As captain of Tyrone Shane was pledged to keep the peace
within the province, to levy no Irish exactions beyond his own
territory, to persuade the other chiefs to take the oath of allegiance, and to attend all hostings of the lord deputy. A year
later he was permitted to assume the tribal sovereignty as 'the
O'Neill'. But his thirst for power being only partially slaked
he now began a campaign of terrorism and conquest, forcing
all the tribes of Ulster, including the Scots of Antrim, to
acknowledge his supremacy. By 1566 he was paramount from
Sligo to Carrickfergus, and began to boast that his ancestors
were kings and that he himself was superior in blood to any earl.
Pope Pius IV addressed him as 'prince of Ulster', commending
him for his constancy in the faith. In April 1566 he appealed
to Charles IX of France and the cardinal of Lorraine for help
to drive the heretical English into the sea. It was now clear
that Shane was aiming at something more than the mastery
of Ulster; for he carried his depredations into the Pale, raided
Meath, made a forcible entry into Newry and Dundrum, and
burnt the metropolitan church of Armagh. Sidney was so
perturbed at the state of affairs that he advised Elizabeth to
take drastic action if she did not wish to lose Ireland as her
sister had lost Calais; and in September of the same year he
conducted a strong punitive expedition into the province for
the purpose of reducing the tyrant to obedience. In a triumphal
march through Ulster, lasting less than two months, Sidney
succeeded in rallying all the tribes to the English side, burnt
Shane's crannog at Benburb, and left the beaten chieftain to
the tender mercies of his enemies, the O'Donnells, who routed
him at the battle of Lough Swilly. In desperation Shane fled
for assistance to the Scots of Clandeboy, who first welcomed
him and then murdered him in drunken brawl (June 1567).

From start to finish, the troubles in Ulster had cost England
no less than £147,000 and 3,500 lives; but the pacification of
the province was as far off as ever, for the tanist Turlogh

Luineach, who now succeeded to the chieftainship of the O'Neills, soon proved himself to be as hostile as Shane to English rule.

During the same period conditions were no better in the south of Ireland, where the earl of Desmond ruled west Munster as tyrannically as O'Neill ruled Ulster, plundering and burning villages and carrying on a frontier warfare against his rival Ormonde. Like Shane he had undertaken, in 1562, to be answerable to the laws, to preserve order within his jurisdiction, and not to make war on the queen's subjects. But in spite of reprimands and warnings to observe his obligations he continued his depredations, and was known to be hand in glove with O'Neill for the overthrow of English sovereignty in the island. 'A man both void of judgement to govern and will to be ruled' was Sidney's summing up of his character, when he visited Munster on his tour of inspection as lord deputy in the spring of 1567. Desmond, in fact, was constitutionally incapable of any sustained policy, either pro-Irish or pro-English: he was loyal when it suited him, disloyal whenever the occasion offered. The lands he governed in the queen's name were among the worst governed in Ireland. If Sidney is to be believed, the queen's name and 'letters of commandment' were no more reverenced in the Desmond country 'than they would be in the kingdom of France'. All the lords and gentry of Cork, except Macarthy More of Muskerry and O'Sullivan Beare of Kerry, complained bitterly of his oppression. 'I never was in a more pleasant country,' wrote Sidney, 'but never saw I a more waste or desolate land,' full of ruined churches and burnt-out villages, and the 'bones and skulls of dead subjects, who, partly by murder and partly by famine, have died in the fields.' Desmond was not the only offender—the earl of Thomond in Clare and the sons of Clanricard in southern Connaught were also disturbers of the peace—but he was the ringleader, and the lord deputy did not hesitate to imprison him in Dublin castle for his misdeeds (1567). Later he was transferred to London, where he remained a prisoner until 1573.

Sidney's analysis of the situation in Ireland, based upon his experience with O'Neill and Desmond, went straight to the root of the matter. He recommended, among other things, the establishment of English presidencies in Munster and Connaught to supersede the palatine jurisdictions of the Anglo-Irish earldoms; the abolition of coign and livery, on which the chiefs depended for the maintenance of their armed bands; the introduction of land tenure by royal patent, by which the chiefs would hold their lands directly from the Crown; the planting of English colonists at points of vantage in the island; and the control of Ulster by the construction of roads, bridges, and fortified castles. Several of these recommendations were adopted, with or without modification, within the next few years.

As a first step Sir Edward Fitton was appointed president of Connaught in 1569, and Sir John Perrot was similarly established in Munster in 1570. At the same time (1569) the Irish parliament passed an act abolishing coign and livery. This was the prelude to a determined attempt to break the military power of the chiefs. It was supplemented by a renewal of the effort to replace the Celtic land system by English feudal tenures. The method which had already been applied by Henry VIII to certain districts is generally referred to as 'surrender and regrant'. In other words, the chiefs were invited to give up their lands to the Crown and then receive them back and hold them in tail male by royal patent. They would thus be placed in practically the same position as the Anglo-Irish earls of Ormonde, Desmond, Kildare, and Clanricard; and if they committed treason their lands could be legally confiscated. Doubtless this latter consideration weighed heavily with the English government. But there is no reason to suppose that the scheme involved injustice to the Irish, if, as actually happened, a serious effort was made to understand the prevailing land system and deal equitably with the respective rights of all parties, chiefs as well as dependants. The best example of what happened is the so-called 'composition' of Connaught, which Sidney proposed in 1571 and Sir John Perrot carried out in 1585. The object to be attained was set down by Walsingham in the following words: 'To give each chief his due, with a *salvo jure* to all others that have right.' With this end in view, inquisitions were held to determine the area of the lands to be distributed and who were their rightful owners according to Irish law; and indentures were drawn up between the Crown and the owners, giving them a legal title, fixing the tribute to be paid to the queen (10s. per 120 acres) and the amount of compensation to be paid to the chiefs in lieu of 'cuttings', 'spendings,' and other Irish exactions. The same method was applied to county Clare and many other Irish districts.

Nor was this all. On the basis of these changes in the land system was to be built up a thorough-going revolution in the whole internal economy of Ireland, involving the introduction of English dress, the county system, sheriffs, justices of the peace and of assize, juries, and in short everything characteristic of English life in the sixteenth century. It was not a question of whether these innovations were suitable for Ireland in the present stage of her development, nor was any consideration given to the fact that political and social institutions cannot be made to order, but are the result of a long process of historical evolution. Ireland was to be remade in the English mould, either with or against her will, for the matter was urgent and there was no other mould available.

The attitude of the government is seen very clearly in relation to costume. Just as in the eighteenth century stress was laid on the necessity of abolishing the kilt as a barbarizing agency in Scotland, so in Ireland in the sixteenth century the wearing of mantles and long hair (glibbs) was regarded as an obstacle to progress. The mantle, wrote Spenser, 'is a fit house for an outlaw, a meet bed for a rebel, and an apt cloak for a thief'. Much vituperation was also vented on the bard who by his glorification of the past threw up a defence round everything that was barbaric and un-English. Irishmen, we are told, had a greater respect for the opinion of the bard than for the thunders of the civil magistrate. Thus Sir John Perrot issued the following regulations for the province of Munster, over which he presided, in 1571:

'The inhabitants of cities and corporate towns shall wear no mantles, shorts, Irish coats, or great shirts, nor suffer their hair to grow long to glibb, but to wear gowns, jerkins, and some civil garments; and no maid or single woman shall put on any great roll or kirchen of linen cloth upon their heads, neither any great smock with great sleeves, but to put on hats, caps, French hoods, tippets or some other civil attire upon their heads: upon pain of £100 to the head officer of the city or corporate town that shall suffer any such Irish garments, and upon forfeiture of the said Irish garments so worn';

and again:

'All carroughs, bards, rhymers, and common idle men or women within this province making rhymes . . . to be spoiled of all their

goods and chattels and to be put in the next stocks, there to remain
till they shall find sufficient surety to leave that wicked "thrade" of
life and fall to other occupation.'

The all-important matter, however, was the general alteration of the social environment. Dress, after all, is merely the
adjustment of a people to its environment: if the environment
be altered, the dress will alter of its own accord. But how was
the Elizabethan government, with its limited resources, to bring
about so vast a change in the economy of a people? Sir John
Davies, writing in the reign of James I, expressed the view that
it could only be done by first breaking the tribes in war and
then 'planting' the country with English colonists. Spenser,
who also had the advantage of seeing the situation in retrospect,
was of the same opinion. 'How then, do you think, is the
reformation thereof to be begun', asks Eudoxius, 'if not by laws
and ordinances?' 'Even by the sword,' replied Irenius; 'for
all these evils must first be cut away by a strong hand before
any good can be planted.' But it was not the queen's policy to
launch out into uncertain financial commitments, nor, perhaps, was she convinced that Ireland was worth such a sacrifice
of blood and treasure. She was willing, however, to experiment
with plantation provided her subjects shouldered the risk in the
event of failure. In this light must be viewed the various
attempts to establish colonies during the reign: they were not
colonies in the strict sense of the word, but simply expedients
whereby what threatened to become a lost cause might be
retrieved by private enterprise. The age was rich in individual
initiative and company-promotion flourished for all manner
of undertakings. Why not a company for the exploitation of
Ireland? There was a surplus population in England who
might be drawn upon for the pioneering work. In fact the
plan was feasible enough provided suitable lands for settlement
could be obtained in Ireland: otherwise the attempt to 'plant'
the country would mean the forcible dispossession of the Irish
—a contingency that must involve the planters in interminable
disputes with the natives.

The first three efforts, those of Sir Peter Carew and other
west-country adventurers in Leinster and Munster (1568–70),
of Thomas Smith, son of Elizabeth's principal secretary of
state, in the district of Ards (1572–3), and of the earl of Essex

in Clandeboy (1573–5), were little better than armed attempts
to expropriate the native Irish from their ancestral lands, and
they ended in bloodshed and disaster. Every tribal chief from
Kilkenny to Kerry, Anglo-Irish as well as pure Irish, including
the Butlers, the southern Geraldines, and the Burkes, forgetting
their own differences, combined in a great confederacy to
defend their patrimonies against the English interlopers; and
for several years intermittent war raged in Munster and Connaught. The Ulster enterprises of Smith and Essex met with
the same resistance from the owners of the soil. Within a year
Smith's colony came to an inglorious end with the murder of
the promoter; and Essex could do nothing to retrieve the situation, although he added to the tale of English atrocities in
Ireland by a massacre of the Scots of Rathlin and the treacherous murder of Brian McPhelim, the lord of south Antrim.
Evidently there was no hope of successful colonization by private
enterprise until some other method was devised for procuring
land than by forcibly dispossessing the native landowners.
For the next ten years interest in the project lapsed; and then
it was revived under happier auspices by Sir Walter Raleigh,
who at this time was dabbling in his Virginian enterprises.
When Raleigh took up the matter the Desmond estates in
Munster had passed into the hands of the Crown, owing to the
treason and death of Desmond (1583), and it was calculated
that half a million acres of the best land in Ireland would be
available for settlement. Plans were drawn up accordingly for
the distribution of the forfeited estates among 4,200 English
settlers, both gentry and peasants, chosen from Cheshire, Lancashire, Somerset, and Dorset. In order to encourage the immigrants and to tide them over the early years of pioneer work,
they were to hold their lands free for three years, and thereafter pay only half rent till 1593, after which full rent would
be charged. Since the purpose was to establish a solid block
of uncontaminated English as a basis for the reorganization of
Ireland, it was carefully provided that no Irish should be
allowed into the settlement, and that no intermarriage should
take place.

No sooner was the scheme launched, however, than difficulties
were encountered. Complaints by the Irish that far too much
land had been pegged off led to judicial inquiries, and three
commissions sat between 1586 and 1593 to investigate the

extent of territory really forfeited by Desmond. The upshot was that in place of 500,000 acres only 202,000 were declared legitimately confiscable. This shrinkage in the amount of land available caused competition among the planters for the better lands, and many families returned to England. Thereupon the native Irish, despite all regulations to the contrary, swarmed into the settlement, attracted by its peace and order; and in the end half the colony consisted of 'rebellious Irish'. The effect of the infiltration was not immediately felt; but when the period of stress and strain arrived in 1598, and all Ireland broke into rebellion, the colony could not muster more than a quarter of the forces it should have had at its disposal. Unable to breast the tide of fanaticism that swept the country from end to end, the colonists fled for safety to the shelter of the cities, and no attempt was made to reinstate them. The Munster colony had vanished like those of Clandeboy and Antrim.

In this brief analysis of Elizabeth's Irish policy we have endeavoured to lay emphasis on principles rather than events, and, for the sake of clearness, have emptied the narrative of much detail. We have also, it will be noted, considered the problem exclusively from the standpoint of Anglo-Irish relations. It is now necessary, if the dramatic occurrences of the latter part of the reign are to be understood, to turn back, change our angle of vision, and look at events from the point of view of the Irish themselves.

The great weakness of Ireland in its resistance to England was its disintegration by clan feuds and consequent lack of a common purpose. Although the aims of the English government were subversive of the whole fabric of traditional Irish life, no leader had as yet arisen capable of uniting all his fellow countrymen in the defence of their heritage, nor had the secret of such a union been discovered. In a general and vague way Shane O'Neill appreciated the needs of the situation, but he had neither the brains nor the character to command the respect of all Ireland. It was James Fitzmaurice, the Geraldine leader in the Munster Confederacy of 1569–73, who first rose to the dimensions of a really great figure in Irish history, and provided his race with a policy and rallying-point. In 1569 he had sent FitzGibbon, the papal archbishop of Cashel, to plead the Irish cause at the court of Spain, and to bring the matter to the notice of the catholic world generally. In Ireland Fitz-maurice constituted himself the apostle of an all-catholic island under the patronage of the pope and the king of Spain. It was the first step towards the foundation of a new Ireland, in which inter-tribal disputes would melt away in a blaze of religious enthusiasm and hatred for the heretical English. For the next ten years, says a modern catholic writer,[1] 'the fretful form of *Eire* stands on the southern shore peering into the mists and gazing across the waters watching for the foreign galleys to bring "Spanish ale" or "wine from the Royal Pope".'

But the time was not yet ripe for foreign intervention. Neither FitzGibbon nor the renegade Sir Thomas Stukeley, who now offered his sword to Spain and became a diligent frequenter of the Spanish court as the *soi-disant* marquess of Leinster, could prevail upon Philip II to take up the Irish cause seriously. Ireland was merely a pawn in the great game of European diplomacy. The eyes of continental catholics were fixed on the main enterprise of England, which was beginning to mono-polize the attention of Spain, rather than upon the Irish, whom the duke de Feria contemptuously described as 'a sort of beg-garly people, great traitors to one another, and of no force'. At length Fitzmaurice, tired of the endless delays and dis-appointments, left Ireland for France in 1575, and eventually made his way to Rome, where he gained the ear of Pope Gregory XIII; and early in 1577 he was rewarded with a brief appointing him leader in a holy war in Ireland against 'that woman who . . . has been cut off from the church'. Dis-appointments, however, dogged Fitzmaurice's footsteps as soon as he tried to furnish out his intended expedition to Ireland; for, although he was able to charter a ship from Lisbon and collect a filibustering force of eighty Spaniards and Portuguese with the money given to him by the pope, he got no farther than St. Malo. The fault was not his but the captain's with whom he sailed, who broke his contract. There was now nothing for it but that the pope himself should organize a fresh expedition from Italy; and this he did in February 1578, placing the egregious Stukeley in command.

Once more the project collapsed, for Stukeley, as we have seen,[2] abandoned his crazy ship at Lisbon and joined Sebastian

[1] M. V. Ronan, *The Reformation in Ireland under Elizabeth, 1558–80* (1930).
[2] p. 143.

of Portugal in his African adventure, to die with him at Alcazar. But Fitzmaurice was not to be daunted. Again he took the matter in hand, collected a motley band of Italians, Spaniards, Portuguese, Flemings, French, and Irish, and along with Dr. Nicholas Sanders and some papal emissaries set sail from Lisbon, in July 1579, to raise a holy war against Elizabeth in Munster. Viscount Baltinglas of Leinster, recently returned from a visit to Rome, responded to the appeal; but with the exception of him and Desmond the Anglo-Irish nobles held aloof from the crusade. Connaught was kept quiet by Sir Nicholas Maltby, who thus prevented the spread of the rebellion to Ulster. Consequently, although Baltinglas cut to pieces an English force at Glenmalure in the Wicklow mountains, and a fresh contingent of Italians and Spaniards arrived at Smerwick in September 1580, the conflagration was soon localized and mastered. Lord Grey and the earl of Ormonde, supported by the English fleet under Winter, compelled the fort at Smerwick to surrender and put the garrison to the sword as pirates (Nov.). Then followed a systematic devastation of Munster by the English in an effort to extinguish the last embers of the revolt—a devastation so ruthless that more than 30,000 native Irish perished in six months of starvation, pestilence, and sudden death, and 'the wolf and the best rebel lodged in one inn, with one diet and one kind of bedding'. The tragic episode was brought to a close with the capture and decapitation of Desmond in the woods of Glanageenty near Tralee (1583), and the forfeiture of his estates to the Crown. Fitzmaurice and Sanders had already died miserably—the former at the hands of the loyal Burkes (1579) and the latter of hunger and dysentery (1581); while Baltinglas fled abroad, leaving his lands like Desmond's to be forfeited for treason.

Although the first attempt to overthrow English rule in Ireland with foreign help thus ended disastrously for the Irish, the fire of religious fanaticism had been kindled in the island, and the counter-reformation, hitherto unable to establish itself in the British isles, at last found a foothold on the tortured soil of the most vulnerable part of Elizabeth's dominions. Increasing numbers of missionary priests poured into Ireland every year, and Irish youths, flocking to the continental seminaries, returned to their native country imbued with a holy zeal for the expulsion of the heretics. It was only a question of time when this widespread propaganda, working on the prevailing discontent, would succeed in fusing all the Irish together, thereby creating the conditions for a general rising, supported by Spain, of which Fitzmaurice had dreamed. For the moment Philip's attitude was not reassuring. He had shown himself sympathetic to the Irish cause to the extent of welcoming refugees and agitators at his court; but his refusal to give official countenance to the Munster rebellion clearly indicated that in his opinion the time was not yet ripe for Spanish intervention. Nor did his attitude of benevolent neutrality undergo any material change until some little time after the disaster to the armada of 1588. The experiences of the luckless Spaniards who fell into the hands of the native Irish, when many of Medina Sidonia's galleons crashed on the Irish rocks, cannot have inspired much confidence in the Spanish government as to the value of Irish friendship. Eight thousand Spaniards, it is said, lost their lives between the Giant's Causeway and Blasquet Sound; and of those who escaped death by drowning far more were murdered by the Irish than were put to death by the English troops. Nevertheless the possibility of Spanish help for Ireland was not destroyed by this untoward event. Spain needed Ireland as a base of operations against England just as much as Ireland needed Spain as an ally. And there were two outstanding Irishmen of the younger generation, Hugh Roe O'Donnell and Hugh O'Neill, earl of Tyrone, who set themselves with resolute will to bring about the desired alliance.

Although the O'Neills and the O'Donnells had been enemies for generations, the common hatred for England acted as a powerful solvent on their intertribal feud, and laid the foundation of an Ulster confederacy which was soon to shake the whole structure of English rule in the island from base to roof. In 1592, when Pope Clement VIII sent James O'Hely, bishop of Tuam, to Ulster with renewed promises of Spanish assistance, O'Donnell at once took the lead in planning a fresh rising, and secured the co-operation of many other chiefs. Tyrone at first held aloof from the movement, being too prudent to commit himself until he was convinced that the rebellion had a fair chance of success. But O'Donnell was the heart and soul of the conspiracy. In 1593 he authorized the dispatch of a letter by

O'Hely to the king of Spain, pleading the cause of the federated chiefs, and offering 7,000 foot and 600 horse for the common enterprise against England. In the margin of the Spanish original of the document Philip scribbled the words: 'If what they say is true, it would be shameful not to help them.' Immediate assistance was, of course, impossible owing to the war in France; but during the year 1594 a Spanish ship reconnoitred the coast from the Shannon to Wexford. More important still: Tyrone now threw off his mask of reserve and entered into direct relations with Spain. On the death of Turlogh Luineach the following year, he publicly assumed the title of 'the O'Neill', and was hailed with almost kingly honours in Ulster.

The new leader was no charlatan. Ever since his return to Ireland in 1576, as baron of Dungannon, he had watched and waited, always preserving a correct attitude to the English, who rewarded him with the earldom of Tyrone in 1585. His ability as a soldier, based upon the training he had received in the English army in Ireland, under Essex (1572–4) and as a cavalry commander in the campaign against Desmond, was only matched by his undoubted genius for unscrupulous intrigue and the plausibility of his tongue. An Irishman at heart, despite his English education, he was intellectually much superior to his compeer, O'Donnell, whose glowing passion left little room for obliquity. Tyrone, in fact, was an accomplished disciple of Machiavelli—flexible, suave, incalculable. When circumstances made it necessary he could appear so reasonable as to deceive even the most hardened negotiators whom Elizabeth sent to deal with him, while all the time his purpose remained fixed and unalterable. The attitude he assumed was that of the extirpator of heresy in Ireland, because he knew that under the banner of religion alone he could win the support not only of his fellow countrymen but also of all foreign powers interested in the suppression of protestantism. But the real object of his crusade was political. It was a question of the preservation of celtic institutions in the island; and to Tyrone there could be no hope for the Irish cause unless Ulster remained sacrosanct to the old order. It is impossible to doubt, however, that his ultimate ambition was to drive the English bag and baggage out of Ireland, to destroy English sovereignty, and to set up a national kingship. He was easily the most dangerous figure who had so far appeared on the stage of Anglo-Irish politics.

Time was no object to Tyrone. He knew that the struggle would be long and arduous, and he proceeded with the utmost caution, keeping alive as long as he could, by dexterous dissimulation, his openness to conciliation and compromise with England, while at the same time plying Spain with urgent requests for substantial help in men and money. Without Spanish assistance he realized that he was powerless; and without the possibility of a settlement with England hovering in the background he could not exert the necessary pressure on Spain. Thus he manœuvred between pro-Spanish and pro-English expressions of sympathy, coercing the English into concessions by his dealings with Philip II, and goading the Spaniard into active intervention by his professions of weakness.

This policy of delay and prevarication on Tyrone's part was possible only because of the inadequacy of the English forces in Ireland, the reluctance of the queen to increase her expenditure on armaments, and the lack of a methodical plan of conquest. It was the inevitable result of a hand-to-mouth statesmanship. But let us not under-estimate the difficulties of a general subjugation of the Irish. Geographically the country was ill adapted for the invader. There was no natural rampart or mountain barrier behind which the aboriginal inhabitants could be penned up until they either succumbed to military pressure or accepted the terms of the conqueror. The mountains, distributed in some five detached groups round the central plain, presented a problem of strategical control which a military race like the Romans might have solved without difficulty, but to the Elizabethans it must have seemed practically insoluble. Moreover, Ireland was as barbarous physically as it was socially. In most parts it consisted of virgin bush, bog, and lake separating settlement from settlement and town from town, without roads or regular means of communication. It was a country, in fact, whose conquest was bound to exact a heavy toll from civilized troops accustomed to operate within reach of material comforts and properly organized military bases. Consequently the Irish wars were exceedingly unpopular in England. In Lancashire and the other recruiting counties it was commonly said that men would rather go to the gallows than to Ireland. Hard knocks and stale victuals if not actual starvation was all they could look for. No doubt the very severity of the service transformed raw recruits quickly into good soldiers; but it was difficult at first to keep them from deserting and selling their weapons to the native Irish; and many captains preferred to hire mercenary Irish who were less expensive and better adapted by their training to the conditions of savage warfare. Indeed the queen's parsimony compelled them to this course. Yet to attempt a subjugation of the Irish by the Irish—for this is what the system involved—was, to say the least, a hazardous and short-sighted policy; for, at a pinch, the mercenary was unreliable. More than once English generals were reduced to inaction by the fear that their Irish auxiliaries might play them false. To make matters worse, the plan was adopted of making the Irish pay for their own conquest. The wars were run on the Irish revenue supplemented by 'cess', an exaction levied on the people of the Pale, who were obliged to furnish the troops with victuals at the queen's price. In practice the system was iniquitous, for it left the peasantry at the mercy of bullying soldiers, who often robbed their homesteads and indemnified them with worthless notes of hand. In 1586, under the lord lieutenancy of Sir John Perrot, 'cess' was abolished in favour of a 'composition' fee of £2,100. Even so, however, it was not until the contest with Tyrone began that the organization of the army and of the method of financing it was placed upon a sound basis. The employment of Irish mercenaries was curtailed as far as possible, and the English exchequer took over the entire responsibility. The arch-rebel could not have been beaten in any other way.

Meanwhile in the spring of 1596 Philip II had committed himself to the Irish enterprise. In April Alonzo de Cobos was dispatched from Santander with sufficient gold and munitions to keep Tyrone and O'Donnell in the field for eighteen months, and several Spanish officers to help in the training of their forces. Captains Cisneros and Medinilla followed with instructions to reach a complete understanding with the chiefs, and to take soundings of the coast in order to select a suitable landing-place. In September Cobos, who had returned to Spain, was sent back to Ulster to make final arrangements for the impending arrival of the Spaniards, and to hearten the Irish. He landed at Killibegs and met the assembled chiefs at a great conference in Donegal, where it was decided that the expedition should disembark at Carlingford or Galway. While these embassies were coming and going Philip had succeeded, despite the dis-

traction caused by Essex's descent on Cadiz (July), in concentrating a powerful armada at Lisbon, consisting of 98 ships
and 10,800 men. By October his preparations were complete,
and the fleet, accompanied by many Irish refugees, moved off
to its rendezvous at Cape Finisterre, wafted by the strains of
the psalm *contra paganos*.

It was not the first time nor was it the last that those who put
their trust in Spanish enterprises on the sea were to be sadly
disillusioned. The ships had hardly sailed when they were
dispersed and partially wrecked by a storm. A second effort a
year later (Oct. 1597) proved to be equally abortive; and then
Philip II, after concluding the peace of Vervins with Henry IV,
died, and was succeeded on the throne by his son Philip III.
The Hispano-Irish alliance had not yet begun to bear fruit,
albeit Spanish diplomacy stood committed to it.

The effect of these disasters to Spain was seriously to endanger Tyrone's position in Ulster; and in December 1597, he
decided to submit, temporarily at least, to the Lieutenant-
General, the earl of Ormonde. Several conferences were held,
during the winter and spring, for the purpose of exploring the
basis of a settlement; but Ormonde was far from confident in
Tyrone's trustworthiness, and the sole result of the discussions
was to reveal the alarming extent of the earl's traffic with Spain,
and the futility of peace negotiations until his power was effectually broken. So far as the campaign in Ireland had gone,
Tyrone had succeeded in maintaining the integrity of his province. The struggle had been waged along the border for the
possession of key positions like the fort on the Blackwater, which
commanded Ulster's communications with the south. The fort
had changed hands several times, and was now occupied by an
English detachment. But soon after hostilities were renewed
in 1598, Tyrone retook it and cut to pieces an English force
which was marching under the command of Sir Henry Bagenal
to its relief (August). The news of this brilliant exploit created
a sensation in Ireland, and coupled with the defeat of Sir
Conyers Clifford by the O'Rourkes of Connaught was instrumental in spreading the rebellion into the other provinces
which had hitherto remained quiescent. Everywhere the native
Irish rose *en masse*, and terrible atrocities took place. In Munster
a new earl of Desmond—the so-called *sugane* (straw rope)
earl—was proclaimed, the old customs were revived, and the

English planters were driven from their lands. In all parts the
chiefs who had conformed to English rule were superseded by
irreconcilables. It was virtually a national revolt. In proportion as the confidence of the Irish increased by their successes
the English lost heart; and the legend grew that the Irish bore
charmed lives. Now was the time for the Spaniard to strike,
for the supreme crisis in Ireland's history had arrived. Accordingly Philip III, who was well informed of the state of affairs,
sent a third armada to Tyrone's assistance in the summer of
1599. Had this expedition reached its destination, it is hard to
believe that English sovereignty in Ireland would have survived
the onslaught. But once more the Spaniard completely failed.
Bad weather forced the fleet to take shelter in Brittany and
eventually to double back to Spain, and the favourable moment
was irretrievably lost. Meanwhile the English grasp on the
situation was gradually restored, first by the young Essex and
later by Lord Mountjoy, the greatest of Elizabethan soldier
administrators.

Although Essex was probably the most finished exponent of
the art of war in England his assumption of the lord deputyship
was a tragic mistake. He was sadly deficient in the patient
organizing genius required for the subjugation of a rebel
country, where all problems are problems of detail, and prudence more important than heroism. Ireland in fact found out
his weaknesses and gave him no scope for the exercise of his
undoubted talents. He was sent to Dublin in April 1599 with
explicit instructions to proceed against Tyrone, crush the rebellion, and pacify the country; and for this purpose a fine army
of 16,000 foot and 1,300 horse was placed at his disposal. But
instead of concentrating at once on the major task, the reduction
of Ulster, he spent his strength in a preliminary attempt to
subdue Munster: with the result that when at length, in response
to urgent messages from the queen, he turned northward, he
found his forces too weak by reason of desertion and sickness
to risk an encounter with Tyrone. He therefore fell into the
trap of beginning peace parleys with the rebel, and ended by
making a treaty, which profoundly irritated the queen and
roused the liveliest apprehension in England. This led, as we
have seen, to his unauthorized and hurried return to London
to justify his conduct; and he fell a victim to the royal displeasure was dismissed the court, and finally deprived of office.

Yet, all things considered, Essex did not entirely fail. The display of military force at a critical moment in Munster probably steadied that province, and his negotiations with Tyrone, albeit they were immediately disowned by the queen, gained time—an important factor at the moment.

Mountjoy, his successor, was a man of an entirely different stamp. Like Sir John Perrot, who to some extent anticipated his methods, he tackled the problem of Ireland's subjugation in the true spirit of a Roman conqueror. Thorough, scientific, and infinitely patient, he knew from prolonged deliberation what the situation demanded, and moved with precision to his goal. He was also fortunate in having for his subordinates Sir George Carew, president of Munster, and Sir Henry Docwra, whose brilliant plan of establishing an English stronghold on Lough Foyle contributed greatly to the final collapse of Tyrone.

The task confronting the new lord deputy and his lieutenants was such as to excite the most dismal forebodings. No part of the country was sincerely loyal to the Crown, not even the Pale, for the papal emissaries had done their work only too thoroughly. 'It is incredible', wrote Carew to Lord Buckhurst (April 1600), 'to see how our nation and religion are maligned and the awful obedience that all the nation stands in to the romish priests, whose excommunications are of greater terror unto them than any earthly horror whatsoever.' In the cities of Munster culprits could not be punished because the mayors and aldermen were disloyal, and juries made no more scruple to 'pass against' an Englishman and the queen than to 'drink milk unstrained'. The gloom was deepened by the constant fear of another Spanish invasion; for Tyrone was moving heaven and earth to bring the Spaniards over, and Philip III, despite his failing finances, was making frantic efforts to launch a fresh armada. Fortunately the year 1600 passed without mishap, and every month saw the work of pacification advance a step farther. By the spring of 1601 the rebellion had been completely stamped out in Munster; the Pale and the central districts were secure; and Connaught and Clare were quiet under the friendly earls of Clanricard and Thomond. Only the power of Ulster remained unbroken, and its subjugation became the capital problem in Irish politics. For rather more than a year Mountjoy had hammered at its defences, constructing blockhouses and planting garrisons at strategic points, and keeping Tyrone continually on the *qui vive* with extensive raids. But in the summer of 1601 he began preparations for invasion on a grand scale, and the rebel chiefs watched them with growing anxiety. Then suddenly the lord deputy was informed by Carew that the Spaniards were on the high seas (29 August), and he abandoned the Ulster enterprise in order to cope with the new danger.

Three weeks later, while the Irish council of state sat at Kilkenny to concert measures of defence, 3,000 Spanish troops under Juan d'Aguila landed at Kinsale, and proceeded to construct a fortified camp. It was fortunate that the blow fell in Munster rather than in Ulster, for the distance of the centre of trouble from Tyrone's country made communication between the Spaniards and the northern Irish, their only possible allies, exceedingly difficult, and at the same time simplified the problem of defence. Moreover, Kinsale was a tiny town of some 200 houses, with no store of provisions or natural defences except on the side of the sea; and as soon as the English fleet under Sir Richard Leveson appeared in Irish waters it was practically impossible for d'Aguila to receive either supplies or reinforcements from Spain. He could not look for assistance to the Munster Irish, who had been thoroughly cowed by Carew. O'Sullivan Beare of Kerry, it is true, offered him 1,000 men if arms and munitions could be provided; but d'Aguila could not do this, because Pedro de Zubiaur's squadron carrying the extra supply of equipment had been compelled by adverse weather to return to Spain before the main body reached Kinsale. Nor again was it possible for the Spaniards to provision themselves by foraging, for they had no horses and the surrounding country was immediately swept bare by Mountjoy's cavalry. On the other hand, ample reinforcements, guns, and stores of all kinds from England soon began to pour into Waterford in response to the lord deputy's appeal; and by 26 October Mountjoy was in a position to invest Kinsale with an army of between 8,000 and 10,000 men.

For two months the siege was prosecuted with vigour, but the wintry weather and the open entrenchments took a heavy toll of the besiegers. On 1 November the fort of Rincurren commanding the entrance to the harbour was captured, and the Spaniards were completely bottled up, being unable either to retire or advance. On 12 November Leveson began the bombardment of Ny Parke, the only other defensive work protecting the town on the seaward side, and it fell on the 20th. A premature assault delivered on 1 December failed to dislodge the Spaniards, but enabled Mountjoy to draw his lines nearer. On the 6th Leveson destroyed Zubiaur's ships, which had just arrived at Castlehaven. It was now evident that the Spaniards of themselves could do nothing to avert the impending disaster: everything depended on Tyrone and O'Donnell, who were now approaching Mountjoy's lines, after vainly trying to raise the country on their march south from Ulster.

On 21 December the English found themselves enveloped by the northern Irish, who cut off their communications with the interior, and threatened to starve them out just as they had hoped to starve out the Spaniards. But d'Aguila could not wait, and on 23 December a joint attack on the English entrenchments was arranged, when Mountjoy's forces caught between two fires would be compelled to abandon the siege and withdraw. By sheer good luck news of the attack reached Mountjoy beforehand, owing to the indiscretion of one of Tyrone's officers; and when the action developed it was the Irish and Spaniards who were surprised. At the cost of a single casualty Mountjoy routed the enemy, driving the Spaniards back into the town and inflicting a loss of 2,000 on the Irish. This, in fact, terminated the fighting, for O'Donnell shortly afterwards escaped to Spain with Zubiaur and Tyrone dejectedly moved off to the north. D'Aguila saw that the time had come to capitulate, and on 2 January he agreed to evacuate Kinsale and Castlehaven, provided the Spaniards were allowed to retain their arms and the English would bear the cost of their repatriation. Mountjoy found no difficulty in accepting these terms, being only too glad to see the Spaniards out of the country. D'Aguila's verdict on the whole episode is tersely expressed in his own words: 'This land seems destined specially for the princes of Hell.'

After the withdrawal of the Spaniards the pacification of Ireland was rapidly concluded. Carew besieged and captured O'Sullivan Beare's stronghold of Dunboy; Mountjoy and Docwra closed in on Tyrone in Ulster, and harried him into the inaccessible wilds of Glenconkein; while Leveson set sail for the coast of Spain to retaliate on the Spaniard by capturing a carrack in Cezimbra roadstead. At the end of March 1603 Tyrone surrendered unconditionally.

The cost of the conquest in blood and treasure cannot be computed; but it is significant that the last four years and a half of the war drained the English treasury to the extent of one and a quarter millions—an enormous sum in those days. And what were the gains? *Pacata Hibernia!* For the first time in its history Ireland was completely subjugated. The clan system was destroyed, and the power of the celtic chief as a war lord was broken beyond hope of recovery. The fall of Ulster proclaimed the end of medieval Ireland and of the elder barbarism which had so long stood like a wall between the Irish race and the cultural life of Europe. Moreover, the triumph of the English had brought with it the establishment of a new land system based upon the principles of the Connaught 'composition', and the reduction of the greater part of the country to shire ground. In 1571 Longford was carved out of the O'Farrell lands and added to Leinster. Five years later (1576) Connaught was subdivided into Galway, Roscommon, Sligo, Mayo, and Clare, the last-named county being subsequently incorporated in Munster. Cavan and Wicklow made their appearance in 1579. Leitrim was created in 1583, and Coleraine, Donegal, Tyrone, Armagh, Fermanagh, and Monaghan in 1586. In fact, with the exception of some minor adjustments, Ireland had acquired by the end of the reign her modern administrative units.

It is harder to determine the economic consequences of the long-drawn-out struggle, for the available evidence is inconclusive. In some respects it was a period of economic retrogression for Ireland. The 'cessing' and billeting of troops, the requisitioning of food-supplies and horses, the destruction of crops by rebels and royalists alike, and all the evils incidental to a military régime reduced many parts of the country to a state of bare subsistence; while the devastation of the five counties of Munster by Lord Grey spread death and starvation over the richest and most productive area in all Ireland. The seaports of the south-west, whose trade with Spain in fish, hides, and beeves depended upon the prosperity of their hinterlands, never recovered from that visitation. Waterford and Galway languished as well as Cork and Limerick. To the devastation of war must be added the deleterious effects produced by the systematic debasement of the coinage—a practice followed by all the Tudors for their own private profit. On the other hand, it may be noted that the increasing demands of the English market for Irish wool, flax, hides, and timber to some extent balanced the account. The rise in importance of Dundalk, Drogheda, Carrickfergus, and the seaports of the north-east, which throve on the English trade, took place side by side with the decay of the Munster seaports. It is probable also that a considerable amount of the money spent by the Elizabethan government on Ireland, during the later phases of the war, remained there to stimulate trade; and the military cantonments of Mountjoy and Carew supplied the peasantry with markets for their produce.

Nor was this all. The cultural life of Ireland benefited by the foundation of Trinity College, Dublin, in 1590, and by the establishment of a printing-press capable of handling texts in Erse. Educationally a new era in Irish history was at hand. But in the matter of religion the population remained overwhelmingly catholic. Indeed the outstanding result of the Elizabethan régime, in the religious sphere, was to deepen the attachment of the Irish race as well as the Anglo-Irish to the church of Rome.

THE END

THE last years of the reign were years of increasing loneliness for the queen. She had outlived her generation. One by one the great men who had served her had passed away. Nicholas Bacon, Lord Clinton, the earl of Bedford, and Sir Thomas Gresham were now mere memories. Leicester died in the year of the armada, Walsingham and Warwick in 1590, Hatton and Shrewsbury in 1591, Hunsdon in 1596, and Burghley in 1598. The latter's death touched Elizabeth profoundly; but being blessed with remarkably good health, by reason, says Camden, of her abstinence from wine and temperate diet, she showed as yet no signs of decrepitude. In the summer she 'progressed', as had always been her habit, when the business of state allowed it; and the winter found her at Whitehall. She continued to celebrate, each year, the anniversary of her accession (Nov. 17) and Twelfth Night, danced galliards *gaîment et de belle disposition,* and went a-maying. Indeed there seems to have been a revival during these later years of the feverish activity and love of amusement that characterized her youth. But it could not be sustained. Essex's death plunged her into a recurrent melancholy from which she never quite recovered. In June 1602 she told de Beaumont, the French ambassador, that nothing now contented her spirit or gave her any enjoyment. 'Her delight', said an observer, 'is to sit in the dark, and sometimes with shedding tears to bewail Essex.' Meanwhile Cecil was in secret correspondence with James of Scotland; and others, including Raleigh, Cobham, and the earl of Northumberland, were said to be working on behalf of the lady Arabella Stuart. Every one was thinking not of the present but of the future—a fact that must have weighed heavily on Elizabeth's mind. Harington, her godson, who visited the court about Christmas 1602, found her in very low spirits, showing 'signs of human infirmity'; and when he tried to divert her by reading some of his verses, she did not respond, but remarked: 'When thou dost feel creeping age at thy gate, these fooleries will please thee less.' She was now in her seventieth or climacteric year. Early in February 1603, at Richmond, she felt well enough to receive the Venetian ambassador, Scaramelli; but before the month was out the death of her cousin, the countess of Nottingham, whom she

dearly loved, caused her fresh grief. Then, in March, the collapse came suddenly. She was really ill; she became sleepless
and could not eat; and in spite of her physicians' advice would
neither take physic nor go to bed. For a fortnight she lay
overcome by a 'settled and unremovable melancholy', hardly
uttering a word. On the 24th, between two and three in the
morning, she died. A few hours later, in accordance with
preparations made by the council, who were fully prepared for
the event, James was proclaimed king, and bonfires were lit
that night in the streets.

A new reign and a new age had begun!

BIBLIOGRAPHY AND WORKS OF REFERENCE

THE standard bibliography for the history of Elizabeth's reign is Conyers Read's *Bibliography of British History: The Tudor Period, 1485–1603* (1933). Without being exhaustive, it is the only systematic survey of the material in print relative to the period. A. W. Pollard and G. R. Redgrave, *A Short Title Catalogue* (Bibliographical Soc., 1926), gives the most complete list of books printed in England, Scotland, and Ireland, as well as of English books printed abroad, before 1640; it includes valuable information as to the libraries and private collections where copies of the books may be found. Reference may also be made to E. Arber, *Transcripts of the Registers of the Stationers' Company* (5 vols., 1875–94), F. Madan, *Early Oxford Press: a Bibliography of Printing and Publishing at Oxford* (1895), and the *British Museum Catalogue of Printed Books*, with its *Supplement* and *Subject Indexes*, which is the best guide to editions, authorship, &c. (new edition in progress, 1931–). Other useful works of general reference are: R. B. McKerrow, *Dictionary of Printers and Booksellers, 1557–1640* (Bibliog. Soc., 1910), the *Catalogue of English Broadsides, 1505–1897* (Bibliotheca Lindesiana, 1898), T. R. Thomson, *Catalogue of British Family Histories* (1928), H. Hall, *List and Index of the Publications of the Royal Historical Society, 1871-1924, and of the Camden Society, 1840-1897* (1925), and *Poole's Index to Periodical Literature*, supplemented by the *Subject Index to Periodicals* (1915–), which is the most important guide to articles in periodicals dealing with English history.

The chief periodicals, journals, and reviews are: *The English Historical Review*, the premier organ of English historical scholarship, *History* (the quarterly journal of the English Historical Association, which also publishes an annual *Bulletin of Historical Literature*), the *Cambridge Historical Journal*, the *Bulletin* of the Institute of Historical Research (useful for corrections to the *Dictionary of National Biography*, summaries of theses, and information regarding the migration of historical documents), the *American Historical Review*, the *Economic History Review*, and the *Mariner's Mirror* (for articles on naval history).

Rymer's *Foedera* (20 vols., 1704–35) is the standard work of reference for treaties and other diplomatic instruments; the

Dictionary of National Biography (22 vols., 1908–9 ed.) is the most valuable authority for biographical matter, and indeed for all other aspects of English history; and the *Victoria County History* (still in progress) is a mine of information on local history. J. Foster, *Alumni Oxonienses, 1500–1714* (1891–2), and J. A. Venn, *Alumni Cantabrigenses*, part i (1922–7), should be used for the lives of university men.

The most important guide to the various repositories of public documents is H. Hall, *Repertory of British Archives*, part i, *England* (1920). For those contained in the Public Record Office consult M. S. Giuseppi, *Guide to the Manuscripts Preserved in the Public Record Office* (2 vols., 1923–4). Detailed inventories and descriptive catalogues of the records are published in the *Reports* of the deputy keeper of the public records, and in the official *Lists and Indexes* of the P.R.O. The British Museum and the Bodleian Library have printed catalogues of their respective manuscript collections, and there is a subject-index in the manuscript room of the British Museum. Manuscript collections in private hands are listed in the *Historical Manuscripts Commission Reports*. The *18th Report* (1917) gives a list of all collections examined to 1916, with a list of materials for diplomatic history in these and the British Museum manuscripts by Frances G. Davenport, and the *19th Report* (1926) carries the former list forward to 1925. For a general account of the Commission's work, and a description of the principal collections, see R. A. Roberts, *The Reports of the Hist. MSS. Comm.* ('Helps for Students of History', No. 22, 1922). For the Scottish records consult M. A. Livingstone, *Guide to the Public Records of Scotland* (1905). Reference should also be made to C. S. Terry, *Index to the Principal Papers relating to Scotland in the H.M.C. Reports* (1908), and C. Matheson's continuation (1928), under the title *Catalogue of the Publications of Scottish Historical Clubs and Societies and of the Volumes relative to Scotland published by the Stationery Office, &c.* Most of the official Irish records preserved in the Public Record Office, Dublin, were destroyed in 1922, and the principal repositories for material on Irish history now available are Trinity College Library and Marsh's Library, Dublin. Useful guidance to source-material on Irish history may be found in Constantia Maxwell, *Irish History from Contemporary Sources, 1509–1610* (1923), and R. H. Murray, *Ireland, 1494–1603* ('Helps for Students of History', No. 33, 1920).

POLITICAL HISTORY

PRINTED SOURCES. The most important collections for the period are the *Calendars of State Papers* published by the Record Office, and the Historical Manuscripts Commission's publications (separate from the *Reports*). For the former, the current *List of Record Publications* issued by His Majesty's Stationery Office should be consulted; for the latter, the *18th* and *19th Reports*, mentioned above, are the best guide. The most valuable of the Commission's publications for the reign is the calendar of the Hatfield (Cecil) manuscripts, 14 vols. now issued (1895–1923). These official publications, however, must be supplemented by such well-known works as J. Strype, *Annals of the Reformation* (4 vols., 1820–40), E. Lodge, *Illustrations of British History* (3 vols., 1838), the *Hardwicke Papers* (ed. Philip Yorke, 2nd earl of Hardwicke, 2 vols., 1778), the *Egerton Papers* (Camd. Soc., xii, 1840), the *Sidney Papers* (ed. A. Collins, 2 vols., 1746), the *Harleian Miscellany* (ed. T. Park, 10 vols., 1808–13; Index, 1885), the *Somers Tracts* (ed. Sir Walter Scott, 13 vols., 1809–15), Murdin's (1759) and Haynes's (1740) collections of the Burghley state papers, F. Peck, *Desiderata Curiosa* (ed. T. Evans, 2 vols., 1779), Sir H. Ellis, *Original Letters* (11 vols., 1824, 1827, and 1846), P. Forbes, *Full View of the Public Transactions in the Reign of Elizabeth* (2 vols., 1740–1), T. Birch, *Memoirs of the Reign of Queen Elizabeth* (2 vols., 1754), T. Wright, *Queen Elizabeth and her Times* (2 vols., 1838), *Cabala* (1691), Harington's *Nugae Antiquae* (ed. T. Park, 2 vols., 1769), Sir R. Naunton, *Fragmenta Regalia* (ed. E. Arber, 1895), *The Losely MSS.* (ed. A. J. Kempe, 1825), *The Sadler Papers* (ed. A. Clifford, 3 vols., 1809), R. Carey's *Memoirs* (ed. Sir W. Scott, 1808), the *Leicester Correspondence, 1585–6* (Camd. Soc., xxvii, 1844), H. Brugmans, *Correspondentie van Robert Dudley, graaf van Leycester* (3 vols. 1931), John Chamberlain's *Letters* (Camd. Soc., lxxix, 1861), *Walsingham's Journal* (Camd. Soc., Misc. x, 1902), N. H. Nicholas, *Memoirs of the Life and Times of Christopher Hatton* (1847), J. Nichols, *Progresses and Public Processions of Queen Elizabeth* (3 vols., 1823), C. Sharpe, *Memorials of the Rebellion of 1569* (1840); the Manningham and Machyn *Diaries* (Camd. Soc., xcix, 1868, and xlii, 1848), the Gawdy *Letters* (ed. I. H. Jeayes, Roxburghe Club, 1906), the *Letters and Epigrams of Sir John Harington* (N. E. McClure, 1930), A. F. Pollard, *Tudor Tracts*

(1903), Fulke Greville, *Life of Sidney* (ed. N. Smith, 1907), and Bacon, *Collected Works* (ed. J. Spedding and others, 1857–74). For useful notes on many of these works consult Read's *Bibliography*.

Of contemporary histories and chronicles, the following are the most important: W. Camden, *Annales Rerum Anglicarum et Hibernicarum Regnante Elizabetha*, 1615 (trans. T. Hearne, 3 vols., 1717); Raphael Holinshed, *Chronicles*, 1577 (ed. H. Ellis, 6 vols., 1807–8); and J. Stow, *Chronicles* or *Annals*, and *Survey of London*. The 1605 edition of Stow's *Annals* is the best, and C. L. Kingsford's edition of the *Survey* (2 vols., 1908) is the standard one. For the Elizabethan taste for history see L. B. Wright's paper in the *Journal of Modern History*, iii. 175–97 (Chicago, 1931).

Among modern works pride of place must be given to Froude's *History*, the most brilliantly written and the most complete for the period it covers, viz. to 1588; but it requires to be checked both in fact and inference by more recent research. M. Creighton, *Elizabeth* (1899), has a certain independent value, and is much better than E. S. Beesley, *Elizabeth* ('Twelve English Statesmen' series, 1892), which is a mere sketch. The standard account of the reign at present is A. F. Pollard, *Political History of England, 1547–1603.* J. E. Neale, *Queen Elizabeth* (1934), may be specially commended as a brilliantly written, accurate, and full account of certain aspects of the period, albeit it does not pretend to be a definitive history. E. P. Cheyney, *History of England from the Defeat of the Armada to the Death of Elizabeth* (2 vols., 1914, 1926), while it has nothing of the literary distinction of Froude's *History*, which it continues, contains some very important chapters on political institutions, piracy, and social and economic affairs. On the catholic side, Lingard's *History*, though still useful, is antiquated, and has been superseded by A. O. Meyer, *England and the Catholic Church under Queen Elizabeth* (trans. from the German by McKee, 1916), which is based on the Vatican archives and carries the *imprimatur* of the Roman church. It is at once scholarly, dispassionate, and reasonable. There is still no adequate life of Lord Burghley, nor of Leicester; but Conyers Read's *Mr. Secretary Walsingham* (3 vols., 1925), one of the most important contributions to Elizabethan history in recent years, helps to fill the gap. The best account of Essex is W. B. Devereux, *Lives and Letters of the Devereux, Earls of Essex* (2 vols., 1853); to which may be added Lytton Strachey's penetrating and suggestive study, *Elizabeth and Essex* (1928). The standard life of Sidney is H. R. Fox Bourne, *Sir Philip Sidney* (1891). On the controversial topic of Elizabeth's relations with Leicester and the Amy Robsart affair, probably the most significant work is Ernst Bekker, *Elisabeth und Leicester* (Giessener Studien, v, 1890).

THE CHURCH SETTLEMENT. For this aspect of the reign, the most important collection of printed sources is the Parker Society publications (54 vols., 1842–55), which contain the works and correspondence of most of the anglican leaders, e.g. Jewel, Pilkington, Grindal, Sandys, Whitgift, &c., and the invaluable *Zurich Letters*. Jewel's *Apology* is indispensable for an understanding of the principles of the settlement: it should be studied in conjunction with Hooker's *Ecclesiastical Polity* (ed. Church and Paget, 1888). Paget's *Introduction to the Fifth Book of Hooker* (separately published) is the best synoptic account of Hooker's 'platform'. The most useful general history of the church for this period is still W. H. Frere, *The English Church in the Reigns of Elizabeth and James* (1904). Details of the settlement should be studied in H. Gee, *The Elizabethan Clergy and the Settlement of Religion* (1898), H. N. Birt, *The Elizabethan Religious Settlement* (1907)—the Roman Catholic standpoint, F. W. Maitland's chapter in the *Cambridge Modern History*, vol. ii, C. G. Bayne, *Anglo-Roman Relations, 1558–65* (1913), the relevant chapters in Pastor's monumental *History of the Papacy* (Eng. trans. by F. I. Antrobus and R. F. Kerr, 22 vols., 1891–1932), F. W. Maitland, *Elizabethan Gleanings* (1900–3), included in the same author's *Collected Papers* (ed. H. A. L. Fisher, 3 vols., 1911), W. H. Frere and W. P. M. Kennedy, *Visitation Articles and Injunctions* (Alcuin Club, 1910), W. P. M. Kennedy, *Elizabethan Episcopal Administration* (Alcuin Club, 1925)—especially valuable for the later history of the church, F. O. White, *Lives of the Elizabethan Bishops* (1898), and W. P. M. Kennedy, *Life of Parker* (1908).

For the detailed study of catholic history recourse should be had to various volumes of the *Catholic Record Society* publications, William Allen's *Letters and Memorials* (ed. T. F. Knox, 1882), T. F. Knox, *The First and Second Douai Diaries* (1878), the 'Memoirs of Robert Persons', ed. J. H. Pollen (*Catholic Record Society Miscellany*, ii (1905), 12–218, and iv (1907), 1–161), the same author's *Unpublished Documents relating to the English*

Martyrs, 1584–1603 (Cath. Rec. Soc. v. xxi), Nicholas Sanders's 'Report to Cardinal Morone' on the Change of Religion in 1558–9 (ed. Pollen in *Cath. Rec. Soc. Misc.* i. 1–46), T. G. Law, *Jesuits and Seculars* (1889), R. Lechat, *Les réfugiés anglais dans les Pays-Bas espagnols* (1914), R. Simpson, *Edmund Campion* (1896), and J. H. Pollen, *The English Catholics in the Reign of Queen Elizabeth* (1920). On the other side, consult Lord Burghley's *Execution of Justice in England*, and *A Declaration of the Honourable Dealing of Her Majesty's Commissioners, &c.* (*Somers Tracts*, i. 189–208 and 209–12).

For the Puritan movement Daniel Neal, *History of the Puritans* (3 vols., 1793), although out of date, chaotic, and prejudiced, is still useful. The most valuable modern book on presbyterial puritanism is A. F. Scott Pearson, *Thomas Cartwright and Elizabethan Puritanism* (1925). It should be read in conjunction with W. H. Frere and C. E. Douglas, *Puritan Manifestoes* (Church Hist. Soc., lxxii, 1907), R. G. Usher, *The Presbyterian Movement* (Camd. Soc., 3rd ser., viii, 1905), and Whitgift's *Works* (ed. J. Ayre, 3 vols., Parker Soc., 1851–3). For the separatists consult C. Burrage, *Early English Dissenters* (2 vols., 1912), the same author's *True Story of Robert Browne* (1906), F. J. Powicke, *Henry Barrow* (1900), and Albert Peel, *The First Congregational Churches* (1920). The Martin Marprelate controversy may be studied in E. Arber, *Introductory Sketch to the Martin Marprelate Controversy* (1895), W. Pierce, *Historical Introduction to the Marprelate Tracts* (1909), *The Marprelate Tracts* (1909), and *John Penry* (1923). The standard work on the High Commission is R. G. Usher, *The Rise and Fall of the High Commission* (1913).

FOREIGN AFFAIRS. *The Calendar of State Papers, Foreign*, whose various volumes cover the period 1558–88, and are still in progress, is the chief authority; but it does not include Anglo-Spanish relations, which are separately treated in the *Calendar of State Papers, Spanish, Elizabeth* (4 vols., 1558–1603). The latter collection is by no means complete, and must be supplemented by material from the great *Colección de Documentos inéditos para la Historia de España* (112 vols., 1842–95), especially vols. 87, 89, 90, 91, 92. There is a catalogue to the documents by J. Paz (1930–1) and a *Nueva Colección* (6 vols., 1892–6). The principal corpus of material for England's relations with the Netherlands is Kervyn de Lettenhove, *Relations politiques des Pays-Bas et de l'Angleterre, 1555–79* (11 vols., 1882–1900). These documents are printed *in extenso,* and although they overlap in some instances those printed in the *Foreign Calendar,* their value is unique. For a fuller understanding of the issues at stake in the Low Countries, recourse should be had to L. P. Gachard's *Correspondance de Philippe II* (5 vols., 1848–79), *de Marguerite d'Autriche* (3 vols., 1867–81), *du Duc d'Albe* (1850), *de Guillaume le Taciturne* (6 vols., 1847–57), and *d'Alexandre Farnèse,* part i (1853), and Groen van Prinsterer's *Archives de la Maison d'Orange-Nassau* (1st ser., 10 vols., 1835–47, and 2nd ser., 6 vols., 1857–62). The following modern works on the Netherlands are very helpful: R. B. Merriman, *The Rise of the Spanish Empire* (vol. 4, 1932); Henri Pirenne, *Histoire de la Belgique* (4 vols., 1902–11); P. J. Blok, *History of the Netherlands* (8 vols., trans. and abridged by Bierstadt and Putnam, 1898–1912); P. Geyl, *The Revolt of the Netherlands* (1932); and, by far the most literary, J. L. Motley, *The Rise of the Dutch Republic* and *The United Netherlands* (1855 and 1860–7). Motley, like Froude, must be read with care because of his strong bias against Philip II. In this place also might be mentioned P. O. de Törne, *Don Juan d'Autriche et les projets de conquête de l'Angleterre* (1928), and J. Kretzchmar, *Die Invasionsprojekte der katholischen Mächte gegen England zur Zeit Elisabeths* (1892). The latter is a valuable study, based on Vatican archives, of the various schemes for the invasion of England. For Anglo-French relations the most important collections of documents are A. Teulet, *Relations politiques de la France . . . avec l'Écosse* (5 vols., 1862), and *Correspondance diplomatique de B. Salaignac de la Mothe Fénelon* (7 vols., Bannatyne Club, 1840). These should be supplemented by H. Ferrière-Percy, *Lettres de Catherine de Médicis* (10 vols., 1880–1909), Sir Dudley Digges, *The Compleat Ambassador*—Walsingham's missions to France, in 1570–3 and 1581 (1655), Xivrey and Guadet, *Lettres Missives d'Henri IV* (9 vols., 1843–76), Le Laboureur, *Mémoires de Michel de Castelnau, seigneur de Mauvissière* (3 vols., 1731), H. Forneron, *Les Ducs de Guise et leur époque* (2 vols., 1877), and the duc d'Aumâle, *Histoire des Princes de Condé.* For the later history of Anglo-French relations the following are indispensable: Laffleur de Kermaingant, *L'Ambassade de France en Angleterre, sous Henri IV, 1598–1605* (4 vols., 1886–95), Henry Unton's *Correspondence, 1591–2* (ed. J. Stevenson, Roxburghe Club, 1847), Ralph Winwood's *Memorials, 1597–1603* (ed. E. Sawyer, 3 vols., 1725), the *Edmondes Papers, 1592–9* (ed. G. G. Butler, Roxburghe Club, 1913). For French public opinion in relation to events in England during the period of the reign consult G. Ascoli, *La Grande Bretagne devant l'opinion française, &c.* (1928). A useful analysis of the various embassies sent to France from England is printed in the *Bulletin of the Institute of Historical Research*, 1926, 1927, 1928, 1929, by F. J. Weaver. Relations of England and Germany, with special reference to Elizabeth's dealings with John Casimir, may be studied in F. von Bezold, *Briefe des Pfalzgrafen Johann Casimir* (2 vols., 1882–4), and V. von Klarwill, *Queen Elizabeth and some Foreigners* (trans. T. N. Nash, 1928), deals with Elizabeth's marriage negotiations with the Archduke Charles. Although the documents quoted in the latter work are from the Vienna archives, the editing is careless. The only papers of value from Italian archives are the *Calendar of State Papers, Rome*, vol. i (1558–71), vol. ii (1572–8), and the *Calendar of State Papers, Venetian*, 9 vols., covering the years 1202–1603.

SCOTLAND AND MARY STUART. The main source for Anglo-Scottish relations is the *Calendar of State Papers relating to Scotland and Mary Queen of Scots, 1547–1603* (ed. J. Bain and W. K. Boyd, 9 vols.): it is complete to 1588. Subsidiary collections of documents are *The Hamilton Papers* (ed. J. Bain, 1890–2), *The Warrender Papers* (ed. A. I. Cameron, Scottish Hist. Soc., 3rd ser., 1931, 1932), The *Letters of Queen Elizabeth and King James VI* (ed. J. Bruce, Camd. Soc., xlvi, 1849), and *The Correspondence of King James VI with Sir Robert Cecil and others, &c.* (ed. J. Bruce, Camd. Soc., lxxviii, 1861). Sir James Melvil's *Memoirs, 1549–93* (ed. T. Thomson, Bannatyne Club, no. 18) are valuable, but should be read with care. Of contemporary histories, G. Buchanan, *Rerum Scoticarum Historia* (trans. Aidman, 1827), is the best; but J. Knox, *The History of the Reformation in Scotland and Correspondence* (vols. i–ii and iii–vi of his collected *Works*, ed. D. Laing, 1846–54), are indispensable. Among modern works, P. F. Tytler, *History of Scotland* (4 vols., 1873–7), may be commended for its fullness: it is entirely political. But P. Hume Brown's (3 vols., 1908–9) is the most dispassionate survey. Both are to be preferred to A. Lang, *History of Scotland* (4 vols., 1900–7), which, though brilliant and suggestive, is not impartial. W. L. Mathieson, *Politics and Religion, 1550–1695* (2 vols., 1902), is also to be commended. The literature dealing with Mary

Stuart is far too voluminous to be detailed here, and much of it is highly controversial. The standard collection of her correspondence is A. Labanoff, *Lettres, Instructions, et Mémoires de Marie Stuart* (7 vols., 1844), but it is incomplete and should be used in conjunction with A. Teulet, *Lettres de Marie Stuart* (1859), and the *Calendar of State Papers, Scotland* (referred to above). Labanoff excludes the 'casket letters' as forgeries: Teulet includes them. For Mary's relations with Rome see J. H. Pollen, *Papal Negotiations with Mary, Queen of Scots, &c.* (Scot. Hist. Soc., xxxvii, 1901), and for her relations with the Guises the same author's *A Letter from Mary, Queen of Scots to the duke of Guise* (ibid., xliii, 1904). The principal material for Mary's period of imprisonment in England is to be found in the *Bardon Papers* (Camd. Soc., 3rd ser., xvii, 1909), the *Letter Books of Amias Poulet* (J. Morris, 1874), and *Mary Queen of Scots and the Babington Plot* (J. H. Pollen, Scot. Hist. Soc., 3rd ser., 1922). J. E. Neale's article on 'Proceedings in Parliament relative to the Sentence on Mary, Queen of Scots' (*Eng. Hist. Rev.*, xxxv, 1920) should also be consulted. Of critical modern biographies of Mary, D. Hay Fleming's (1898), T. F. Henderson's (1905), and F. A. M. Mignet's (1851) are the most scholarly; but S. Zweig, *The Queen of Scots* (1935), is a valuable interpretative study. Fleming stops at 1567, and has invaluable appendixes: Henderson's and Mignet's volumes are complete lives. J. D. Leader, *Mary, Queen of Scots in Captivity* (1880), should be consulted for a detailed account of the imprisonment. S. Cowan, *The Last Days of Mary Stuart* (1907), is useful for its translation of Burgoyne's Journal—originally printed in M. R. Chantelauze, *Marie Stuart, son procès et son exécution* (1876). Another useful book is A. Chéruel, *Marie Stuart et Catherine de Médicis* (1858). *King James's Secret* (R. S. Rait and A. I. Cameron, 1927) gives an account of the king's attitude to his mother's trial and death, with illustrative matter from the *Warrender Papers*. The celebrated 'casket letter' controversy may be followed in W. Goodall, *Examination, &c.* (2 vols., 1754), J. Hosack, *Mary Queen of Scots and her Accusers* (1870–4), T. F. Henderson, *The Casket Letters* (1890), and A. Lang, *The Mystery of Mary Stuart* (1904); but see also footnote to p. 84. For contemporary controversial literature dealing with Mary, consult J. Scott's *Bibliography of Works relating to Mary, Queen of Scots* (Edin. Bibliog. Soc., no. 2, 1890).

IRELAND. The most important collections of printed documents are: the *Calendar of State Papers, Ireland* (ed. Hamilton and Atkinson, 11 vols.), which are complete for the reign; the *Calendar of Carew MSS.* (ed. Brewer and Bullen, 6 vols.); the *Calendar of Fiants* (published by the deputy keeper of public records in Ireland, 1875–90, 6 vols.); the *Statutes of Ireland* (8 vols., 1765); the *Acts of the Privy Council in Ireland* (J. T. Gilbert in *H.M.C. 15th Report*, 45, and App. iii, 1–296); the *Calendar of Patent and Close Rolls of Chancery in Ireland* (ed. J. Morrin, 3 vols., 1861–3); R. Lascelles, *Liber Munerum Publicorum Hiberniae*—a record of all official appointments (2 vols., 1852) with Index in App. iii to *9th Report* of the deputy keeper of public records, Ireland (1877); J. Lodge, *Desiderata Curiosa Hibernica* (2 vols., 1772); *The Irish Correspondence of James Fitzmaurice of Desmond* (ed. J. O'Donovan, Royal Soc. of Antiquaries of Ireland, 2nd ser., 1859, 354–69); *Letters from Sir Robert Cecil to Sir George Carew* (ed. J. Maclean, Camd. Soc. lxxxviii); *The Social State of the Southern and Eastern Counties of Ireland in the Sixteenth Century* (ed. H. F. Hore and J. Graves, Kilkenny Arch. Soc., 1868–9); T. Stafford, *Pacata Hibernia*—important for Tyrone's rebellion (ed. S. H. O'Grady, 2 vols., 1896); *Ireland under Elizabeth and James I*, described by Spenser, Moryson, and Davies (ed. H. Morley, Carisbrooke Lib., ser. x, 1890); and O'Clery's *Life of Hugh Roe O'Donnell* (ed. D. Murphy, 1893). For the ecclesiastical history of the period recourse should be had to the following: H. Cotton, *Fasti Ecclesiae Hibernicae* (6 vols., 1848–78), which gives brief biographies and details of appointments of prelates of the anglican church; W. M. Brady, *The Episcopal Succession in England, Scotland, and Ireland* (3 vols., 1876–7), and P. F. Moran, *The Episcopal Succession in Ireland during the Reign of Elizabeth* (1866), which give similar information for the Roman Church; E. P. Shirley, *Original Letters and Papers* (1851); *Ibernia Ignatiana* (ed. E. Hogan, 1880), which gives letters and papers from jesuit archives dealing with the jesuit activities in Ireland; the *Spicilegium Ossoriense* (ed. P. F. Moran, 1874–84, 3 vols.), valuable for letters of Irish catholics to Philip II; and W. M. Brady, *State Papers concerning the Irish Church in the time of Elizabeth* (1868).

The standard modern history of Ireland in the Tudor period is R. Bagwell, *Ireland under the Tudors* (3 vols., 1890). M. Bonn, *Die englische Kolonisation in Irland* (2 vols., Stuttgart, 1906), is an analytical work of high quality, on the sociological side, but unfortunately it still lacks a translator. Of the shorter histories, W. O'Connor Morris, *Ireland, 1494–1868* (1909), is the best. The ablest ecclesiastical histories are J. T. Ball, *The Reformed Church of Ireland, 1537–1886* (1890), on the anglican side, and Belles-heim, *Geschichte der katholischen Kirche in Irland* (3 vols., Mainz, 1890–1), for the catholic standpoint. Other works on special subjects are: E. W. L. Hamilton, *Elizabethan Ulster* (1919), H. Wood, 'The Court of Castle Chamber, or Star Chamber of Ireland' (*Proc. Royal Irish Academy*, xxxii, sect. c, no. 90), C. I. Falkiner, 'The Parliament of Ireland under the Tudors' (*Proc. R.I.A.* xxv), A. K. Longfield, *Anglo-Irish Trade in the Sixteenth Century* (1930), M. V. Ronan, *The Reformation in Ireland under Elizabeth* (1930)—informative but not well arranged, J. B. Kelso, *Die Spanier in Irland* (Leipzig Diss., 1902), W. F. Butler, *Gleanings from Irish History* (1925), specially important for the English land policy in Ireland, and the same author's *Confiscation in Irish History* (1917). The 'Journal of the Irish House of Lords in Sir John Perrot's Parliament', ed. F. J. Routledge in *E.H.R.* xxix. 104–17, is important for parliamentary history, and J. H. Pollen, 'The Irish Expedition of 1579' (*The Month*, ci, 1903, 69–85), for Thomas Stukely's enterprise.

The Irish chronicles of greatest value are the *Annals of the Four Masters* (ed. J. O'Donovan, 7 vols., 1856) and the *Annals of Loch Cé* (ed. W. Hennessy, Rolls Series, 2 vols., 1871).

CONSTITUTIONAL

The most important collections of printed documents are the following: *The Statutes of the Realm* (ed. Luders, Tomlins, and Raithby, 11 vols., 1810–28), J. R. Tanner, *Tudor Constitutional Documents* (1922), G. W. Prothero, *Select Statutes and Documents* (1898), and R. Steele, *Tudor and Stuart Proclamations* (2 vols., 1910). Tanner, Prothero, and Steele have valuable introductions; but there is no adequate history of the constitution for the period. H. Hallam, *The Constitutional History of England from the Accession of Henry VII to the Death of George II* (1846) is antiquated, and F. W. Maitland, *The Constitutional History of England* (1908) is too brief; but both are helpful. The best account is to be found in Sir W. S. Holdsworth, *A History of English Law* (vol. iv), 9 vols., 1922–6. Cheyney's *History*, referred to above, ought also to be consulted. The only contemporary survey of the constitution is Thomas Smith, *De Republica Anglorum* (ed. L. Alston, 1906).

For parliament the authorities are Sir Simonds D'Ewes, *Journals of the Parliaments during Elizabeth's Reign* (1693), H. Townshend, *The Last Four Parliaments of Elizabeth* (1680), the *Journals of the House of Lords* (1846), and the *Journals of the House of Commons* (ed. Vardon and May, 1803). In connexion with the last two works, reference should be made to articles by A. F. Pollard on 'The Authenticity of the Lords' Journals in the Sixteenth Century' (*Trans. Royal Hist. Soc.*, 3rd ser., viii, 1914, 17–40) and by J. E. Neale on 'The Commons' Journal of the Tudor Period' (*Trans. R.H.S.*, 4th ser., iii, 1920, 136–70). Of modern works on parliament, A. F. Pollard, *The Evolution of Parliament* (1926), and C. H. McIlwain, *The High Court of Parliament* (1910), may be singled out as specially worthy of attention; and Neale's articles in the *E.H.R.*, vols. xxxiv (1919) and xxxvi (1921), are valuable elucidations of particular points connected with Elizabeth's parliamentary tactics. The same author's papers on 'The Commons' Privilege of Free Speech' (*Tudor Studies*, ed. R. W. Seton-Watson, 1924), on 'Peter Wentworth' (*E.H.R.* xxxix, 1924), and on 'Three Elizabethan Elections' (*E.H.R.* xlvi, 1931) are important contributions to parliamentary history. C. G. Bayne's article on 'The First House of Commons of Queen Elizabeth' (*E.H.R.* xxiii, 1908) is also valuable.

The best guide to the printed and unprinted sources for the history of the council is E. R. Adair, *Sources for the History of the Council in the Sixteenth and Seventeenth Centuries* ('Helps for Students of History', no. 51, 1924). The standard work of reference for the council is, of course, J. R. Dasent, *Acts of the Privy Council of England* (32 vols., 1890–1907), but it has many gaps. On this subject consult Adair's articles in the *E.H.R.*, vols. xxx and xxxviii. The following modern works will be found helpful: A. V. Dicey, *The Privy Council* (1887), Lord Eustace Percy, *The Privy Council under the Tudors* (1907), Dorothy Gladish, *The Tudor Privy Council* (1915), Rachel R. Reid, *The King's Council in the North* (1921)—an exhaustive study, and Caroline A. J. Skeel, *The Council in the Marches of Wales* (1904). Other useful works on the administrative aspect of government are Florence M. G. Evans, *The Principal Secretary of State, &c.* (1923), Gladys Scott Thomson, *Lords Lieutenant in the Sixteenth Century* (1923), and C. H. Beard, *The Office of Justice of the Peace* (1904). But reference should also be made, as in the case of constitutional history generally, to Holdsworth and Cheyney.

For the public finance of the period consult S. Dowell, *A History of Taxation and Taxes in England* (4 vols., 1888)—the only, though far from adequate, general account of the system of taxation, H. Hall, *A History of the Customs revenue in England* (2 vols., 1885), F. C. Dietz, *The Exchequer in Elizabeth's Reign* (Smith Coll. Studies in History, iii, 1918, no. 2), and an article by A. P. Newton on 'The Establishment of the Great Farm of the Customs' (*Trans. R.H.S.*, 4th ser., i, 1918).

MILITARY

M. J. W. Cockle, *A Bibliography of English Military Books to 1642* (1900), is the best guide to literature on the art of war; but there is no comprehensive bibliography dealing with campaigns. Information on this subject must be sought for in Froude, Cheyney, Motley, Bagwell, and other historians of the period. E. van Meteren, *A True Discourse Historical* (trans. T. Churchyard and R. Robinson, 1602), gives an account of Norreys's campaigns, 1577–89, and afterwards in Portugal and elsewhere; and C. R. Markham, *The Fighting Veres* (1888), is important for the military history of the last ten years of the reign. For the armour of the period consult C. J. Ffoulkes, *Armour and Weapons* (1909) and *The Armourer and his Craft* (1912). Contemporary books on war that may be consulted by the specialist are: Whithorne's translation of Machiavelli's *Arte della Guerra* (1588), Barnaby Rich's *Pathway to Military Practice* (1587), Sir John Smith's *Discourses* (1590), Roger Williams, *A Brief Discourse of War* (1590), and Sir H. Knyvett's *Defence of the Realm* (ed. 1906). The standard account of the army is, of course, J. H. Fortescue, *History of the British Army* (13 vols., 1899–1930): vol. i, bk. ii, chaps. 2–5 deal with our period.

NAVAL

Naval bibliographies are best represented by W. G. Perrin, *Admiralty Library*, a subject-catalogue of printed books in the admiralty library (1912), part i, G. E. Mainwaring, *A Bibliography of British Naval History* (1930), and the bibliographies appended to J. S. Corbett's standard volumes on *Drake and the Tudor Navy* (2 vols., 1898–9) and *The Successors of Drake* (1900), which are both indispensable. For the administration of the admiralty consult M. Oppenheim, *History of the Administration* of the Royal Navy (1896), and R. G. Marsden, *Select Cases from the Admiralty Courts* (Selden Soc., vol. ii, 1897). The Navy Record Society publications contain much material of first-rate importance, notably vols. i and ii (*State Papers relating to the Defeat of the Armada*, by J. A. K. Laughton), vol. xi (*Papers relating to the Spanish War, 1585–7*, by J. S. Corbett), vols. xxii, xxiii, xlvi, and xlvii (W. Monson's *Naval Tracts*, ed. M. Oppenheim, with a valuable introduction), and xi, app. A ('Guns and Gunnery in the Tudor Navy', by J. S. Corbett). The story of the armada from the Spanish side is given in Fernandez Duro, *La Armada Invincible* (2 vols., 1884–5), and J. A. Froude, *The Spanish Story of the Armada* (1882), which is also based on the Spanish authorities.

ECONOMIC

There is no general or complete bibliography at present available, but many of the works referred to below contain useful bibliographies on aspects of the subject. R. H. Tawney and Eileen Power, *Tudor Economic Documents* (3 vols., 1924), is the best selection of material. E. Lipson, *The Economic History of England* (3 vols., 1915–31), is the most systematic account of the whole subject; but W. Cunningham, *Growth of English Industry and Commerce* (2 vols., 1903–5), which is arranged on an entirely different plan, is also to be commended. For the agrarian revolution the best book is R. H. Tawney, *Agrarian Problem in the Sixteenth Century* (1912), but the following should also be consulted: R. W. Prothero, *English Farming Past and Present* (1919), A. H. Johnson, *The Disappearance of the Small Landowner* (Ford Lectures, Oxford, 1909), E. F. Gay, 'Inclosures in England in the Sixteenth Century' (*Quar. Jour. Econ.* xvii. 576–97), A. Savine, 'Bondmen under the Tudors' (*Trans. R.H.S.*, 2nd ser., xvii. 235–89) and 'English Customary Tenure in the Tudor Period' (*Quar. Jour. Econ.* xix. 33–80). Lipson uses most of the conclusions arrived at by Gay, Savine, and Johnson in his *History*. Relevant contemporary material will be found in *A Discourse of the Common Weal of this Realm of England* (ed. Elizabeth Lamond, 1929), P. Stubbes's *Anatomy of Abuses* (ed. Furnivall, New Shakspere Society, 1877–9), and W. Harrison's *Description of England* (ed. Furnivall, ibid., 1908).

For industry and commerce the most important works are: G. Unwin, *Industrial Organization in the Sixteenth and Seventeenth*

Centuries (1904), and his collected papers (*Studies in Economic History*, ed. R. H. Tawney, 1927). The history of the joint stock companies should be studied in W. R. Scott, *English, Scottish, and Irish Joint Stock Companies* (3 vols., 1911–12). R. H. Tawney, *Religion and the Rise of Capitalism* (1926), is as important as his *Agrarian Problem* for a proper understanding of the economic tendencies of the age. G. Unwin, *The Guilds and Companies of London* (1908), should also be consulted. For monetary theory and practice consult Thomas Wilson, *A Discourse upon Usury* (ed. R. H. Tawney, 1925): it contains a valuable introduction.

It is impossible to give a list of the books dealing with the different industries, but the following represent a choice of the most significant: J. U. Nef, *The Rise of the British Coal Industry* (2 vols., 1932), H. Heaton, *The Yorkshire Woollen and Worsted Industries* (1920), E. Lipson, *British Woollen and Worsted Industries* (1921), G. R. Lewis, *The Stannaries* (Harv. Econ. Ser., iii, 1908), H. Hamilton, *The English Brass and Copper Industries to 1800* (1926), J. T. Jenkins, *The Herring and Herring Fisheries* (1927).

For prices consult J. E. T. Rogers, *A History of Agriculture and Prices in England* (7 vols., 1866–1900), supplemented by N. S. B. Gras, *The Evolution of the English Corn-market from the Twelfth to the Eighteenth Century* (Harv. Econ. Stud. xiii), and A. P. Usher, 'Prices of Wheat and Commodity Price Indexes for England, 1259–1930' (*Harv. Rev. of Econ. Statistics*, xiii. 103–13). The history of poor relief is dealt with by E. M. Leonard, *The Early History of English Poor Relief* (1900), and S. and B. Webb, *English Local Government: English Poor Law History*, part i, *The Old Poor Law* (1927); and there is a good account of monopolies in W. H. Price, *English Patents of Monopoly* (Harv. Econ. Stud. i, 1906).

For England's foreign trade, reference should be made to C. Read's article in *E.H.R.* xxix. 515–24, R. Ehrenberg, *Hamburg und England im Zeitalter der Königin Elisabeth* (1896)—the main authority for the struggle between the Merchant Adventurers and the Hanse, N. R. Deardorff, *English Trade in the Baltic during the Reign of Elizabeth* (1912), M. Wretts-Smith, 'The English in Russia during the Second Half of the Sixteenth Century' (*Trans. R.H.S.*, 4th ser., iii. 72–102), H. Stevens, *The Dawn of British Trade to the East Indies, 1599–1603* (1886), A. C. Wood, *The History of the Levant Company* (1935), and H. G. Rawlinson, 'Embassy of William Harborne to Constantinople,

1583–8' (*Trans. R.H.S.*, 4th ser., v, 1922, 1–27). J. W. Burgon, *The Life and Times of Sir Thomas Gresham* (2 vols., 1839), is valuable for England's borrowings abroad; and R. Ehrenberg, *Das Zeitalter der Fugger* (2 vols., 1896), is the last general account of the sixteenth-century money market. The latter work has been translated and abridged by H. M. Lucas under the title *Capital and Finance in the Age of the Renaissance* (1928).

For voyages, discoveries, and navigation the important repository of material is Richard Hakluyt's *Principal Navigations* &c. (ed. W. Raleigh, 12 vols., 1903–5), and *Hakluytus Posthumus or Purchas his Pilgrims* (ed. W. Raleigh, 20 vols., 1905–7), together with the various publications of the Hakluyt Society. The following books will also be found valuable: Elizabeth G. R. Taylor, *Tudor Geography* (1930) and *Late Tudor and Early Stuart Geography* (1934), E. Arber, *The First Three English Books on America* (1895), G. B. Parks, *Richard Hakluyt and the English Voyagers* (Amer. Geog. Soc. Pub., 1928), J. A. Williamson, *Sir John Hawkins* (1927), J. S. Corbett, *Drake and the Tudor Navy* (2 vols., 1899)—the standard life, H. R. Wagner, *Sir Francis Drake's Voyage Around the World* (1925), J. A. Williamson's article on 'Books on Drake' (in *History*, xii. 310–21), W. Stebbing, *Sir Walter Raleigh* (1899), or E. Thompson, *Sir Walter Raleigh* (1935), C. R. Markham, *A Life of Sir John Davis*, F. Jones, *The Life of Sir Martin Frobisher* (1878), W. G. Gosling, *The Life of Sir Humphrey Gilbert* (1911), and F. E. Dyer, 'The Elizabethan Sailorman' (*Mariner's Mirror*, x. 133–46).

LITERATURE, THOUGHT, AND SCIENCE

The most valuable guide to the literature of the period is the *Cambridge History of English Literature* (A. W. Ward and A. R. Waller, 14 vols., 1907–16): the chapters are by specialists, and there are full bibliographies. For books published since 1916 consult the bibliographies contained in the March issue of the publications of the Modern Language Association each year, and *The Year's Work in English Studies* (published by the Clarendon Press annually for the English Association). Valuable material will be found in the publications of the Shakespeare Society (1841–53) and the New Shakspere Society (1874–86), the Spenser Society (1867–94), E. Arber's English Reprints (7 vols., 1868–70), the Ballad Society (1868–73), and the Malone Society reprints of Elizabethan plays. E. K. Chambers,

The Elizabethan Stage (4 vols., 1923), is an extensive treatment of this subject, and contains a copious bibliography. F. E. Schelling, *The Elizabethan Drama* (2 vols., 1908), and W. M. A. Creizenach, *The English Drama in the Age of Shakespeare* (1916), should also be consulted: together with *Henslowe's Diary* (ed. W. W. Greg, 2 vols., 1904–8).

The best book on Shakespeare is E. K. Chambers, *William Shakespeare* (1930); but the Shakespeare literature is voluminous, and should be approached under the guidance of Chambers's bibliography, attached to the aforesaid volume. The literature on Spenser is also considerable, and guidance should be sought for in F. I. Carpenter, *A Reference Guide to Edmund Spenser* (1923). For special studies of aspects of Elizabethan literature the following may be singled out: J. Erskine, *The Elizabethan Lyric* (1905), J. J. Jusserand, *The English Novel in the Time of Shakespeare* (trans. E. Lee, 1890), P. Sheavyn, *The Literary Profession in the Elizabethan Age* (1909), G. Gregory Smith, *Elizabethan Critical Essays* (2 vols., 1904), J. T. Murray, *English Dramatic Companies* (2 vols., 1910), and R. B. McKerrow's edition of the *Works of Thomas Nashe* (5 vols., 1904–10) with its valuable introduction on pamphleteering.

The political thought of the period may be studied in Lord Acton's *History of Freedom and other Essays* (ed. J. N. Figgis, 1909), J. N. Figgis, *The Theory of the Divine Right of Kings* (1896) and *From Gerson to Grotius* (1907), H. O. Taylor, *Thought and Expression in the Sixteenth Century* (2 vols., 1920), L. Einstein, *The Italian Renaissance in England* (1902) and *Tudor Ideals* (1921), and, above all, J. W. Allen, *History of Political Thought in the Sixteenth Century*. For Machiavelli's influence (or lack of influence) on England consult M. Praz, 'Machiavelli and the Elizabethans' (*Proc. Brit. Acad.* xiii, 1928), and L. A. Weissberger, 'Machiavelli and Tudor England' (*Pol. Sci. Quar.* xlii, 1927).

For Elizabethan education the best guides are: J. B. Mullinger, *The University of Cambridge* (2 vols., 1873–84), and C. E. Mallet, *A History of the University of Oxford* (3 vols., 1924–7), A. R. M. Stowe, *English Grammar Schools in the reign of Queen Elizabeth* (1908), Foster Watson, *The English Grammar Schools to 1660, their Curriculum and Practice* (1908), Roger Ascham, *English Works* (ed. W. A. Wright, 1904), Richard Mulcaster, *Positions* (abridged ed. by J. Oliphant, 1903), and Foster Watson's translation of Vives' *Linguae Latinae Exercitatio* (in *Tudor School*

Boy Life, 1908). For a general discussion of Renaissance education consult W. H. Woodward, *Studies in Education during the Age of the Renaissance* (1906).

MUSIC, THE ARTS, AND SOCIAL LIFE

The most satisfactory history of Elizabethan music is E. Walker, *A History of Music in England,* chs. iii and iv (1907). Ecclesiastical music is very fully dealt with in P. C. Buck, *Tudor Church Music* (10 vols., 1923-30). E. H. Fellowes, *The English Madrigal Composers* (1921), is valuable for this aspect of the subject, and F. W. Galpin, *Old English Instruments of Music,* &c. (1911), is necessary for an understanding of the various instruments and their uses. G. Grove's *Dictionary of Music and Musicians* (ed. H. Colles, 1927-8) is the best authority for biographical detail.

The authorities on architecture for the period are R. Blomfield, *History of Renaissance Architecture in England* (2 vols., 1897), and T. Garner and A. Stratton, *The Domestic Architecture of England during the Tudor Period* (2 vols., 1911, 1929). For painting and engraving consult C. H. C. Baker and W. G. Constable, *English Painting of the Sixteenth and Seventeenth Centuries* (1930), and S. Colvin, *Early Engraving and Engravers in England* (1905).

The best historical account of gardening is R. Blomfield, *The Formal Garden in England* (8 vols., 1901). F. W. Fairholt, *Costume in England* (4th ed. by H. A. Dillon, 1896), is the standard work of reference for the dress of the period; and manners and customs may be studied in G. Harris, 'Domestic Everyday Life', &c. (*Trans. R.H.S.* ix, 1881), and F. J. Furnivall, *Manners and Meals in Olden Time* (Early Eng. Text Soc. xxxii, 1868). For sports and pastimes consult Mr. Justice Madden, *The Diary of Master William Silence* (1907), Ascham, *Toxophilus,* Egerton Castle, *Schools and Masters of Fence* (1893), J. Strutt, *Sports and Pastimes of the People of England* (1801)—for bull- and bear-baiting, and R. Laneham, *Description of the Entertainment at Kenilworth* (ed. Furnivall, New Shakspere Soc., 1890).

P. Smith, *History of Modern Culture,* vol. i (1930), is the best general guide to the intellectual progress of the age, and the following books should be used for special subjects: W. Notestein, *A History of Witchcraft in England* (1911), W. W. Bryant, *A History of Astronomy* (1907), C. F. Smith, *John Dee* (1909), J. N. Stillman, *The Story of Early Chemistry* (1924), W. W. R.

Ball, *A Short Account of the History of Mathematics* (1915), and, above all, William Gilbert, *De Magnete, &c.* (trans. Gilbert Club, 1900), the most striking contribution of the age to natural science. For medicine and surgery consult W. Osler, *The Evolution of Modern Medicine* (1921), A. H. Buck, *The Growth of Medicine* (1917), and J. F. South, *Memorials of the Craft of Surgery in England* (1886).

Information about social health and disease may be found in C. Creighton, *A History of Epidemics in Britain* (1891), and F. P. Wilson, *The Plague in Shakespeare's London* (1927).

LISTS OF THE HOLDERS OF CERTAIN OFFICES

Archbishops of Canterbury

1558   Matthew Parker.
1575   Edmund Grindal (suspended in 1577; performed only spiritual
        functions till his death in 1583).
1583   John Whitgift.

Lord Chancellors and Keepers of the Great Seal

1558   Sir Nicholas Bacon, Lord Keeper.
1579   Sir Thomas Bromley, Lord Chancellor.
1587   Sir Christopher Hatton, Lord Chancellor.
1591   Sir John Puckering, Lord Keeper.
1596   Sir Thomas Egerton, Lord Keeper.

Lord Treasurers

1558   William Paulet, Marquess of Winchester.
1572   William Cecil, Lord Burghley.
1599   Thomas Sackville, Lord Buckhurst.

Secretaries of State

1558   Sir William Cecil.
1572   Sir Thomas Smith (d. 1576).
1573   Sir Francis Walsingham (d. 1590).
1577   Thomas Wilson (till 1581).
1586   William Davison (till 1587).
1590-6 No principal secretary.
1596   Robert Cecil.
1600   John Herbert.

Lord Chief Justices of the King's Bench

1558   Sir Robert Catlin.
1574   Sir Christopher Wray.

Lord Chief Justices of the Common Pleas

1558   Anthony Brown.
1559   Sir James Dyer.
1582   Edmund Anderson.
1592   Sir John Popham.

Lord High Admirals

1558   Edward, Lord Clinton, afterwards earl of Lincoln.
1585   Charles, Lord Howard of Effingham, afterwards earl of Nottingham.

Chief Governors of Ireland

1560   Thomas Radcliffe, earl of Sussex (Lord Lieutenant).
1565–7
1568–71 } Sir Henry Sidney (Lord Deputy).
1575–8
1580   Arthur, Lord Grey de Wilton (Lord Deputy).
1584–8   Sir John Perrot (Lord Deputy).
1588–94 Sir William Fitzwilliam (Lord Deputy).
1594–7   Sir William Russell (Lord Deputy).
1597   Thomas, Lord Burgh (Lord Deputy).
1599   Robert Devereux, earl of Essex (Lord Lieutenant).
1600   Charles Blount, Lord Mountjoy (Lord Deputy).

N.B. Lord Lieutenant was a title of higher distinction than Lord Deputy, but both offices carried the same, or much the same, powers.

Speakers of the House of Commons

1558   Sir Thomas Gargrave.
1562–3   Thomas Williams.
1566   Richard Onslow.
1571   Christopher Wray.
1572   Robert Bell.
1575   Robert Bell.
1580   John Popham.
1584–5   John Puckering.
1586   John Puckering.
1588–9   George Snagg.
1592–3   Edward Coke.
1597   Christopher Yelverton.
1601   John Crooke.

INDEX

F f

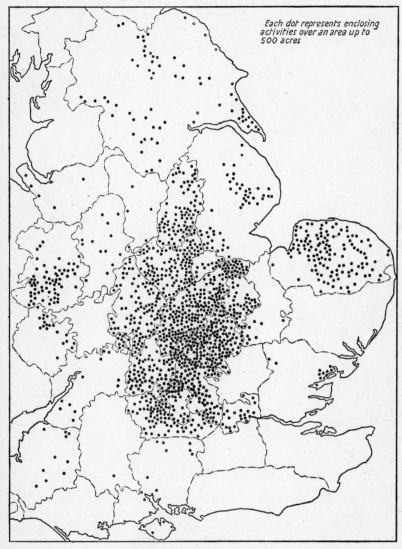

THE 'INCLOSURES', 1485–1607
Adapted from E. F. Gay's map in *Quarterly Journal of Economics*, vol. xvii (1903), p. 576

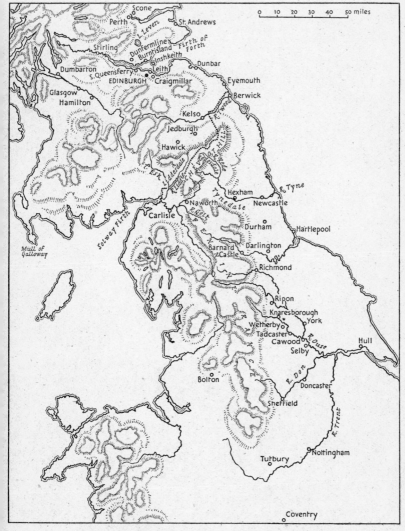

ENGLAND AND SCOTLAND
To illustrate Anglo-Scottish relations and the rebellion of the Earls

G g 2

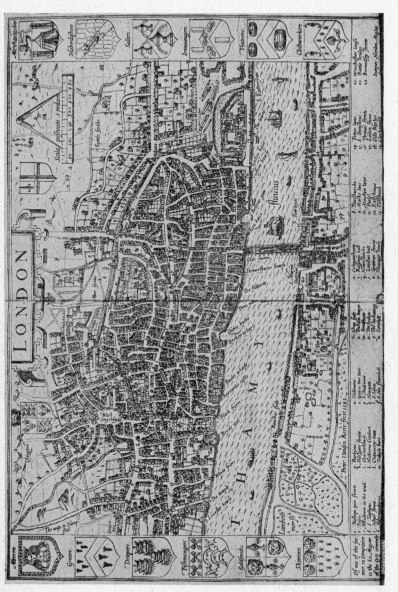

NORDEN'S MAP OF LONDON, 1593

The United Provinces

FRIESLAND

GRONINGEN

Enckhuizen

OVERYSSEL

Amsterdam

UTRECHT

GELDER

Zutphen
ZUTPHEN
Doesburg

Gouda
Rotterdam
Schiedam
Brill

Schouwen
Duiveland
ZEELAND
Walcheren
Flushing
Middleburg
Axel
Sluys

B R A B A N T

R. Maas

R. Rhine

Ostende
Nieuport
Bruges
Ghent
ANTWERP
Dunkirk
Dixmude
R. Scheldt
MALINES
Aerschot

LIMBURG

FLANDERS
Ypres
Oudenarde
Alost
Louvain
St. Omer
Courtrai
Brussels
Tirlemont
A R T O I S
R. Lys
Lille
Tournai
Gembloux
Namur

Douai
Mons
HAINAULT
Valenciennes
NAMUR
Marche
Arras
Cateau-Cambrésis

Le Catelet

LUXEMBURG

0   20   40   60 miles

THE NETHERLANDS

FLORES

TERCEIRA
Angra
S. MICHAEL

Miles (approx)
0   50   100

VillaFranca

K. OF
SCOTLAND

Firth of Forth

IRELAND

KINGDOM

OF

ENGLAND

R. Thames
LONDON   Queenborough
Plymouth   Portsmouth   Margate
Falmouth   Chatham   Blankenberg
Scilly   Dover   Nieuport   ANTWERP
Is.   Isle of Wight   Strait of Dover   Dunkirk
The Lizard   Gravelines
ENGLISH   CHANNEL   Doullens   Cambrai   NETHERLANDS
Ushant   Brest   Le Havre   Dieppe   Amiens
Paimpol   Rouen   Noyon   Vervins
Lower   Dreux   Seine   Château-
BRITTANY   Crozon   St. Malo   Ivry   PARIS   Thierry
Hennebon   Upper   Craon   Vassy
Blavet   NORMANDY   Orleans   Troyes

R. Loire   Amboise   Sancerre

KINGDOM

OF

FRANCE

La Rochelle

C. Finisterre   Montauban
Corunna   Bayonne   Nimes
Ferrol
Vigo   Bilbao
PYRENEES
KINGDOM

K. OF PORTUGAL TO SPAIN 1580-1640

Peniche   OF
LISBON   R. Tagus   MADRID
Cezimbra
SPAIN
Loule
C. St. Vincent   Faro   Seville

Cadiz

→ Course of the Spanish Armada
0   50   100   150   200   250 miles

Events in France and the War with Spain

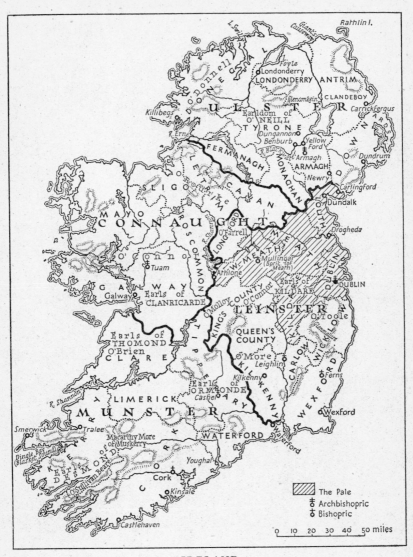

IRELAND

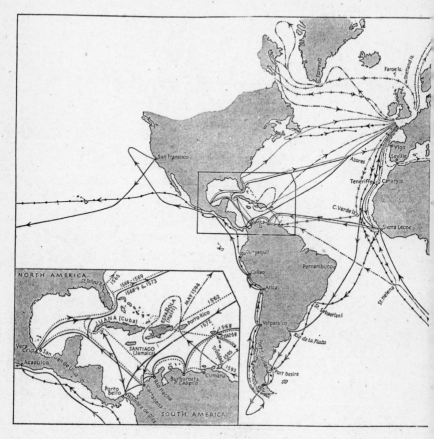

NORTH AMERICA

St. John's   1565   1568-1569
              1568-9 & 1573

JUANA (Cuba)   ESPAÑOLA (Haiti)   MAY 1594
                          Porto Rico   1562
Vera
Cruz   San Juan de Ulúa   SANTIAGO   1572
Acapulco   (Jamaica)   1568
                          1562-58
                          Dominica
                   Trinidad   1593
           Porto   Burboroata   Cumana
           Bello   Rio Hacha   P. Cabello
                 Cartagena
                 Nombre de Dios

SOUTH AMERICA

San Fransisco

Torrelt   Faroe Is.   Shetland Is.

Vigo
Seville
Azores   Teneriffe   Canary Is.
C. Verde Is.   Sierra Leone

Gu*yaquil   Pernambuco
Callao
Arica   St. Helena

Valparaiso   Sr. Sebastian   St. Helena
         Rio de la Plata
Port S. Julian
Port Desire

VOYAGES AND

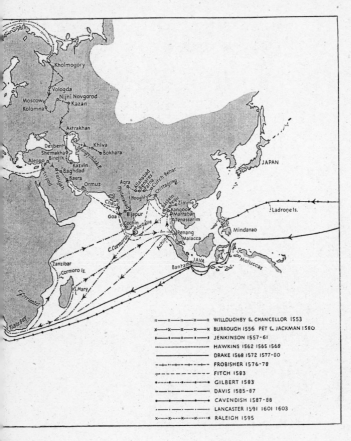

GENEALOGICAL TABLE I

QUEEN ELIZABETH AND HER RELATIONS

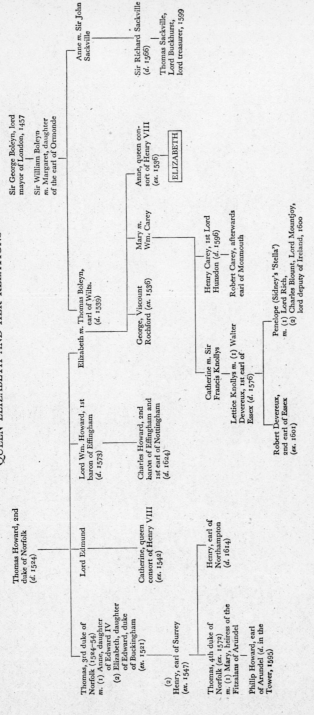

THE HOUSES OF TUDOR, STUART, AND SUFFOLK

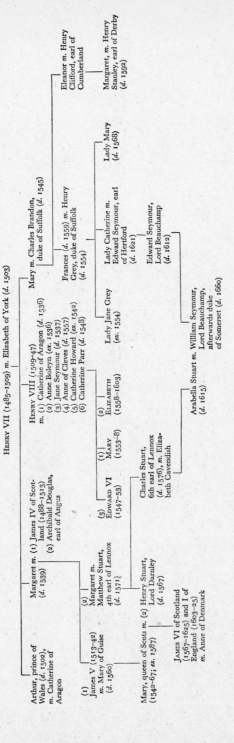

GENEALOGICAL TABLE III

THE DE LA POLE, POLE, AND HUNTINGDON FAMILIES

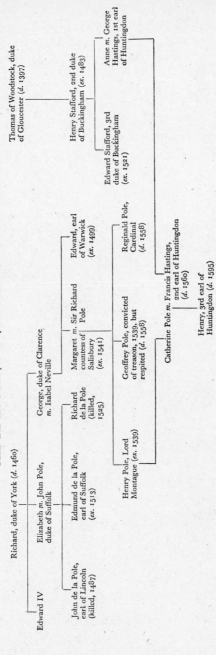

Richard, duke of York (d. 1460) — Thomas of Woodstock, duke of Gloucester (d. 1397)

Edward IV

Elizabeth m. John Pole, duke of Suffolk

George, duke of Clarence m. Isabel Neville

Henry Stafford, 2nd duke of Buckingham (ex. 1483)

John de la Pole, earl of Lincoln (killed, 1487)

Edmund de la Pole, earl of Suffolk (ex. 1513)

Richard de la Pole (killed, 1525)

Margaret m. Sir Richard Pole, countess of Salisbury (ex. 1541)

Edward, earl of Warwick (ex. 1499)

Edward Stafford, 3rd duke of Buckingham (ex. 1521)

Anne m. George Hastings, 1st earl of Huntingdon

Henry Pole, Lord Montague (ex. 1539)

Geoffrey Pole, convicted of treason, 1539, but respited (d. 1558)

Reginald Pole, Cardinal (d. 1558)

Catherine Pole m. Francis Hastings, 2nd earl of Huntingdon (d. 1560)

Henry, 3rd earl of Huntingdon (d. 1595)